Sebastian Wolff

# Asynchronous variational integration of structural collision dynamics

Sebastian Wolff

# Asynchronous variational integration of structural collision dynamics

## Numerical methods in structural mechanics

Südwestdeutscher Verlag für Hochschulschriften

**Impressum/Imprint (nur für Deutschland/only for Germany)**
Bibliografische Information der Deutschen Nationalbibliothek: Die Deutsche Nationalbibliothek verzeichnet diese Publikation in der Deutschen Nationalbibliografie; detaillierte bibliografische Daten sind im Internet über http://dnb.d-nb.de abrufbar.
Alle in diesem Buch genannten Marken und Produktnamen unterliegen warenzeichen-, marken- oder patentrechtlichem Schutz bzw. sind Warenzeichen oder eingetragene Warenzeichen der jeweiligen Inhaber. Die Wiedergabe von Marken, Produktnamen, Gebrauchsnamen, Handelsnamen, Warenbezeichnungen u.s.w. in diesem Werk berechtigt auch ohne besondere Kennzeichnung nicht zu der Annahme, dass solche Namen im Sinne der Warenzeichen- und Markenschutzgesetzgebung als frei zu betrachten wären und daher von jedermann benutzt werden dürften.

Verlag: Südwestdeutscher Verlag für Hochschulschriften GmbH & Co. KG
Heinrich-Böcking-Str. 6-8, 66121 Saarbrücken, Deutschland
Telefon +49 681 37 20 271-1, Telefax +49 681 37 20 271-0
Email: info@svh-verlag.de

Approved by: Wien, TU, Diss., 2011

Herstellung in Deutschland:
Schaltungsdienst Lange o.H.G., Berlin
Books on Demand GmbH, Norderstedt
Reha GmbH, Saarbrücken
Amazon Distribution GmbH, Leipzig
**ISBN: 978-3-8381-3046-0**

**Imprint (only for USA, GB)**
Bibliographic information published by the Deutsche Nationalbibliothek: The Deutsche Nationalbibliothek lists this publication in the Deutsche Nationalbibliografie; detailed bibliographic data are available in the Internet at http://dnb.d-nb.de.
Any brand names and product names mentioned in this book are subject to trademark, brand or patent protection and are trademarks or registered trademarks of their respective holders. The use of brand names, product names, common names, trade names, product descriptions etc. even without a particular marking in this works is in no way to be construed to mean that such names may be regarded as unrestricted in respect of trademark and brand protection legislation and could thus be used by anyone.

Publisher: Südwestdeutscher Verlag für Hochschulschriften GmbH & Co. KG
Heinrich-Böcking-Str. 6-8, 66121 Saarbrücken, Germany
Phone +49 681 37 20 271-1, Fax +49 681 37 20 271-0
Email: info@svh-verlag.de

Printed in the U.S.A.
Printed in the U.K. by (see last page)
**ISBN: 978-3-8381-3046-0**

Copyright © 2011 by the author and Südwestdeutscher Verlag für Hochschulschriften GmbH & Co. KG and licensors
All rights reserved. Saarbrücken 2011

Für

Stefanie, Johannes

und

meine Eltern Gudrun und Hans-Martin

# Danksagung

Die vorliegende Arbeit entstand während meiner Arbeit als wissenschaftlicher Mitarbeiter am Forschungsbereich für Baumechanik und Baudynamik der Technischen Universität Wien. Vorausgegangen ist die Mitarbeit in der Dynardo GmbH, wo ich Ideen für mögliche Zielstellungen dieser Arbeit gefunden habe.

Mein besonderer Dank gilt Herrn Professor Christian Bucher für die hervorragende Betreuung dieser Arbeit. Hervorzuheben ist die gewährte Freiheit in der Entwicklung der Forschungsthemen, das Vertrauen in meine Fähigkeiten und seinen Optimismus nach Fehlschlägen, seine Geduld, das genaue Zuhören und Diskutieren meiner neuen und häufig noch nicht ausgereiften Ideen. Herrn Professor Erwin Stein danke ich für die Übernahme des Gutachtens und für seine Hinweise zur Ausarbeitung.

Herrn Dr. David Schneider danke ich für die wertvollen Ratschläge, die er mir beim Lesen der ersten Entwürfe gegeben hat. Weiters danke ich allen Mitarbeitern des Forschungsbereiches für die angenehme Arbeitsatmosphäre. Die Pausengespräche waren immer erfrischend und brachten die gesuchte Ablenkung vom oft monotonen Forschungsalltag.

Besonderer Dank gebührt meiner Familie Stefanie und Johannes, die mich insbesondere in den letzten Monaten eher zu ertragen als zu genießen hatten. Viel Verständnis und Motivation sind in diese Arbeit geflossen.

# Abstract

The growing power of modern workstations enables engineers to simulate more and more complex mechanical models by computers. In particular, nonlinear problems from structural dynamics are computationally intensive. Hence, there is ongoing demand in the development of new and improvement of existing algorithms. The present thesis deals with the simulation of the dynamics of flexible bodies subject to material and geometrical nonlinearities, as well as discontinuous phenomena arising from collisions.

The equation of motion is discretized following the principle of variational integration, by what conservation laws of the continuous problem are valid in the discrete model. Existing approaches are presented. By combination of different procedures a mollified implicit-explicit algorithm is developed. It allows larger critical time steps and is particular suited for problems with non-dominant nonlinearities. The presentation of variational integrators includes the temporal discretization of holonomic and unilateral constraints. The spatial discretization is performed by a modified finite element method. The accuracy of isoparametric elements is increased by enforcing stress continuity locally. This happens by the assumption of a continuously interpolated deformation gradient. The stability of the formulation is discussed in detail. For the temporal discretization an asynchronous strategy is employed. The equation of motion is integrated explicitly, whereby some critical time step length must not be exceeded. Asynchronous methods apply individual time steps to each spatial domain. Substructures with softer material behaviour or larger finite elements can, therefore, be integrated by a larger time step. The thesis develops strategies to estimate the local time step size for the new element formulation and to efficiently treat nodal restraint conditions. It studies, how temporally-adaptive step sizes influence stability and accuracy. Furthermore, this work presents procedures for spatial discretization and detection of collision problems. In particular, the concept of distance fields is enhanced in this respect. The contact conditions from impenetrability and friction are enforced by discontinuous velocity changes in a spatially asynchronous and temporally adaptive manner.

# Contents

**Symbols and abbreviations**     1

**1 Introduction**     7
    1.1 Motivation . . . . . . . . . . . . . . . . . . . . . . . . . . . . 7
    1.2 Related approaches . . . . . . . . . . . . . . . . . . . . . . . 11
       1.2.1 Variational integrators . . . . . . . . . . . . . . . . . . 11
       1.2.2 Continuous assumed gradient elements . . . . . . . . 17
       1.2.3 Contact . . . . . . . . . . . . . . . . . . . . . . . . . . 21
    1.3 Contributions and outline . . . . . . . . . . . . . . . . . . . . 24

**2 Variational mechanics**     29
    2.1 Introduction . . . . . . . . . . . . . . . . . . . . . . . . . . . . 29
    2.2 Preliminaries . . . . . . . . . . . . . . . . . . . . . . . . . . . 30
    2.3 Principle of Hamilton . . . . . . . . . . . . . . . . . . . . . . 30
       2.3.1 Lagrangian dynamics . . . . . . . . . . . . . . . . . . 32
       2.3.2 Hamiltonian dynamics . . . . . . . . . . . . . . . . . 32
    2.4 Preserved quantities . . . . . . . . . . . . . . . . . . . . . . . 33
       2.4.1 Noether's theorem . . . . . . . . . . . . . . . . . . . . 34
       2.4.2 Conservation of linear momentum . . . . . . . . . . . 35
       2.4.3 Conservation of angular momentum . . . . . . . . . . 35
       2.4.4 Conservation of energy . . . . . . . . . . . . . . . . . 35
       2.4.5 Symplecticity . . . . . . . . . . . . . . . . . . . . . . . 36
       2.4.6 Liouville's theorem . . . . . . . . . . . . . . . . . . . 38
    2.5 Example: linear SDOF system . . . . . . . . . . . . . . . . . 38

**3 Variational integrators**     41
    3.1 Introduction . . . . . . . . . . . . . . . . . . . . . . . . . . . . 41

| | | |
|---|---|---|
| 3.2 | Geometric integrators | 43 |
| 3.3 | Discrete Euler-Lagrange equation | 44 |
| 3.4 | Preserved discrete quantities | 45 |
| | 3.4.1 Discrete time Noether's theorem | 45 |
| | 3.4.2 Conservation of linear momentum | 45 |
| | 3.4.3 Conservation of angular momentum | 46 |
| | 3.4.4 Preservation of the symplectic structure | 46 |
| | 3.4.5 Discrete energy | 47 |
| 3.5 | Error analysis | 47 |
| 3.6 | Linear stability analysis | 49 |
| | 3.6.1 Fundamental matrix | 49 |
| | 3.6.2 Combining different integrators | 50 |
| | 3.6.3 Lyapunov stability | 51 |
| 3.7 | Example integrators | 52 |
| | 3.7.1 Symplectic Euler | 52 |
| | 3.7.2 Velocity Verlet | 53 |
| | 3.7.3 Midpoint | 54 |
| | 3.7.4 Multiple time stepping | 54 |
| | 3.7.5 Exponential integrators | 57 |
| | 3.7.6 Implicit-explicit integrators | 59 |
| | 3.7.7 Rowlands's method | 60 |
| 3.8 | Constraints | 61 |
| | 3.8.1 Reduction of degrees of freedom by projection | 61 |
| | 3.8.2 Lagrange-d'Alembert principle | 63 |
| | 3.8.3 SHAKE | 64 |
| | 3.8.4 RATTLE | 65 |
| | 3.8.5 Unilateral constraints | 66 |
| 3.9 | Examples | 69 |
| | 3.9.1 A perturbed linear oscillator | 69 |
| | 3.9.2 A geometrically nonlinear oscillator | 73 |
| | 3.9.3 Nonlinear vibration of a cantilever beam | 76 |
| | 3.9.4 Linear vibration of a beam | 79 |
| | 3.9.5 Nonlinear pendulum: Instabilities in implicit midpoint and trapezoidal Newmark | 80 |

|  | 3.9.6 | Nonlinear pendulum: Stable discretizations | 84 |
|---|---|---|---|
|  | 3.9.7 | Double pendulum with attached spring and mass | 86 |
|  | 3.9.8 | Collision of two linked linear springs | 88 |

# 4 Continuous assumed gradient method    93

4.1 Introduction . . . . . . . . . . . . . . . . . . . . . . . . . . 93
4.2 Fundamentals of continuum mechanics . . . . . . . . . . . . . 94
    4.2.1 Kinematics . . . . . . . . . . . . . . . . . . . . . . . 94
    4.2.2 Kinetics . . . . . . . . . . . . . . . . . . . . . . . . 97
    4.2.3 Constitutive equations . . . . . . . . . . . . . . . . . 98
    4.2.4 Equilibrium equations . . . . . . . . . . . . . . . . . 100
4.3 Finite element interpolation of the continuum . . . . . . . . . 101
4.4 Assumed gradient field . . . . . . . . . . . . . . . . . . . . . 102
    4.4.1 Formulation . . . . . . . . . . . . . . . . . . . . . . 102
    4.4.2 Principle of Hu-Washizu . . . . . . . . . . . . . . . . 102
    4.4.3 Solution strategy: Dual multiplier space . . . . . . . . 103
    4.4.4 Strain energy and numerical integration . . . . . . . . 105
    4.4.5 Linearization . . . . . . . . . . . . . . . . . . . . . . 106
4.5 Regular mesh generation . . . . . . . . . . . . . . . . . . . . 107
4.6 Nodal integration . . . . . . . . . . . . . . . . . . . . . . . . 108
    4.6.1 Nodal averages . . . . . . . . . . . . . . . . . . . . . 108
    4.6.2 Analysis of instabilities . . . . . . . . . . . . . . . . . 109
    4.6.3 Penalty regularization . . . . . . . . . . . . . . . . . 110
    4.6.4 Conforming regularization . . . . . . . . . . . . . . . 112
4.7 Smoothed Finite Element Method . . . . . . . . . . . . . . . 114
    4.7.1 Smoothing operator . . . . . . . . . . . . . . . . . . 114
    4.7.2 Smoothing cells . . . . . . . . . . . . . . . . . . . . 114
    4.7.3 Relation . . . . . . . . . . . . . . . . . . . . . . . . 115
4.8 Stable interpolation schemes . . . . . . . . . . . . . . . . . . 115
    4.8.1 Nodal support with bubble stabilization . . . . . . . . 116
    4.8.2 Edge support . . . . . . . . . . . . . . . . . . . . . . 117
    4.8.3 Face support . . . . . . . . . . . . . . . . . . . . . . 118
    4.8.4 Assumed higher order gradient . . . . . . . . . . . . . 119
    4.8.5 Deriving continuum elements . . . . . . . . . . . . . 119

4.9 Implementation . . . . . . . . . . . . . . . . . . . . . . . . . . . 122
    4.9.1 Simplified dual mapping . . . . . . . . . . . . . . . . . 122
    4.9.2 Integration of volume integrals . . . . . . . . . . . . . 122
    4.9.3 Constitutive laws . . . . . . . . . . . . . . . . . . . . . 124
4.10 Error analysis . . . . . . . . . . . . . . . . . . . . . . . . . . . 125
4.11 Examples . . . . . . . . . . . . . . . . . . . . . . . . . . . . . 126
    4.11.1 Vibration analysis of a two-dimensional cantilever . . . . . . 126
    4.11.2 Spurious modes in nodal integration . . . . . . . . . . . 128
    4.11.3 Forced vibration of a cantilever beam . . . . . . . . . . 128
    4.11.4 Forced vibration of a geometrically nonlinear cantilever beam 131

# 5 Asynchronous variational integration    133
5.1 Introduction . . . . . . . . . . . . . . . . . . . . . . . . . . . . 133
5.2 Asynchronous Euler scheme . . . . . . . . . . . . . . . . . . . 134
5.3 Discretization of the space-time integral . . . . . . . . . . . . 136
5.4 Nodal restraints . . . . . . . . . . . . . . . . . . . . . . . . . . 140
5.5 Estimating the time step length . . . . . . . . . . . . . . . . . 141
5.6 Example: Asynchronous integration of a cantilever beam . . . . 145
    5.6.1 Model problem . . . . . . . . . . . . . . . . . . . . . . 145
    5.6.2 Benchmark against synchronous time stepping . . . . . . 146
    5.6.3 Equally distant nodes . . . . . . . . . . . . . . . . . . . 148

# 6 Variable step size integration    151
6.1 Introduction . . . . . . . . . . . . . . . . . . . . . . . . . . . . 151
6.2 Explicit symplectic energy momentum integration . . . . . . . . 152
6.3 Time transformations . . . . . . . . . . . . . . . . . . . . . . . 154
6.4 Variational kick and drift operators . . . . . . . . . . . . . . . 156
6.5 Asynchronous variable time steps . . . . . . . . . . . . . . . . 157
6.6 Example time step functions . . . . . . . . . . . . . . . . . . . 161
6.7 Time step selection and solution . . . . . . . . . . . . . . . . . 164
6.8 Examples . . . . . . . . . . . . . . . . . . . . . . . . . . . . . 166
    6.8.1 Synchronous integration of a single degree of freedom system with variable step sizes . . . . . . . . . . . . . . . . 166
    6.8.2 Asynchronous variable time steps applied to a linear oscillator 169
    6.8.3 Limiting cases of variable time steps . . . . . . . . . . . 171

|        |       | 6.8.4 Variable step size integration of a cantilever beam | 174 |
|---|---|---|---|

# 7 Collision dynamics — 179

- 7.1 Introduction . . . . . . . . . . . . . . . . . . . . . . . . . . . . 179
- 7.2 Contact mechanics . . . . . . . . . . . . . . . . . . . . . . . . 180
    - 7.2.1 Problem description . . . . . . . . . . . . . . . . . . 180
    - 7.2.2 Kinematics . . . . . . . . . . . . . . . . . . . . . . . 181
    - 7.2.3 Gap function . . . . . . . . . . . . . . . . . . . . . . 183
    - 7.2.4 Contact integral . . . . . . . . . . . . . . . . . . . . 184
    - 7.2.5 Constraints . . . . . . . . . . . . . . . . . . . . . . . 185
- 7.3 Distance field . . . . . . . . . . . . . . . . . . . . . . . . . . . 186
    - 7.3.1 Level sets . . . . . . . . . . . . . . . . . . . . . . . . 186
    - 7.3.2 Discrete distances . . . . . . . . . . . . . . . . . . . 188
    - 7.3.3 Computing discrete distances . . . . . . . . . . . . . 188
    - 7.3.4 Stable interpolation and assumed distance gradients . . . . 190
    - 7.3.5 Computing the closest point projection . . . . . . . . . . . 194
    - 7.3.6 Replacing the gap function . . . . . . . . . . . . . . . 195
- 7.4 Asynchronous collisions . . . . . . . . . . . . . . . . . . . . . 198
    - 7.4.1 Asynchronous collision detection . . . . . . . . . . . . 198
    - 7.4.2 Normal contact . . . . . . . . . . . . . . . . . . . . . 200
    - 7.4.3 Normal contact with nodal restraints . . . . . . . . . . 200
    - 7.4.4 Coulomb friction . . . . . . . . . . . . . . . . . . . . 203
    - 7.4.5 Time step selection . . . . . . . . . . . . . . . . . . . 204
- 7.5 Examples . . . . . . . . . . . . . . . . . . . . . . . . . . . . . 205
    - 7.5.1 Two elastic bars . . . . . . . . . . . . . . . . . . . . . 205
    - 7.5.2 Elastic block sliding on rigid obstacle . . . . . . . . . . 208
    - 7.5.3 Block assembly . . . . . . . . . . . . . . . . . . . . . 209

# 8 Summary — 213

- 8.1 Longer time steps in explicit dynamics . . . . . . . . . . . . . 213
- 8.2 Time discretization of nonlinear constraints . . . . . . . . . . 215
- 8.3 Continuous assumed gradient method . . . . . . . . . . . . . 216
- 8.4 Asynchronous variational integration . . . . . . . . . . . . . . 219
- 8.5 Variable time steps . . . . . . . . . . . . . . . . . . . . . . . . 220

|     | 8.6   | Asynchronous collisions | 221 |
|-----|-------|-------------------------|-----|

## A  Verification of CAG elements — 225

- A.1 Patch test … 225
- A.2 Cantilever beam … 226
- A.3 Lamé problem … 230
- A.4 Cook's tapered panel … 231
- A.5 Compressed block … 234
- A.6 Performance of distorted elements … 236
- A.7 Convergence of natural frequencies … 237
- A.8 Numerical efficiency … 239

## B  Variations on the contact interface — 243

## C  Spatial discretization of the contact boundary — 245

- C.1 Exact integration domains … 245
- C.2 Integration strategies … 246
- C.3 Mortar method … 247
- C.4 Node-to-element integration … 249

## D  Collision detection — 251

- D.1 Global contact search … 252
  - D.1.1 Bounding volume hierarchies … 252
  - D.1.2 Position codes … 253
  - D.1.3 Application in dynamics … 255
- D.2 Local contact search … 256

## Bibliography — 257

## Index — 280

# Symbols and abbreviations

The subsequent list describes frequently used symbols and abbreviations. In elasticity, it is common to indicate matrix and vector quantities by bold letters. This notation is used in chapters 4 and 7.

| | |
|---|---|
| $i, j, k, \ldots, z$ | indices; in continuum mechanics in $\mathbb{R}^3$ |
| $\alpha, \beta, \gamma, \ldots, \omega$ | dimensional indices ($1 \ldots 3$ in $\mathbb{R}^3$, $1, 2$ in $\mathbb{R}^2$) |
| $A, B, C, \ldots, Z$ | indices of spatial support points or nodes |
| $(\circ)_k^-$ | discrete quantity directly prior (left to) time node $k$ |
| $(\circ)_k^+$ | discrete quantity directly after (right to) time node $k$ |
| $D_i \circ$ | partial derivative for the $i$th argument of $\circ$ |
| $(\circ)^{pred}$ | quantity at predictor 'time' |
| $(\hat{\circ})$ | quantity in contact frame |
| $(\circ)_i$ | $i$th component of vector $(\circ)$ |
| $(\circ)_{ij}$ | component of matrix $(\circ)$ at row $i$, column $j$ |
| | |
| $\alpha$ | fictive time variable |
| $\gamma_A$ | assumed distance gradient |
| $\delta_{ij}$ | Kronecker delta |
| $\epsilon$ | strain vector |
| $\theta_k$ | discrete system times in AVI |
| $\kappa$ | coefficient of restitution |
| $\lambda$ | Lyapunov exponent |
| $\lambda$ | Lagrange multiplier |
| $\mu$ | Coulomb coefficient of friction |
| $\nu$ | Poisson's ratio |
| $\xi$ | parameterized material coordinate (eg. element coordinate system) |
| $\Pi$ | potential function |
| $\rho$ | mass density |
| $\hat{\tau}$ | tangential basis in contact frame |
| $\phi_h$ | continuous time-$h$ flow |
| $\Phi_h$ | discrete time-$h$ flow |
| $\Omega$ | spatial domain in $\mathbb{R}^3$ |
| $\partial\Omega$ | boundary of $\Omega$ ($\mathbb{R}^2$) |

# Symbols and abbreviations

| | |
|---|---|
| $c$ | velocity of wave propagation |
| $d$ | distance function |
| $d^h$ | interpolated distance function |
| $g(q)$ | vector of constraint equations |
| $\hat{g}$ | gap function |
| $h$ | time step size |
| $h_{crit}$ | critical time step |
| $h_i^j$ | time step size at step $j$ of the $i$th potential |
| $j$ | discrete momentum |
| $n$ | dimension |
| $n$ | normal vector |
| $p$ | canonical momentum |
| $q$ | vector of generalized coordinates |
| $q^k(t)$ | interpolated $q(t)$ in $k$th time element |
| $q_i^k$ | sicrete generalized coordinate vector at $i$th support point in $k$th time element for $q^k(t)$ |
| $q_k$ | discrete coordinate at time node $k$ |
| $s$ | time step function |
| $t$ | real stress vector |
| $t$ | time |
| $t_i^j$ | $j$th time step of $i$th potential energy |
| $u$ | vector of displacements |
| $v$ | velocity vector |
| $w$ | conjugate momentum of time $t$ (negative energy) |
| $x$ | spatial coordinate in deformed configuration |
| $z$ | phase space |
| | |
| $\mathcal{A}$ | averaging operator |
| $C$ | contact frame |
| $C$ | damping matrix |
| $E$ | Young's modulus |
| $E_d$ | discrete energy |
| $\mathbb{F}_{ij}$ | Jacobian in phase space |
| $F_{\alpha\beta}$ | deformation gradient |
| $F(q)$ | restoring force vector |
| $F(t)$ | external force vector |
| $G$ | gradient of linear constraints |
| $G^h$ | natural distance gradient |
| $G_{dyn}(u, \delta u)$ | virtual Lagrangian |
| $H(q,p)$ | Hamiltonian |
| $\bar{H}$ | shadow Hamiltonian |
| $H_k^d$ | discrete Hamiltonian |
| $\mathcal{I}$ | preserved quantity (Noether's theorem) |
| $J$ | Jacobian (volumetric strain) |

# Symbols and abbreviations

| | |
|---|---|
| $\mathbb{J}$ | symplectic matrix of Hamilton's equations |
| $K$ | stiffness matrix |
| $L(q,\dot{q},t)$ | Lagrangian |
| $L_d$ | discrete Lagrangian |
| $L_A$ | interpolation function of multipliers |
| $M$ | mass matrix |
| $M_A$ | interpolation function, i.e. of assumed gradients |
| $N_A$ | finite element shape function |
| $P$ | 1st Piola-Kirchoff stress |
| $\mathbb{P}$ | RATTLE projection matrix |
| $\mathbb{P}^c$ | collision projection matrix |
| $S$ | 2nd Piola-Kirchoff stress |
| $S$ | action |
| $S_k^d$ | discrete elemental action |
| $T$ | Cauchy stress |
| $T(\dot{q}), T(p)$ | kinetic energy |
| $T_d$ | discrete kinetic action |
| $U$ | total strain energy |
| $U^d$ | strain energy density function |
| $V$ | volume |
| $V(q,t)$ | potential energy |
| $V_d$ | discrete potential action |
| $V_i(q,t)$ | asynchronous potential energy |
| $W_A$ | spatial weight |
| $X$ | material point coordinate (spatial coordinate in reference configuration) |
| $Z = (Q, J)$ | extended phase space |

# Symbols and abbreviations

| | |
|---|---|
| **AABB** | axis aligned bounding box |
| **AVI** | asynchronous variational integration (integrator) |
| **C3D_$x$N** | $x$-noded three-dimensional continuum element |
| **C3D_$x$N_NI** | $x$-noded three-dimensional continuum element, nodal integration |
| **C3D_$x$N_$y$I** | $x$-noded three-dimensional continuum element, nodal integration with $y$ interior points |
| **C3D_$x$N_$y$E** | $x$-noded three-dimensional continuum element, edge integration with $y$ points per edge |
| **C3D_$x$N_$y$C** | $x$-noded three-dimensional continuum element, assumed gradient interpolation with $y$ support points |
| **CAG** | continuous assumed gradient |
| **CFFM** | closest feature front marching |
| **DCR** | decomposition contact response |
| **FEM** | finite element method |
| **IMEX** | implicit-explicit (integrator) |
| **KAM** | Kolmogorow-Arnold-Moser |
| **MDOF** | multiple degrees of freedom |
| **MOLLY** | mollified impulse method |
| **NI** | nodal integration |
| **NICE** | nodally integrated continuum element |
| **r-RESPA** | reversible reference system propagator algorithm |
| **SDOF** | single degree of freedom |
| **SEM** | symplectic energy momentum integration |
| **SFEM** | smoothed finite element method |
| **VV** | Velocity Verlet |

## Differentiation and variations

Differentiation with respect to time is denoted by putting a superimposed dot on the corresponding function. Thus, the material velocity of a point is given by

$$\mathbf{v}^{(i)} = \dot{\phi}^{(i)}(\mathbf{X}^{(i)}) = \frac{d}{dt}\phi^{(i)}(\mathbf{X}^{(i)})$$

Partial derivatives are denoted $(\cdot)_{,i}$ or $(\cdot)_{,\alpha}$.

A functional is a function which takes functions as arguments. The variational derivative of a functional is a directional derivative for a variational parameter $\epsilon$. A perturbation of a function $\phi$ in direction of a function $\eta_\phi$ is denoted by

$$\phi_\epsilon = \phi + \epsilon \overset{\star}{\phi}$$

The directional derivative (Gateaux derivative) of this function defines its variation $\delta\phi$

$$\delta\phi := D_{\eta_\phi}[\phi] = \frac{d}{d\epsilon}\bigg|_{\epsilon=0}[\phi_\epsilon] = \overset{\star}{\phi}$$

# Symbols and abbreviations

where $D_\mathbf{a}(g(\mathbf{x}))$ denotes the derivative of a function $g$ in direction $\mathbf{a}$ at point $\mathbf{x}$. The variational derivative of a functional is then

$$\delta f := D_{\eta_\phi}[f(\phi_\epsilon)] = \frac{d}{d\epsilon}\bigg|_{\epsilon=0} [f(\phi + \epsilon \overset{\star}{\phi})]$$

Applying the chain rule to a functional with $N$ parameters yields

$$\delta f = \sum_i^N \frac{\partial f}{\partial \phi_i} \frac{\partial \phi_i}{\partial \epsilon}\bigg|_{\epsilon=0} = \sum_i^N \frac{\partial f}{\partial \phi_i} \delta\phi_i$$

## Summation indices

The Einstein convention is applied in this work (without considering the location of indices - subscripts vs. superscripts). For example, the following terms are equivalent:

$$[AB]_{ik} = A_{ij}B_{jk} = \sum_j A_{ij}B_{jk}$$

# Chapter 1

# Introduction

## 1.1 Motivation

In many engineering tasks, the distribution of stresses, strains and deformation of structures subject to dynamic loading are of interest. Dynamic simulations, however, need a lot of computational resources. They are only possible because the numerical power of today's computers is much higher than a few decades ago. At the same time, however, the level of detail of mechanical models is improving by refining the approximation of geometrical features and by a better representation of the true material behaviour. In particular, the number of unknown variables and the degree of nonlinearities have been increased: nonlinear elastic material laws, plasticity and viscoelastoplasticity, geometric nonlinearities, structural instabilities with snap-through and bifurcation points, contact/impact with nonlinear friction laws, crack propagation, automatic mesh refinement etc. are issues in what engineers are interested in and what increases the numerical effort.

The right choice of the finite element formulation and of the solution method can help to reduce the numerical complexity. For example, specific finite element types were developed for thin-walled structures introducing structural elements for beams and shells and helping to reduce the number of unknowns. When considering the solution procedure, it is important how many degrees of freedom are involved and how they are coupled, whether an iterative or non-iterative solution is required, whether the response can be obtained explicitly or implicitly, how fast the system of equation can be generated, how certain components, for example the contact formulation, interact with each other, etc.

Implicit methods often lead to iterative methods. Linear systems may be solved non-iteratively, but the sparsity of the matrix defining the equations gets lost due to an increased number of couplings among the degrees of freedom. The treatment of nonlinear constraints becomes rather complex in implicit methods. Their advantage, however, is the stability and accuracy for relatively large time steps. Even for very large time steps implicit time stepping methods may be stable. This may compensate the numerical effort of the solution procedure because less time steps are required in order to simulate a certain time interval.

On the other hand, explicit schemes lead to very simple systems of equations which generally can be solved efficiently, are non-iterative and are easy to implement. Explicit methods are, however, subject to a critical time step which decreases the overall efficiency. The critical time step becomes smaller with stiffer materials and with finer meshes.

Methods were developed which try to combine advantages of both, explicit and implicit methods: mixed methods where explicit and implicit schemes are applied to different regions in the same structure, multiscale methods such as implicit-explicit or exponential integrators which increase the critical time step and improve the accuracy of the linear response, modal superposition methods which reduce the number of unknowns and increase the critical time step, other model reduction techniques, etc. Generally, these methods aim at finding a compromise between implicit and explicit methods - a larger time step of the explicit part and a simplified solution of the implicit part. The additional numerical effort must be balanced by the larger time step. For reduced order modeling, the loss of accuracy must be considered.

When concentrating on explicit methods, three factors are generally used to improve the efficency: the number of unknowns (more accurate finite element formulations may reduce the required number of degrees of freedom), the number of integration points (the less material law evaluations per force vector, the less computing time is needed) and the critical time step. Therefore, finite elements are often secured against locking behavior limiting the time step. Material laws are usually expressed in an updated Lagrangian framework which reduces the effort in evaluating geometrically nonlinear strains. Constitutive relations are solved by explicit methods, in particular the enforcement of yield surfaces. Reduced integration schemes in conjunction with hourglass control decrease the number of integration points. Finite elements with extraordinarily small dimensions are often neglected in the simulation such that they can not destroy the critical time step length. The time step can be further imcreased by mass scaling, i.e. increasing the mass of the smallest elements.

Recently, a new philosophy was introduced to explicit schemes: asynchronous integration. Therein, every part of a structure is analyzed with an individual time step size. Thereby, the total number of material law evaluations is reduced. In standard schemes, the smallest or, respectively, the 'stiffest' finite element in the model governs the time step being applied to the whole structure. In asynchronous integration, the 'small' and 'stiff' elements are integrated with a small time step while a larger time step is applied to 'soft' and 'large' finite elements. The differences of the time steps within the considered model must be large enough in order to compensate the additional effort which originates from maintaining the asynchronicity. That means, asynchronous procedures may improve the numer-

## 1.1. Motivation

ical efficiency if there exist some regions in the considered finite element mesh with much smaller element sizes than in other regions. This assumption applies to many engineering problems. One often refines a finite element mesh in regions where an accurate stress evaluation takes place, for example in case of notch stresses. Furthermore, asynchronous integration could be an ideal companion for automatic mesh refinement and specific crack propagation strategies, where the mesh is adopted during the simulation.

On the other hand, improving the formulation of finite elements may help to reduce the number of equations and the number of integration points in order to obtain a desired accuracy at the same time. When considering an isoparametric finite element approximation, one observes that strains and stresses are continuous functions in the interior of finite elements, but are discontinuous along finite element interfaces. This is in contrast to local balance equations, i.e. the equilibrium of forces which is only satisfied in its weak form. Assuming homogenious material behaviour, the strains should be continuous at element interfaces as well. A finite element formulation satisfying this constraint may drastically improve accuracy. An example is isogeometric analysis. It implements smooth interpolation functions (and, hence, continuous strains), but does not provide the flexibility of FEM with respect to mesh generation. Efficient and stable finite element formulations with continuous strains are still missing.

Contact and impact problems are another component of the simulation procedure which is numerically expensive. The correct and robust detection of interpenetrations must be ensured. Complex contact laws (nonlinear friction, inelastic impacts) increase the degree of nonlinearity. When large time steps are employed, a consistent determination of glide path and interpenetration are important to avoid locking effects and divergence. Various types of geometries must be considered being arbitrarily complex, i.e. concave vs. convex, smooth vs. sharp features, corners vs. curvatures and plane surfaces, etc. The algorithm must treat the contact problem with as little user input as possible. Additionally, the contact search alone puts high demands on memory consumption and computing time. The complexity of search algorithms with respect to the number of surface patches is decisive. First implementations tested all surfaces segments among each other on intersection leading to $n \times n$ complexity which is not suitable for large-scale problems. The issue of complexity continues in the choice of the interpolation of contact tractions: The better the approximation of contact tractions, the more accurate but also the larger the numerical expenses.

In particular the latter can be easily answered in explicit integration. When using very small time steps the contact search and solution must be reasonable fast. Furthermore, the contact formulation should not influence the length of the critical time step. This is, for example, the case in penalty formulations. Therein, penetrations are avoided by the application of additional forces. These contact forces are proportional to some parameter. The larger the parameter, the smaller are the allowed penetrations. But at the same time the critical time step due to the contact forces is reduced. Recently, collision integrators were proposed which apply a non-smooth update to the solution trajectory and which do not affect the step size.

The question is, now, whether the principle of asynchronicity can be applied to impact problems and whether it may help to improve the efficiency. The idea behind asynchronous collisions is that the frequency of contact evaluations depends on local mesh properties. The larger the velocity and the smaller the finite elements, the more frequent one has to test on intersection in order to stay accurate. Other elements may be evaluated less frequent.

Generally, the choice of a procedure by which time stepping schemes can be derived, is of rather philosophical nature. For example, in many text books the central difference method is derived from a geometric consideration, i.e. by the assumption of a piecewise quadratic interpolation of the displacements in time. Properties of conservative mechanical systems (energy, momentum, symplecticity, phase space volume) are often missing in engineering education. Research has shown, however, that numerical schemes which preserve invariants of mechanical systems are generally more accurate and more robust in long-term simulation than numerical schemes which do not preserve these quantities. Procedures were developed which ensure the preservation of quantities. This is independent from the choice of the detailed solution procedure (implicit, explicit, etc.).

The class of geometric or, respectively, mechanical integrators can be distinguished into energy-momentum, energy-symplectic and momentum-symplectic preserving algorithms. It is generally not possible to preserve all of the three quantities. In this work, symplectic-momentum integrators are favored. They do not preserve the energy, but vector- and matrix-valued invariants. The energy can be computed easily and may serve as an error indicator. The error in symplecticity is, for example, difficult to obtain. Longterm stability and approximate preservation of energy are ensured because a symplectic method provides the exact solution to an alternative model which often is only a small perturbation to the original system. Variational integrators provide a simple and strict procedure to derive symplectic-momentum methods. Furthermore, their formulation fits perfectly into elasticity problems which can be expressed in terms of Hamilton's variational principle as well.

All these considerations lead to the following observations: contact algorithms, the treatment of linear and nonlinear constraints, for example arising in domain decomposition, and the finite element formulation should be harmonized with the chosen time integration method. At the same time, it would be interesting to learn how the numerical efficiency can be improved by the mentioned issues. These statements are the motivation of the present work.

## 1.2 Related approaches

### 1.2.1 Variational integrators

Numerical integration schemes derived from variational principles were not uncommon in the past. Nevertheless, one of the first integrators which were consequently derived from a discretized version of Hamilton's priniple is the Moser-Veselov algorithm [195] for the integration of the rigid body motion. It was shown in [263] that the used methodology preserves a symplectic form. The same procedure was derived in [5] and the existence of a discrete version of Noether's theorem was shown. The term 'mechanical integrator' first appears in [266]. In the given sense, mechanical integrators denote either energy-momentum or symplectic-momentum schemes since energy-momentum-symplectic schemes are not possible when using constant time steps [281]. Recently, a new class of mechanical integrators with variable time steps was presented [75, 76] being symplectic-energy preserving. The discrete mapping [266] is given in Lagrangian form, i.e. it computes the coordinates at the next time step from the coordinates at the current and the previous time step. Holonomic constraints are embedded and the SHAKE algorithm [229] is recovered. One can show that the explicit Newmark algorithm [200] can be derived from a variational principle [119]. The trapezoidal Newmark method [200] can be obtained by a variable transformation starting at the variational implicit midpoint rule. Certain symplectic Runge-Kutta methods can be identified as variational integrators [23]. A detailed discussion on variational integrators is presented in [185]. Popular methods like Newmark, Verlet [262], SHAKE [229] or RATTLE [3] are derived, external forces and constraints are explained. [156] gives a survey on variational integrators including collisions, asynchronous integration, space-time methods, etc.

The Hamiltonian viewpoint, defining the state by coordinates and momentum, can be derived using the discrete principle of Hamilton. An earlier work [236] discretizes the principle of least action using independent interpolations of coordinates and momentum. [137, 153] use the discrete Legrende transformation to introduce discrete momenta from a Lagrangian formulation. Related is the Hamilton-Pontryagin principle [24, 137, 251]. Therein, individual interpolations of velocity and momentum are used. The discrete momenta are determined as Lagrange multipliers which weakly enforce the equality between velocity and the time derivatives of the coordinates.

**Multiple time stepping**

Explicit time integration schemes may become inefficient in the presence of fast oscillators in the system. For example, finite elements of very small size and/or with stiff material properties reduce the critical time step of the whole system. An approach to overcome this problem are mixed methods using explicit integrators for one domain and implicit methods for another mesh region. [10, 12] present a mixed method with nodal partition while [105, 106] use an element partition. Multiple time stepping algorithms were developed to integrate different parts of structures with different step sizes. This strategy is also known as subcycling. The first subcycling approach in structural dynamics is given in [11]. The time step ratios were obtained by bisection. An approach using non-integer ratios of the time steps was presented [199]. A clock is introduced which counts the smallest time step in the system. All nodes are thus updated which are behind the clock. The Verlet-I/r-RESPA schemes for molecular dynamics are presented by [79, 258]. Both are extensions of Velocity Verlet, but only the latter is symplectic. Therein, the potential function is interpreted as a sum of slow and fast potentials, each integrated with different time steps. The time step ratio is an integer value. [248] presents a similar method applied to the context of finite element meshes. [247] extends the algorithm to non-integer time step ratio and gives stability criteria. The results are affirmed by [43] comparing different subcycling algorithms and proving the Smolinski and Sleith algorithm [248] as the only stable one. Belytschko's method [11] is shown to be only statistically stable [43], improving stability with growing number of degrees of freedom. [44] improves accuracy by reestablishing momentum conservation at element interfaces with different time steps. The method is only statistically stable. Stability can be enforced by damping high-frequent oscillations using the generalized-alpha method [35, 109]. Another approach [78] enforces continuity of velocities at interfaces and is stable but dissipative.

On the other front, development in molecular dynamics algorithms focused on symplectic stabilizations, i.e. by filtering those forces which lead to instabilities. MOLLY is presented in [68], a mollified impulse method which filters momentum changes in a symplectic transformation by evaluating forces at averaged positions. [113] uses a different averaging operator suitable for molecular dynamics, i.e. it eliminates completely the directions in slow impulses which excite fast forces. [182] combines MOLLY with a B-Spline weighting function and adds Langevin damping. Using nonsymplectic averaging methods, stability criteria can be tied to the slow forces [149]. The approach was extended [148] to adaptive time steps and smooth force decompositions with time-interpolated forces.

**Asynchronous variational integrators**

A generalization of r-RESPA are asynchronous variational integrators (AVIs) [154, 155, 157, 158]. Therein, time step sizes are individually assigned to each finite element at arbitrary ratios. The method is variational and implements a priority queue which decides on the sequence of drift phases and velocity kicks. The method can be extended to parallel implementations [117]. Convergence can be proved for linear elasticity [63]. Reliable stability criteria are difficult to find [64, 65] which was exemplified for a two-potential problem leading to similar results as for

## 1.2. Related approaches

r-RESPA. AVIs were successfully employed to improve accuracy and efficiency in domain decomposition [14, 69] where each domain is assigned to individual time step sizes. They managed to develop implicit AVIs, but require synchronization times and lead to full-implicit couplings in between. The benefits of AVIs in efficiency were applied to contact/impact problems [91] using a quadratic penalty formulation to compute contact forces.

There exist other asynchronous approaches though not being variational. [74] applies energy-symplectic methods to the context of asynchronous integration. Another notable approach to asynchronous integration is presented by [282] providing the tent-pitcher algorithm which meshes space and time asynchronously. The method is based on a space-time discontinuous Galerkin method and creates linear tetrahedra in two-dimensional space and time. The asynchronous discontinuities lead to very small systems of equations being solved efficiently. The method was improved by several authors: [56] finds a local condition for time step size selection. [1] provides theoretical background and presents another time step function. [256] further simplifies the time step selection strategy.

**Exponential integrators**

When multiple time stepping schemes are applied to a system where the fast forces are linear then more complex algorithms can be used to integrate the linear part of the response. A family of symplectic exponential integrators (or integrating factor methods) was established by fitting Verlet's method to the accurate solution of the linear forces [70] involving matrix exponentials for the linear parts and Verlet's method for the nonlinear parts. Another popular exponential integrator was derived by a variable transformation and applying Euler's method to the nonlinear forces [146]. The latter is not symplectic, but can be applied to 1st and 2nd order differential equations. Other Gautschi-type methods were presented [47, 87] which differ in the solution of the nonlinear parts. [47] applies Verlet to the space of generalized coordinates and not to the exponentially transformed space as in [70]. [87] perform a frequency expansion of the solution and are interested in near-conservation of the energy. [259] presents an error analysis, a generalized scheme to construct Gautschi-type integrators and a new method which is not symplectic, but reduces the integration error. A practical comparison of existing schemes is given in [41] illustrating different viewpoints on Gautschi's method and discussing adaptive matrix exponentials and the application of MOLLY [68] in the context of exponential integration.

Due to the use of the matrix exponential these methods are recommended for problems of small to moderate size. An approach to large scale problems was presented in [98] where processed integrators involving matrix-vector products are exponentially fitted and the matrix exponential itself is approximated by a Krylov subspace method. Instead of approximating the matrix exponential one could instead integrate the linear forces using an implicit method, for example the implicit midpoint rule [252]. Exponential integrators can be further combined with methods of model order reduction which reduce the number of variables and share the requirement that the nonlinear forces are small compared with the linear ones. Strategies to model reduction are based on modal truncation using an eigenvector basis [29, 111, 206, 261], Ritz vectors or other basis vectors which are easier to construct [2, 33, 66, 226, 270], combined eigenvectors and Ritz vectors for nonlinearities [122, 132], balancing [231], proper orthogonal decomposition which finds the optimal basis to given snapshots [123, 124, 132, 136, 176, 188], combining POD with wavelets [135], etc.

**Processed methods**

Various approaches have been developed to generate higher order schemes by processing low-order methods. One branch of development in this regard are composition methods [278] which allow to combine symmetric methods of even order, for example Verlet, to create higher order schemes. This is done by chaining the base integrator forward and backward in time using different step sizes such that the sum of all steps equals the desired time step length. Composition methods can be applied to constrained dynamical systems, i.e. composing SHAKE and RATTLE [224].

Another approach is presented in [228]. Instead of additional integration points Rowlands uses derivatives of higher order while remaining explicit. The method was revisited by several authors establishing the class of processed integrators [19, 178, 242]. Processing the Verlet method was used to maximize the stability intervals of the central difference method [177], basically by changing the weights in Rowlands' method.

**Variable time steps**

Variable step sizes offer the opportunity to adapt the time step to local error estimators and, thus, to improve numerical efficiency. Simply varying the time step length does, however, reduce favourable properties of symplectic methods such as long-term stability as noted in [31]. This article compares various Runge-Kutta methods where constant step size symplectic methods outperformed adaptive symplectic and non-symplectic schemes. Symplectic integration with adaptive time step was presented in [81]. Therein, the reason for the bad performance is identified as a violation of the symplectic form. Adaptation of the time step introduces the time as an additional coordinate for which symplecticity is not preserved in general. Symplecticity is restored by additive terms which may lead to iterative schemes even for explicit methods.

In [189] an explicit symplectic variable time step method is obtained through replacing the Hamiltonian by another energy function which includes time and

## 1.2. Related approaches

energy as additional coordinates and momentum. The latter has been applied to the N-body problem and cometary orbits [54, 190]. Generalized versions of Mikkola's method are provided in [20, 214].

Adaptive step sizes with arc-length parametrization as in [81] are realized through the Adaptive Verlet method [100] which is explicit and time-reversible, but not symplectic. An asymptotic error analysis giving hints on step size selection is presented in [37].

One of the first variational approaches to variable time steps is presented in [237] where the implicit midpoint rule is combined with variable time steps which enforce energy preservation. The same method is derived by [118] as a variational integrator. A generic scheme for deriving variational symplectic-energy-momentum schemes is given in [153]. These schemes may lead to unsolvable equations. An approach to eliminate this problem was presented in [238] for the implicit midpoint rule involving additional inequality constraints. An alternative approach to variational energy-symplectic integrators is presented in [28] where the integration rule is changed in every time step in order to preserve the energy.

### Error analysis

Backward error analysis [80, 86, 242] is a useful tool providing deeper insight into symplectic geometry. A simplified methodology was presented in [225]. Backward error analysis was compared with global error expansions in [85] and found to be more meaningful with respect to long term stability, energy drift, etc. Asymptotic expansions and backward error analysis were used in [32] to prove that symplectic one-step methods are more accurate than others. It was applied to explicit multiple time stepping [161] in order to give hints on suitable time step sizes. Backward error analysis was used in conjunction with asymptotic expansions to explain the numerical stability of the adaptive Verlet method [37]. [244] analyses stability, provides an error analysis and gives closed-form solutions of shadow Hamiltonians for selected symplectic integrators. A methodology to approximate shadow Hamiltonians in general systems is provided in [243] and improved in [55] being able to filter energy oscillations due to the discrete nature of the numerical method and, therefore, to monitor energy drifts arising from numerical instabilities.

### Stability

Linear stability analysis of numerical integrators is often based on eigenvalue analysis of the constant propagation matrix and can be found in standard text books, for example [8].

The methodology can be used to prove linear stability of multi time stepping algorithms [43, 65, 232, 247]. Multi time stepping algorithms are subject to resonances which lead to energy drifts and, therefore, long-term instability. Those were first reported in [17] where nonsymplectic multi time stepping methods were shown to exhibit bad long-term accuracy and stability, but their symplectic counterparts were shown to exhibit stronger instabilities due to resonances. [18] analyzes instabilities in r-RESPA which were identified to appear in narrow time step intervals such that time steps being slightly longer than a critical step size render the scheme stable. Stability was analyzed using eigenvalue analysis combined with backward error analysis in [232]. Stable and unstable intervals were identified using the shadow Hamiltonian. Thresholds were given for selected schemes. The same methodology was applied to linear stability analysis of asynchronous variational integrators in [65].

Backward error analysis can be used to analyze the stability of time stepping schemes when applied to nonlinear problems. Stability may be defined by compactness of the shadow Hamiltonian which was exemplified for different integrators in [227]. An alternative approach was presented earlier in [245] using KAM theory to analyze nonlinear stability of members of the Newmark family.

A basic concept of stability analysis is Lyapunov's theory [181]. It provides quantities to measure the stability of a trajectory subject to small perturbations. Applications are, for example, given in [27, 121, 134, 198].

**Holonomic constraints**

In Lagrangian dynamics, one of the first methods to deal with nonlinear holonomic constraints is SHAKE [229] in combination with Verlet. The constraint discretization of SHAKE was shown to be variational in [185, 266]. The application of the variational principle gives rise to alternative discretizations. An enforcement at the midpoint is presented in [152] and applied to rigid body motion and contact of rigid bodies. Spurious high frequency modes were observed in case of ill-conditioned mass matrices and were damped out. Furthermore, spurious oscillations in the end-point coordinates of each time step were observed by [115] which grow over time and render the scheme unstable.

In Hamiltonian dynamics, RATTLE [3] was applied to Velocity Verlet with nonlinear holonomic constraints. The scheme was later shown to be equivalent to SHAKE and to be symplectic [150]. An application to the dynamics of rigid bodies was presented in [6]. A modification of RATTLE which is more efficient, but not symplectic, was presented by [82]. RATTLE was embedded into the concept of variational integrators [185]. A backward error analysis and approximate shadow Hamiltonian are given in [88]. In [159] the variational discretizations of constraints are extended to the Legrende transformation and the discrete Nullspace method is introduced which reduces the number of variables by projection onto the constraint manifold.

## 1.2. Related approaches

**Unilateral constraints and collisions**

Unilateral constraints often arise in contact/impact problems where an impenetrability condition [239] must be satisfied which is represented by an inequality equation. There are generally two approaches to dynamic discretization of the constraint: enforcing impenetrability at discrete points in time or enforcing the time derivative of impenetrability (persistency condition) being zero.

The enforcement of the persistency condition is known as Laursen-Chawla algorithm [142]. Therein, an analysis of the generalized-alpha and Newmark methods leads to the observation that energy conservation is tied to the persistency condition. An application to the augmented Lagrangian and penalty method is presented which are either conservative or dissipative and allow small penetrations. The algorithm was later combined with the impenetrability condition whereby energy conservation is restored by an additional velocity update [143, 179]. A variational treatment based on a discrete version of the classical principle of Hamilton is given by [58] who introduce the collision time as additional degree of freedom in explicit integrators and perform a velocity jump which is equivalent to enforcing the persistency condition. This leads to a non-smooth trajectory. The approach was further simplified in [36] introducing Decomposition Contact Response which is a non-iterative treatment at discrete points in time extending to inelastic and frictional contact problems. The enforcement of the persistency condition is related to variational inequalities [193, 194] involving a different treatment of Hamilton's principle. Extensions of Hamilton's principle to variational inequalities are presented in [72, 151].

The enforcement of the impenetrability condition is generally solved using Lagrange multipliers [34] or the penalty method [275]. Issues were reported regarding energy conservation [34] and spurious oscillations in surface velocities when applied to trapezoidal Newmark. The generalized alpha method was used to damp the spurious frequencies arising in frictional and non-frictional contact problems in [39,40]. An alternative approach to eliminating spurious oscillations is based on additional projections [48]. A similar modification to Newmark's method was presented earlier in [120, 213] where instabilities due to non-smooth geometries are eliminated. Energy preservation and oscillating post-contact velocities are identified as improper treatment of inertia in [125] proposing a redistribution of mass on the contact surface to overcome these problems.

### 1.2.2 Continuous assumed gradient elements

Nonlinear structural analysis using the finite element method frequently involves isoparametric low-order continuum element types. The design of such elements typically aims at considering numerical efficiency, accuracy and locking behavior, in particular volumetric or incompressible locking which led to various modifications. In the recent years, assumed $C0$-continuous fields of strain quantities have been investigated. By using these techniques, most of the mentioned design constraints can be addressed.

**Numerical efficiency**

In static and implicit analysis the numerical effort is closely related the factorization of the stiffness matrix. The structure of the stiffness matrix, i.e. band structure and bandwidth, are decisive for the duration of its factorization. In explicit dynamic analysis the only matrix to be factorized is constant and narrow banded (often diagonal). Then the numerical costs mainly depend on the effort for computing the strain energy and restoring force vector which in turn is proportional to the number of integration points. Therefore, reducing the number of integration points can improve the numerical efficiency of the method. Basically, there exist two approaches: Either reduced order numerical integration of the strain energy [223], or integration points which are shared by more than one element. For example, the integration points can be located in the finite element nodes involving nodal averages of the strains in order to compute a unique nodal value of the strain energy and its derivatives [22]. Both approaches to numerical integration yield an overly-soft behavior: Reduced order integration leads to non-physical zero-energy modes (hourglassing) and nodal integration leads to spurious low-energy modes. Therefore, both integration methods must be stabilized.

**Accuracy**

It is well known that standard finite elements tend to be too stiff. Besides mesh refinement and application of constant strain operators [51, 164], it can be shown that this effect can be reduced by eliminating the discontinuity of the strains among finite element interfaces. By application of nodal averages of strains or of the deformation gradient several authors were able to improve the accuracy of finite elements [22, 133, 280].

**Incompressible locking**

Standard finite elements applied to the analysis of (nearly) incompressible media tend to artificially stiffen the structure. Incompressibility appears as a constraint on the available deformation modes. The result is a response that is exceedingly stiff encountering volumetric locking. Among a lot of approaches to treat volumetric locking are reduced integration, mixed variational and projection methods, mixed displacement/pressure formulations, assumed strain, incompatible-mode, enhanced assumed strain methods, etc.

Improvements to standard elements can be obtained by using mixed formulations with interpolation of two or more field variables being treated individually, i.e.

## 1.2. Related approaches

interpolations of pressure and displacements. When dealing with mixed formulations, the order of approximation of pressure and displacement cannot be chosen arbitrarily. It must satisfy the LBB condition (inf-sup condition) to ensure stability and optimal convergence or must be used together with stabilization methods. Some of the stabilization techniques are based on hourglass control in reduced integration, see [13, 60, 222, 223].

A different approach to handle incompressibility constraints is the incompatible mode technique proposed by Wilson [271] for quadrilateral elements and later generalized to the assumed strain method (EAS) [241]. The basic idea is to add additional degrees of freedom to the element, i.e. displacement degrees of freedom in case of incompatible modes, for example bubble modes or some discontinuous function, or strain degrees of freedom in case of EAS. The method exhibited hourglass instabilities in some cases, but a few improvements were proposed to overcome these effects.

Pure displacement formulations go back to reduced and selective integration methods and the B-bar projection technique [108], wherein different interpolations for the volumetric and deviatoric components of deformation are applied. For some of these methods equivalence with mixed methods could be proved [103]. Interpreted in terms of a projection technique [108] one can include within the same B-bar framework the mean dilatational formulation [197]. The B-bar method was extended to nonlinear elasticity in [192]. For large deformations, the so-called $\bar{F}$ scheme has been developed in [45, 107, 240]. Its idea is based on the B-bar method which is directly applied to the deformation gradient instead of the symmetric strain tensor. It involves a product decomposition into volumetric and deviatoric components of the deformation gradient.

A specific assumed strain method is the nodal-averaged pressure formulation for tetrahedra [21]. The idea was extended to uniform-strain tetrahedral elements which correspond to displacement-based nodal-average operators for the deformation gradient [22, 51, 71, 116]. Nodal integration was generalized to other finite element types [133, 169] and found to eliminate incompressible locking. A simple explanation for this behavior is that the number of incompressibility constraints is related to the number of integration points and, in the case of nodal integration, even coincides with the number of nodes. The number of independent incompressibility constraints and the degrees of freedom must be balanced. This is usually not satisfied by isoparametric finite elements and, therefore, leads to artificial stiffening effects on isochoric deformation shapes.

**Nodal integration**

Nodal integration of the strain energy originates from the class of meshfree methods. Using smooth shape functions nodal strain measures are naturally given and can be used for a nodal integration rule. However, it suffers from spurious zero energy modes. The matter of singular modes is addressed by adding an artificial potential function to the strain energy that penalizes the norm of the pointwise stress residual being integrated over the volume covered by the described material [9]. The pointwise stress residual is measured as the error of the strong equilibrium conditions. The first nodal integrated approach in FEM is found in [22]. Therein, a tetrahedral element is nodally integrated. Earlier works were interested in reducing volumetric locking by applying nodally averaged pressure fields or averaged determinants of the deformation gradient [21, 51, 116]. The deformation gradient itself was averaged at the nodes according to the relative volume of the surrounding elements [22]. The authors found the new elements to be very efficient, but observed nonphysical low-energy modes, though they regarded these effects as little important since they appeared in a very limited number of use cases. In [22], it is suggested to stabilize the elements by using information on the deformation gradient from the last time step in explicit simulations. The stabilization of spurious modes was discussed in detail [217]. The suggested stabilization method adds an artificial energy to the strain energy, similar to [9]. The new aspect of the contribution was that the modified tetrahedral elements exactly behave like linear elastic finite elements with standard Gaussian quadrature for small strains. This is done by penalizing the difference of the averaged nodal strain and the natural strain. A norm of this difference can be obtained by assembling an energy using a constant material tensor. This stabilization scheme was generalized and applied to nodal integration of meshfree methods [215].

Using nodal integration, an assumed strain method is derived using the method of weighted residuals [133]. The approach consistently extends to tetrahedral and hexahedral finite elements of linear and quadratic shape functions. Volumetric locking is eliminated and illustrated by examples. The authors do not mention instabilities, but for linear tetrahedral elements it corresponds to the method of [22]. The weighted residual approach was later discussed in detail [26]. Extensions to higher order hexahedra are presented and the appearance of spurious modes eliminated using a penalty method.

**Smoothed finite element method**

A related approach is the smoothed finite element method (SFEM). Herein, the integration domain of each finite element is subdivided into smoothing cells. For each cell, a constant strain tensor is computed as a "smoothed" representation of the strain field.

SFEM is a variant of point interpolation methods (PIM). A strain-smoothing technique applied to meshfree method was used to stabilize in nodal integration [114]. The technique was extended to a generalized gradient smoothing technique allowing discontinuous functions [162]. The approach provides a theoretical foundation to the so-called weakened weak formulation [163]. This formulation includes linearly conforming point interpolation methods (LC-PIM) [175, 279], the

## 1.2. Related approaches

linearly conforming radial point interpolation method [165] and the least-square fitted point interpolation method PIM-LSS [276] that use incompatible shape functions. When applying the same ideas to FEM, one obtains the element-based smoothed FEM (SFEM) [164, 166, 170]. The SFEM has been applied to two-dimensional $n$-sided polygonal elements [42] and extended for plate and shell analysis [202, 205].

Nodal integration was successfully implemented in this framework (NS-FEM) [169]. In case of linear triangles and tetrahedra, NS-FEM is identical with the methods of [22, 51] or the LC-PIM [165] using linear interpolation. A penalty method to stabilize NS-FEM was presented [280].

LC-PIM provides an upper-bound to the exact solution of the strain energy as shown numerically [173]. Both, upper and lower bounds in the strain energy can be obtained by combining the 'overly-soft' NS-FEM with 'overly-stiff' FEM or SFEM. The overly-soft nature of NS-FEM was balanced by superposing the FEM solution yielding the $\alpha$-FEM [168]. Therein, $\alpha$ is a problem-dependent parameter which scales the contributions of NS-FEM and FEM, respectively. $\alpha$ can be chosen such that the strain energy error norm becomes zero. There exists, however, no verification if the optimally chosen parameter reintroduces spurious non-zero energy modes of NS-FEM. The idea of superposing two solutions was, however, further applied to SFEM [171] and LC-PIM [172].

An edge-based smoothed FEM (ES-FEM) was proposed for two-dimensional problems [167]. The ES-FEM uses triangular elements. The stiffness matrix is computed using strains averaged over smoothing cells associated with the triangle edges. The construction of smoothing cells was extended to quadrilateral elements [204]. ES-FEM was later extended to three-dimensional elasticity problems [93, 203]. Therein, the smoothing cells are created around the faces of linear tetrahedra. The two-dimensional ES-FEM exhibits very good properties: (1) solutions are between the bounds given by NS-FEM and FEM. (2) The results are much more accurate than those of FEM. (3) There exist no spurious non-zero energy modes. (4) Numerical efficiency is better than FEM when using the same set of nodes. The same is true when applied to three dimensions.

### 1.2.3 Contact

The literature on algorithms treating contact-impact problems is numerous. An attempt to a categorizing overview on the variety of procedures is given by [201] citing more than 600 publications. At this point, only those being related to the contact formulation used in this thesis shall be named.

**Distance fields**

An alternative approach to contact detection relies on the evaluation of distance fields. Distance fields provide an implicit representation of the closest point projection. Algorithms regarding distance fields go back to the level set equation. The level set method was presented by [209] who described the temporal propagation of moving interfaces by numerical methods solving the Hamilton-Jacobi equation. This is performed by a finite difference scheme working on a rectangular grid in two or three dimensions. Information on normal vectors and curvature can be obtained. The fast marching method [234] provides an efficient numerical scheme of complexity $n \log n$. The algorithm computing the levels is a reinterpretation of the propagation process, i.e. the time where the interface passes a certain grid point is influenced only by those neighboring grid points which are previously passed by the interface. An overview on the theory of level set and fast marching methods and their applications to problems of various areas are given in [233,235], for example shape offsetting, computing distances, photolithography development, seismic travel times, etc. Distance fields are a special case of the level set equation where the absolute value of the advection velocity is 1.

The concept of distance fields was introduced to contact problems in [96] using first order tetrahedral meshes. The distance field is generated on a supplementary grid and evaluated at the finite element nodes. Simplicity and robustness compared with closest point projection is emphasized, in particular not needed smoothness conditions on the shape of the boundary. Self-contact, large deformations and deep interpenetrations may be treated. Exact intersection polygons are determined on which contact forces are computed by the penalty method. More details on the employment of the distance field are provided in [59]. It focuses on the precomputation of the distance field by fast marching. A simple partial update strategy during a time integration is proposed for regions where intersections actually occur. More details of the approach are presented in [95].

A supplementary grid is not required if the distance field is interpolated on the finite element mesh. This is constricted by the lack of efficient level set methods on unstructured meshes. A fast marching method is adopted to acute triangle meshes in [7]. The basic problem are instabilities which arise by propagating approximate levels along arbitrarily changing directions. Instead of propagating the approximate distance, [186] computes accurate distances of grid points to the initial interface, but propagates a reference to the surface patches to which the closest point projection refers to. The idea was adopted to tetrahedral finite element meshes [184] with application to collision detection eliminating the supplementary grid in [59]. A partial distance field update strategy is provided for simplex meshes. Although not related to distance fields, a partial update strategy is presented by [94] which improves the robustness of closest point projection approximation in two dimensions. By identifying the actually intersected boundary as initial interface, the distances will only be computed with respect to the

## 1.2. Related approaches 23

selected surface patches and the partial update is restricted to the finite element nodes which are actually in contact. The approximate distances and normal vectors are propagated similar to the original fast marching approach.

In contact algorithms based on closest point projection, it is not uncommon to assume the existence of normal vectors being averaged on the boundary nodes. This strategy was proposed [264] to improve the robustness and to simplify implementations. A continuous normal vector field is created being equivalent to a smooth gap function. The deadzone problem at corners and edges on the boundary is eliminated and the number of iterations is reduced. It still needs the creation of a halo, i.e. a volume around a surface segment, during the global search phase. Spatial discretization methods of the contact tractions were developed using the concept of averaged normals, for example the mortar method implementing a segment-to-segment integration for nonconfoming domains [218]. In the context of collision detection based on distance fields, the approach of assumed distance gradients is very similar to the concept of averaged normal vectors. Assumed distance gradients were developed [196] to eliminate spurious oscillations in levelset methods and were shown to be sufficiently accurate when applied to triangular and quadrilateral meshes.

**Mortar method**

The mortar method was originally developed for weakly coupling fields in non-conforming domain decomposition problems [16]. The optimality of the spatial discretization is proven, where optimality contends a solution error limited to the sum of the approximation errors of each bonded domain. The mortar method was applied in the context of contact problems [187] where the contact tractions must be mapped onto the finite element mesh boundary being nonconform with the frame where the contact constraints are enforced. [144] briefly discusses issues of the implementation including aspects of numerical integration and curved interfaces. These topics are explained in detail in [218] where the mortar method is applied to contact in large-deformation analysis without the need of an intermediate surface. An extension to frictional problems with a focus on a non-locking glide path estimation and its linearization are presented in [219]. The implementation of friction contact is further explained in [145] and summarized by [216]. The ideas are extended to an interelementary contact interface [127] which is represented by enriching the finite elements' shape function space.

Mortar methods generate discrete constraint equations which can be used in conjunction with Lagrange multipliers or, due to the specific structure of the mortar conditions, can be used to directly eliminate degrees of freedom. This involves the inversion of a sparse square matrix. By the introduction of dual Lagrange multiplier spaces [272] this matrix becomes diagonal. Dual multiplier spaces were employed in conjunction with multigrid solvers to improve the efficiency of the numerical solution [129]. [211] presents dual shape function spaces for linear and quadratic elements. Example shape functions in two and three dimensions for surface and volume interpolations and stability conditions are presented in [126]. A discussion on quadratic dual shape functions is presented in [138] leading to the result that a simple strategy to construct stable and optimal dual higher order function spaces is hard to find. As a result, an optimal quasi-dual function space is introduced [139] being not diagonal, but still simplifies the inversion of the quadratic mortar matrix. Dual Lagrange multiplier spaces combined with an algebraic multigrid solver as solution procedure are successfully applied to contact problems in [102]. The approach is extended to contact of thin-walled structures with shell elements [92]. The same article presents a generic algorithm to construct linear dual shape function spaces. Dual multiplier spaces may lead to hourglass-like solutions in presence of curved interfaces. Stable schemes for curved quadrilateral elements are presented in [62].

The application of the mortar method involves the choice of a domain in which the Lagrange multipliers must be interpolated. Various approaches exist on the choice of this surface which in turn serves as domain for numerical integration. An intermediate surface is employed in [187]. The mortar segments are constructed in two dimensions by projecting the finite element nodes of both surfaces onto the other. An intermediate surface with a discretization that is completely independent of the finite element meshes being in contact and fixed in time is presented for two dimensions [221]. The approach was extended to three dimensions [220]. The same article proposes a contact patch test in 3D. Details on the structure of equations using this approach in frictional contact are presented in [73]. When interpreted in terms of the mortar method, the involving localized Lagrange multipliers are, however, not interpolated by continuous functions, but are applied to individual discrete points on the interface.

## 1.3 Contributions and outline

The major motivation behind this thesis is to deal with algorithms that improve the efficiency of explicit simulation of collision dynamics. The philosophy behind asynchronous schemes, which adopt the numerical expenses to spatially local properties such as mesh fineness, is an elegant concept in the author's optinion. By chosing the numerical effort to be minimal per spatial domain it may help to reduce numerical cost while obtaining a reasonable good accuracy. This is, for example, in contrast to model order reduction techniques which may fail to represent important components of the response. The concept of explicit asynchronous integration interacts with all algorithmic parts which treat a dynamic contact problem, i.e. the finite element formulation, the time stepping scheme, the treatment of constraints, the contact detection and contact solution.

When the thesis was started, the original idea behind an optimal combination of asynchronous integration with finite elements was to minimize the number of

## 1.3. Contributions and outline

stress evaluations by nodal integration in space and time. Within this concept, the stresses would be evaluated at the finite element nodes only. The time stepping scheme would propagate an asynchronous space-time front through the time where the integration domains are defined such that only a single finite element node is propagated at one time step. When implementing nodal integration, however, it turns out to be unstable. As a result, a stable modification must be found which in turn affects the formulation of the asynchronous integrator.

Nodal integration of low-order finite elements is analyzed in depth. The target is the identification of the reasons behind the instabilities. Helpful is the comparison of existing approaches to stabilization which are based on penalty methods. This work introduces a continuous assumed interpolation of the deformation gradient inducing continuous strains and stresses. Nodal integration is interpreted in terms of the assumed gradient field. The result of the analysis is that the interpolation of the deformation gradient by nodal averages is not suitable to approximate the natural spatial distribution of strains. The polynomial degree of the interpolation must be increased which leads to additional support points in the finite elements' interior. The choice of the number and location of the new supports is trivial for one-dimensional elements, but difficult for higher dimensions. The number and position of additional support points and the associated polynomials must be balanced with numerical efficiency: A minimal number of support points should be chosen in order to be stable and efficient while an even larger number may significantly affect the accuracy. Interpolation schemes are proposed for first order, two and three dimensional isoparametric finite elements. Furthermore, a strategy is proposed how interpolation functions can be derived for arbitrary finite element types. This is demonstrated by the isoparametric 10-noded tetrahedron. The derived elements can be interpreted as a new methodology to finite elements based on assumed strains. The discussion of the new method is extended to requirements on mesh generation. Compared with FEM, additional conditions must be satisfied if different finite element types are present. The effects of a violation of these conditions are explained sketchily. Numerous examples from static elasticity and dynamics verify the new method with respect to stability, accuracy, convergence and numerical efficiency.

The foundation of asynchronous integration are variational integrators. They preserve certain invariants of the continuous system. As a consequence, methods which are systematically derived from this principle are often better than time stepping schemes which are derived by less strict rules and, therefore, do not preserve important invariants. The knowledge of most engineers regarding properties of conservative systems is often restricted to the balance of energy and momentum. To begin with, continuum dynamics of conservative mechanical systems is presented by means of the principle of Hamilton. The explanation includes the definition of invariants and Noether's theorem, the preservation of linear and angular momentum, energy and the symplectic form. The exposure follows standard text books, but adopts the nomenclature used by variational integrators.

The derivation of variational integrators will be explained using different existing approaches, i.e. Lagrangian dynamics versus Hamiltonian dynamics by Legrende transformation and the Hamilton-Pontryagin principle. It is followed by a summary of explicit, symplectic-momentum preserving, one-step integrators which can be interpreted being variational. Subsequently, the presented time stepping schemes are compared by example problems. The target of the comparison is to find out, how these methods can be used to improve numerical efficiency by increasing the critical time step in structural dynamics assuming certain conditions. Stability is measured in terms of Lyapunov's condition. By combination and reinterpretation of existing methods, a new time stepping scheme (the mollified implicit-explicit integrator) is found which obeys very large stability intervals for systems with dominant linear response and rather small nonlinearities. This is verified using model problems from structural dynamics. The stabilization implicates the replacement of the original discrete model by another which becomes unneglibably inaccurate in case of rather large nonlinear effects and large time steps. As a consequence, it was not possible to incorporate it within an asynchronous procedure and is not employed in later chapters.

The presentation of variational integrators includes the treatment of nonlinear constraint equations. A correct and efficient treatment is important when collisions and restraints are incorporated into the asynchronous procedure. Existing methods will be presented in the variational context. Examples compare the robustness of various approaches. They also explain the appearance of spurious oscillations in the implicit midpoint scheme and trapezodial Newmark when applied to nonlinear constraints. The collision integrator for explicit treatment of inequality constraints is then applied as a standalone procedure to nonlinear holonomic constraints and as a stabilization to the midpoint scheme with promising results.

Asynchronous integration is presented in a general form, which is based on a generalized potential energy and which includes other multiscale methods as special cases. The formulation is then applied to structural elasticity. Previous approaches deal with isoparametric finite elements. Here, the asynchronous space-time discretization focuses on finite element formulations with continuous properties, for example the new finite element formulation based on continuous assumed gradients. As a result, the finite elements can not be considered individually. An elemental stiffness matrix does not exist and, therefore, elemental natural frequencies and wave speeds can not be obtained easily. Thus, an estimation of the critical time step can not be found using classical methods. Nevertheless, two approaches are developed based on the CFL condition which relates the critical time step to the wave propagation speed. Only one is found to be robust in the examples. Although previous approaches to asynchronous integration include models with fixed nodes, a consideration of general nodal restraints and their efficient numerical treatment is missing. This is provided in this thesis. The

## 1.3. Contributions and outline

consistent treatment of nodal restraint conditions becomes important when addtional constraints arise, for example in contact situations.

An attempt to improve accuracy and stability of time stepping schemes are variable time steps. Therein, the time step size is adopted during the simulation. Two existing approaches are presented, symplectic-energy-momentum integration and time transformations. The latter introduces a fictitious time variable to which a fixed step size method can be applied. This is usually done by replacing the Hamiltonian. With an application to asynchronous discretizations in mind, the time transformations will be applied directly to the discrete principle of Hamilton in this work. The principle of Hamilton-Pontryagin can be used to simplify the equations by eliminating the time step length from denominators. The strategy is further generalized to include synchronous and asynchronous integration schemes. New limitations on the choice of suitable time step functions become obvious. The discretization leads to local asynchronous systems of equations which are linear and explicit in most variables, but quadratic with respect to the time step size. An efficient solution procedure is developed. Examples illustrate fields of applications in principle. Single degree of freedom systems serve to verify stability, convergence, efficiency and accuracy. When relating the accuracy to the computing time, no benefit can be identified when applied to linear oscillators. Applied to asynchronous elasticity of finite element structures, no convergence can be found.

In the final part, the thesis presents asynchronous collision integrators. It starts with the presentation of contact mechanics of a continuum following standard text books. It further explains existing methods of collision detection based on spatial partitioning and discretization methods of the contacting boundaries, in particular the mortar method. The algorithmic treatment requires a simple and robust procedure in order to perform an accurate local contact detection and computation of the closest point projection. The implementation is based on distance fields. Previous approaches to contact algorithms using distance fields are limited to penalty methods, first order tetrahedron finite elements and level sets being interpolated on a rectangular grid. The idea of discrete distances is extended to arbitrary interpolations based on the existing finite element geometry. A consistent framework is created which replaces the contact constraints arising from closest point projection. The discussion extends to possible instabilities and their elimination. Finally, the presented algorithms regarding contact search, distance fields, collision integrators and restraints are combined and applied to asynchronous integration. The contact laws under consideration are elastic normal contact, inelastic normal contact and Coulomb friction. The advantage of asynchronous collision treatment is that very small systems of equations are to be solved such that the contact conditions can be enforced efficiently. In opposite to penalty methods, the used collision integrator does not affect the critical time step. Hence, a time step adaptation only based on local accuracy conditions is possible.

The algorithms and examples presented in this thesis are implemented in the software package SLangTNG [30]. Its source code is distributed under the BSD license.

This work is organized is as follows: This chapter contains a discussion of related research in the fields of variational integration, temporal discretization of constraints, finite element formulation, distance fields applied to contact and mortar method. The related bibliography with respect to spatial contact algorithms is presented in the respective sections. Chapter 2 presents the principle of Hamilton and its properties. Section 3 explains variational integrators, it presents various numerical schemes and the treatment of constraints. The continuous assumed gradient method is derived in chapter 4. Asynchronous variational integration is illustrated in section 5 being applied to continuous assumed gradient elements and nodal restraints. Variable time steps are studied in chapter 6. Section 7 presents the application of collision integrators in the asynchronous context including details regarding contact mechanics, distance fields and collision detection.

# Chapter 2

# Variational mechanics

## 2.1 Introduction

This section presents the classical principle of Hamilton [90]. Using variational calculus, it can be used to describe the solution trajectory of conservative systems and to prove properties of such systems, i.e. preservation of energy, momentum and the symplectic form.

The aim of this section is two-fold: First, the reader should understand the theoretical foundations of variational dynamics and the mathematical beauty behind Hamilton's variational principle. This includes the derivation of the equation of motion in the sense of Euler and Lagrange (1755; second order ordinary differential equation) and of Hamilton (1833; two first order partial differential equations). Hamiltonian systems preserve various invariants, but only a few, i.e. energy and momentum, are known to most engineers. The matter of invariants is explained through definitions and proofs. The latter are presented in a form which only needs knowledge on matrix algebra and variational calculus for understanding. It should be noted that differential geometry and Lie algebra are common to simplify notation and proofs.

As a second intent, the connection to the derivation of numerical methods shall be understood. Although the general formulation of Hamilton's principle is discussed, it is expressed in terms of discrete time intervals which could be interpreted in terms of finite elements in time and may serve as a basic tool for discretization. To derive numerical methods, the equations of motion were often used as a basis. Then standard methods for the solution of differential equations can be applied. Interestingly, only a few of them exhibit good accuracy and long-term stability. These properties can be regarded to invariants being preserved (or not preserved) by the original system and the discretized model.

The subsequent paragraphs follow books and lecture notes given in [84, 86, 97, 242] including theorems and proofs. For the separation into discrete time intervals the notation of [153] is used, more details in [137, 267].

## 2.2 Preliminaries

**Theorem 1.** *For holonomic constraints and forces, which may be derived from a potential function $V(q, \dot{q}, t)$, one defines the Lagrange function*

$$L(q, \dot{q}, t) := T - V, \quad T = \frac{1}{2} \dot{q}^T M \dot{q} \tag{2.1}$$

*where $T$ is the sum of the kinetic energies of all system particles and $V$ the sum of all potential energies of the particles in the system. The Lagrange function is a function of the generalized coordinates $q$, their time derivatives $\dot{q}$ and of the time.*

**Theorem 2.** *The action $S$ is defined as the integral of the Lagrange function over the time interval $[a, b]$*

$$S = \int_a^b L(q, \dot{q}, t) dt \tag{2.2}$$

**Theorem 3.** *The solution trajectories $q(t)$ of a dynamical system given through a Lagrange function $L$ are determined as stationary points of the action $S$, i.e.*

$$\delta S = 0 \tag{2.3}$$

*which is Hamilton's principle.*

**Theorem 4.** *A coordinate $q_i$ is called cyclic, if only its time derivative is contained in the Lagrangian function, i.e. if*

$$\frac{\partial}{\partial q_i} L(q, \dot{q}, t) = 0 \tag{2.4}$$

**Theorem 5.** *The canonical momentum (generalized momentum) $p_i$ to a generalized coordinate $q_i$ is defined as*

$$p_i := \frac{\partial L(q, \dot{q}, t)}{\partial \dot{q}} \tag{2.5}$$

## 2.3 Principle of Hamilton

Without loss of generality, the action integral is written as a sequence of mutually exclusive finite time intervals,

$$S = \sum_{k=0}^{N-1} \int_{t_k}^{t_{k+1}} L(q, \dot{q}, t) dt \tag{2.6}$$

This formulation allows the application of the stationarity principle to the exact - and - to numerical solutions. It is based on the concept of generalized variational integrators presented by [153]. Therein, the chosen intervals denote finite time elements. Each time element is defined by some time increment length $h^k$, where a local time coordinate $t^k \in [0 \ldots h^k]$ is used. Boundary conditions are given through $q_0^k$ and $q_1^k$ denoting the generalized coordinates at the beginning and the

## 2.3. Principle of Hamilton

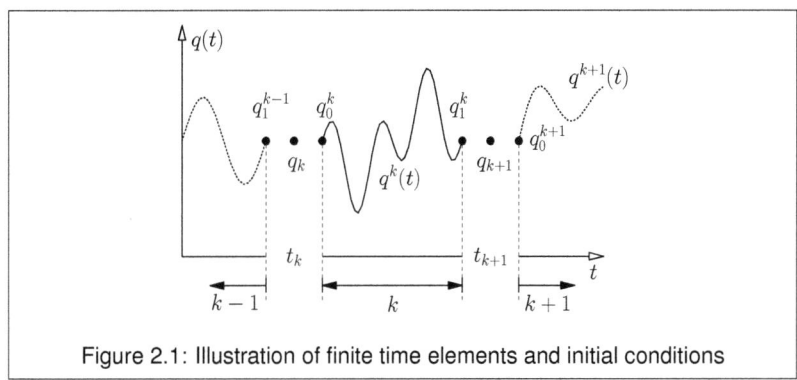

Figure 2.1: Illustration of finite time elements and initial conditions

end of the element, see figure 2.1. The discrete Lagrangian of the $k$th element is introduced,

$$L_d^k = \int_0^{h^k} L(q^k, \dot{q}^k, t) dt \qquad (2.7)$$

with Lagrange function $L$. Then Hamilton's principle of least action declares the stationarity of the $k$th finite element action

$$S^k = L_d^k + j_0^k(q^k(0) - q_0^k) + j_1^k(q_1^k - q^k(h^k)) \qquad (2.8)$$

inserted in the sum of equation (2.6). The quantities $j_0^k$ and $j_1^k$ are Lagrange multipliers which enforce the boundary conditions $q_0^k = q^k(0)$ and $q_1^k = q^k(h^k)$.

The elemental action integrals will be assembled in the global action $S$. Since each time element is adjacent to others, $C^0$-continuity of the generalized coordinates must be ensured by inter-element conditions. Therefore, global Lagrange multipliers $j_{k-}$ and $j_{k+}$ are introduced in order to enforce the local boundary conditions, i.e.

$$S = \sum_{k=0}^{N-1} S^k + \sum_{k=1}^{N} j_{k-}(q_k - q_1^{k-1}) + \sum_{k=0}^{N-1} j_{k+}(q_0^k - q_k) \qquad (2.9)$$

This action sum adds up all contributions of the time elements and some global constraints, whereby $q_k$ are the vectors of generalized coordinates $q$ at the nodes $k$. $j$ are Lagrange multipliers which enforce $C^0$-continuity between the elements. The multiplier $j_{k-}$ enforces the coordinates of the right element border of the element $(k-1)$ left from node $k$ being identical with the nodal coordinates at node $k$. The multiplier $j_{k+}$ enforces the coordinates of the left element edge of element $(k)$ being identical to the nodal coordinates of node $k$.

The total variation of the global action sum yields $q_k = q_1^{k-1} = q_0^k$ and $j_{k-} = j_{k+} = j_1^{k-1} = j_0^k$. By introducing the global constraints, no additional knowledge is obtained. But it allows the individual consideration of the local action integrals of each element. From now on, the index of the element $k$ will be omitted when considering a single time element $k$.

## 2.3.1 Lagrangian dynamics

Consider the action integral of a finite time element

$$L_d = \int_0^h L(q,\dot{q})dt, \quad S^k = L_d + j_0(q(0) - q_0) + j_1(q_1 - q(h)) \quad (2.10)$$

The variation of $L_d$ is performed using integration by parts, i.e.

$$\begin{aligned}\delta L_d &= \int_0^h (L_q \delta q + L_{\dot{q}} \delta \dot{q}) dt \\ &= \int_0^h (L_q - \frac{d}{dt}L_{\dot{q}}) \delta q dt + [L_{\dot{q}} \delta q]_0^h \end{aligned} \quad (2.11)$$

Inserting $\delta L_d$ into $\delta S^k$, one obtains the boundary conditions

$$\begin{aligned} 0 &= q(0) - q_0 \\ 0 &= q_1 - q(h) \\ 0 &= j_0 - L_{\dot{q}}(q(0), \dot{q}(0)) \\ 0 &= L_{\dot{q}}(q(h), \dot{q}(h)) - j_1 \end{aligned} \quad (2.12)$$

and the Euler-Lagrange equation

$$0 = L_q - \frac{d}{dt}L_{\dot{q}} \quad (2.13)$$

The Euler-Lagrange equation defines a system of $n$ differential equations of 2nd order and is used to set up the equations of motion. The $4n$ boundary conditions require $2n$ initial conditions and provide $2n$ conditions to determine the variables enforcing continuity to adjacent time elements. Typically, $(q_0, j_0)$ or $(q_0, q_1)$ are used as initial conditions.

## 2.3.2 Hamiltonian dynamics

The multipliers $j_0$ and $j_1$ may be interpreted in terms of momentum. To prove this, the action integral is transformed in terms of the Hamilton-Pontryagin principle. Herein, the time derivative $\dot{q}$ is replaced by the velocity function $v$; the identity of both quantities is enforced by Lagrange multipliers $p$, such that

$$L_d = \int_0^h (L(q,v) + p(\dot{q} - v)) dt \quad (2.14)$$

The variation of the local action with fixed end points $q_0$ and $q_1$ yields:

$$\begin{aligned} 0 &= \delta S^k \\ &= \delta(j_0(q(0) - q_0)) + \delta(j_1(q_1 - q(h))) \\ &\quad + \int_0^h ((L_v(q,v) - p)\delta v + (\dot{q} - v)\delta p + L_q(q,v)\delta q + p\delta \dot{q}) dt \end{aligned} \quad (2.15)$$

## 2.4. Preserved quantities

After applying integration by parts one obtains:

$$\begin{aligned}
0 =\ & \delta\left(j_0(q(0) - q_0)\right) + \delta\left(j_1(q_1 - q(h))\right) + [p\delta q]\big|_0^h \\
& + \int_0^h \left((L_v(q,v) - p)\delta v + (\dot{q} - v)\delta p + L_q(q,v)\delta q - \dot{p}\delta q\right) dt
\end{aligned} \quad (2.16)$$

$$\begin{aligned}
0 =\ & \delta j_0(q(0) - q_0) + \delta j_1(q_1 - q(h)) \\
& + \delta q(0)(j_0 - p(0)) + \delta q(h)(p(h) - j_1) \\
& + \int_0^h \left(\delta v(T_v(v) - p) + \delta p(\dot{q} - v) + \delta q(-V_q(q) - \dot{p})\right) dt
\end{aligned} \quad (2.17)$$

As a result, the Lagrange multiplier $p$ turns out to be the momentum of the system, defined by $p = T_v(v)$. The multipliers $j$, enforcing the continuity conditions between the time elements, are the momenta at the element's boundaries, i.e. $j_0 = p(0)$ and $j_1 = p(h)$. Another result are Hamilton's canonical equations. Defining the Hamiltonian by application of the Legrende transformation,

$$H(q,p) = pv - L(q,v) \leftrightarrow L_d = \int_0^h (p\dot{q} - H(q,p))\, dt \quad (2.18)$$

one obtains the variational principle

$$\begin{aligned}
0 =\ & \delta j_0(q(0) - q_0) + \delta j_1(q_1 - q(h)) \\
& + \delta q(0)(j_0 - p(0)) + \delta q(h)(p(h) - j_1) \\
& + \int_0^h \left(\delta p(\dot{q} - H_p) + \delta q(-H_q - \dot{p})\right) dt
\end{aligned} \quad (2.19)$$

and from that the Hamiltonian equations:

$$\dot{q} = H_p \quad (2.20)$$

$$\dot{p} = -H_q \quad (2.21)$$

## 2.4 Preserved quantities

**Theorem 6.** *A quantity $f$ is preserved (is a first integral or is a constant of motion) if the total derivative for the time of all solution trajectories is zero, i.e.*

$$\dot{f} = \frac{d}{dt} f(q, \dot{q}, t) = 0 \quad (2.22)$$

**Theorem 7.** *The conjugated momentum of a cyclic coordinate is preserved.*

*Proof.* This immediately follows from the Euler-Lagrange equation, i.e.

$$0 = \frac{d}{dt} L_{\dot{q}_i} - L_{q_i} = \frac{d}{dt} L_{\dot{q}_i} = \frac{d}{dt} p_i$$

□

## 2.4.1 Noether's theorem

**Theorem 8.** *Let a transformation of variables $q \to q'$ be diffeomorph, i.e. the transformation and its inverse transformation are differentiable and the determinant of the transformation matrix is not zero. Let $q'$ be dependent on a single scalar parameter $\epsilon$, i.e. $q \to q'_\epsilon(q)$ such that $q'_{\epsilon=0}(q) = q$. Then $q'(q)$ is called a symmetric transformation if*

$$L(q'_\epsilon(q), \dot{q}'_\epsilon(q), t) = L(q, \dot{q}, t) + \frac{d}{dt} f_\epsilon(q, t) \qquad (2.23)$$

**Theorem 9.** *Let $q'_\epsilon$ a single parametric symmetric transformation with $q'_{\epsilon=0}(q) = q$, then Noether's theorem [207] states that the quantity*

$$\mathcal{I}(q, \dot{q}, t) = \frac{\partial L}{\partial \dot{q}_i} \left. \frac{\partial q'_{\epsilon i}}{\partial \epsilon} \right|_{\epsilon=0} - \left. \frac{df_\epsilon}{d\epsilon} \right|_{\epsilon=0} \qquad (2.24)$$

*is preserved.*

---

*Proof.* Define $L = L(q, \dot{q}, t)$ and $L' = L(q'_\epsilon(q), \dot{q}'_\epsilon(q), t)$. First, it is proven that the transformed system has the identical Euler-Lagrange equation

$$0 = \frac{d}{dt} \frac{\partial L}{\partial \dot{q}_i} - \frac{\partial L}{\partial q_i} = \frac{d}{dt} \frac{\partial L'}{\partial \dot{q}'_i} - \frac{\partial L'}{\partial q'_i}$$

which is obtained by deriving the Euler-Lagrange equations individually for both coordinates from Hamilton's principle. Then the symmetry condition is used. Thus, first deriving using the chain rule, then applying the transformed Euler-Lagrange equation, applying the identity $q' = q|_{\epsilon=0}$ and finally setting new brackets,

$$\begin{aligned}
\left. \frac{\partial L'}{\partial \epsilon} \right|_{\epsilon=0} &= \left. \left( \frac{\partial L'}{\partial q'_i} \frac{dq'}{d\epsilon} + \frac{\partial L'}{\partial \dot{q}'_i} \frac{d\dot{q}'}{d\epsilon} \right) \right|_{\epsilon=0} \\
&= \left. \left( \frac{d}{dt} \frac{\partial L'}{\partial \dot{q}'_i} \frac{dq'}{d\epsilon} + \frac{\partial L'}{\partial \dot{q}'_i} \frac{d}{dt} \frac{dq'}{d\epsilon} \right) \right|_{\epsilon=0} \\
&= \left. \left( \frac{d}{dt} \frac{\partial L}{\partial \dot{q}_i} \frac{dq'}{d\epsilon} + \frac{\partial L}{\partial \dot{q}_i} \frac{d}{dt} \frac{dq'}{d\epsilon} \right) \right|_{\epsilon=0} \\
&= \left. \frac{d}{dt} \left( \frac{\partial L}{\partial \dot{q}_i} \frac{dq'}{d\epsilon} \right) \right|_{\epsilon=0} = \frac{d}{dt} \left( \left. \frac{\partial L}{\partial \dot{q}_i} \frac{dq'}{d\epsilon} \right|_{\epsilon=0} \right)
\end{aligned}$$

Additionally, the symmetry condition may be derived directly. Applying the identity $q' = q|_{\epsilon=0}$ one obtains

$$\begin{aligned}
\left. \frac{\partial}{\partial \epsilon} L' \right|_{\epsilon=0} &= \left. \frac{\partial}{\partial \epsilon} \left( L(q, \dot{q}, t) + \frac{d}{dt} f_\epsilon(q, t) \right) \right|_{\epsilon=0} \\
&= 0 + \left. \frac{d}{dt} \frac{\partial f_\epsilon(q, t)}{\partial \epsilon} \right|_{\epsilon=0}
\end{aligned}$$

From that,

$$\frac{d}{dt} \left( \left. \frac{\partial L}{\partial \dot{q}_i} \frac{dq'}{d\epsilon} \right|_{\epsilon=0} \right) = \left. \frac{d}{dt} \frac{\partial f_\epsilon(q, t)}{\partial \epsilon} \right|_{\epsilon=0}$$

and, therefore,

$$\frac{d}{dt} \mathcal{I}(q, \dot{q}, t) = 0$$

□

## 2.4.2 Conservation of linear momentum

Apply a translation $q'(t) = q(t) + \epsilon q_0$. The Lagrangian is invariant under translation, i.e. $L(q, \dot{q}, t) = L(q', \dot{q}', t)$ and $f_\epsilon(q,t) = 0$. Inserting the transformation into Noether's theorem, one obtains the preserved quantity

$$\mathcal{I} = \frac{\partial L}{\partial \dot{q}_i} \frac{dq'_i}{d\epsilon}\bigg|_{\epsilon=0} = p_i q_{0,i} \tag{2.25}$$

Therefore, the momentum $p$ in direction of $q_0$ is preserved. Since $q_0$ is arbitrary, the global momentum is preserved.

## 2.4.3 Conservation of angular momentum

Let $L$ be invariant under the rotation $q' = R_\epsilon q$ ($q'_i = R_{\epsilon,ik} q_k$) with rotation matrix $R$ and $f_\epsilon = 0$. Using Noether's theorem one obtains the preserved quantity

$$\mathcal{I} = \frac{\partial L}{\partial \dot{q}_i} \left( \frac{dR_{\epsilon,ik}}{d\epsilon} \bigg|_{\epsilon=0} \right) q_k \tag{2.26}$$

The derivative of the rotation matrix for the scalar parameter may be written in terms of the vector product of a linearized rotation transform given by $\omega_\epsilon$, i.e.

$$\mathcal{I} = p_i(\bar{\omega}_{\epsilon,ik} q_k) = p \cdot (\omega_\epsilon \times q) = \omega_\epsilon \cdot (q \times p) = \omega_\epsilon \cdot L \tag{2.27}$$

Therefore, the angular momentum $L$ is preserved for the given direction $\omega$. Since $R$, respectively, $\omega$ may be arbitrary, the angular momentum is preserved in general.

## 2.4.4 Conservation of energy

The conservation of energy originates from the invariance regarding temporal translations. To do so, the vector of generalized coordinates is extended by the time $t$, such that $Q \to (q,t)(\alpha)$ and $t'(\alpha) = t(\alpha) + \epsilon$ with new time variable $\alpha$. Define the Lagrangian

$$\mathcal{L}\left(Q, \frac{dQ}{d\alpha}\right) = L(q, \dot{q}, t) \frac{dt}{d\alpha}, \quad \dot{q} = \frac{dq}{d\alpha}\left(\frac{dt}{d\alpha}\right)^{-1} \tag{2.28}$$

and rewrite Hamilton's principle regarding the new time variable $\alpha$

$$S = \int_a^b \mathcal{L} \, d\alpha = \int_{a(\alpha)}^{b(\alpha)} L \, dt \tag{2.29}$$

The reformulation of Hamilton's principle states that the solution trajectories are determined as stationary points of both integrals and that the solutions are identical for both. Therefore, Noether's theorem may be applied to the transformed system, i.e.

$$\mathcal{I} = \frac{\partial \mathcal{L}}{\partial(\partial Q_i / \partial \alpha)} \frac{dQ'_i}{d\epsilon}\bigg|_{\epsilon=0} \tag{2.30}$$

Herein, one is interested in the last component of $Q$, i.e. in the time $Q_\mu = t$. Therefore,

$$\mathcal{I} = \frac{\partial \mathcal{L}}{\partial(\partial Q_\mu/\partial\alpha)} \frac{dQ'_\mu}{d\epsilon}\bigg|_{\epsilon=0} = \frac{\partial \mathcal{L}}{\partial(dt/d\alpha)} \qquad (2.31)$$

by using the definition of $\mathcal{L}$ one obtains

$$\begin{aligned}\mathcal{I} &= \left(\frac{\partial}{\partial(dt/d\alpha)} L\left(q, \frac{dq}{d\alpha}\left(\frac{dt}{d\alpha}\right)^{-1}, t\right)\right) \cdot \left(\frac{dt}{d\alpha}\right) + L \cdot 1 \\ &= \left(-\frac{\partial L}{\partial \dot{q}} \frac{dq}{d\alpha}\left(\frac{dt}{d\alpha}\right)^{-2}\right) \cdot \left(\frac{dt}{d\alpha}\right) + L \\ &= L - \frac{\partial L}{\partial \dot{q}} \dot{q} \qquad (2.32)\end{aligned}$$

which is the negative energy of the system. Therefore, the energy is a preserved quantity.

### 2.4.5 Symplecticity

**Theorem 10.** *A symplectic matrix is a $2n \times 2n$ matrix $\mathbb{F}$ satisfying the condition*

$$\mathbb{F}^T \mathbb{J} \mathbb{F} = \mathbb{J} \qquad (2.33)$$

*where $\mathbb{J}$ is a fixed nonsingular, skew-symmetric matrix.*

In the context of Hamilton's principle, $\mathbb{J}$ is chosen to be

$$\mathbb{J} = \begin{pmatrix} 0 & I_n \\ -I_n & 0 \end{pmatrix} \qquad (2.34)$$

where $I_n$ is the $n \times n$ unity matrix. Hamilton's equations may be rewritten such that

$$\begin{pmatrix} \dot{q} \\ \dot{p} \end{pmatrix} = \mathbb{J} \begin{pmatrix} H_q \\ H_p \end{pmatrix} \qquad (2.35)$$

and, thus, introducing the phase space $z$,

$$\dot{z} = \mathbb{J} H_z, \quad z = \begin{pmatrix} q \\ p \end{pmatrix} \qquad (2.36)$$

**Theorem 11.** *The smooth coordinate transformation $\psi : \mathbb{R}^{2n} \to \mathbb{R}^{2n}$, $(\tilde{q}, \tilde{p}) \leftrightarrow (q, p)$ is said to be canonical if for any Hamiltonian $H$ Hamilton's equations are equivalent to*

$$\dot{\tilde{q}} = \frac{\partial \tilde{H}}{\partial \tilde{p}}, \quad \dot{\tilde{p}} = -\frac{\partial \tilde{H}}{\partial \tilde{q}}$$

*where $\tilde{H} = H \circ \psi$.*

## 2.4. Preserved quantities

**Theorem 12.** *The transformation* $\psi : \mathbb{R}^{2n} \to \mathbb{R}^{2n}$, $(q,p) \leftrightarrow (Q,P)$ *is canonical (or symplectic) if and only if its Jacobian* $\mathbb{F}_{ij} = \partial Z_i / \partial z_j$ *satisfies the relation*

$$\mathbb{F}^T \mathbb{J} \mathbb{F} = \mathbb{J}$$

*Proof.* Assume a coordinate transformation $Z = Z(z)$ with Jacobian matrix $\mathbb{F}$,

$$\mathbb{F}_{ij} = \frac{\partial Z_i}{\partial z_j}$$

which is smooth, i.e.

$$\det(\mathbb{F}) \neq 0$$

Further assume the satisfaction of the Hamiltonian equations

$$\dot{Z} = \mathbb{J} H_Z$$

They may be transformed into

$$\begin{aligned}
\dot{Z}_h &= \mathbb{F}_{hi}\dot{z}_i = \mathbb{F}_{hi}\mathbb{J}_{ij}\frac{\partial H}{\partial z_j} = \mathbb{F}_{hi}\mathbb{J}_{ij}\frac{\partial H}{\partial Z_k}\frac{\partial Z_k}{\partial z_j} \\
&= \mathbb{F}_{hi}\mathbb{J}_{ij}H_{Z,k}\mathbb{F}_{kj} = \mathbb{F}_{hi}\mathbb{J}_{ij}[\mathbb{F}^T]_{jk}H_{Z,k} \\
\dot{Z} &= \mathbb{F}\mathbb{J}\mathbb{F}^T H_Z
\end{aligned}$$

leading to

$$\mathbb{F}\mathbb{J}\mathbb{F}^T = \mathbb{J}$$

From that one obtains $\mathbb{J}\mathbb{F}^T = \mathbb{F}^{-1}\mathbb{J}$. Transposing it and using $\mathbb{J}\mathbb{J} = -\mathbb{J}\mathbb{J}^T = -\mathbb{I}_{2n}$ leads to the condition (12). □

**Theorem 13.** *For fixed* $t \in \mathbb{R}$ *the flow* $\phi_t^H : \mathbb{R}^{2n} \to \mathbb{R}^{2n}$ *of Hamilton's equations* $\dot{z} = \mathbb{J}\nabla_z H(z)$ *are a canonical transformation.*

*Proof.* Since $H$ is assumed smooth the flow $\phi_t^H(z)$ is a smooth function of $t$ and $z$. Since $(\phi_t^H)^{-1} = \phi_{-t}^H$ it defines a smooth coordinate transformation.

$$\nabla\left(\frac{d}{dt}\phi_t^H(z)\right) = \nabla\left(\mathbb{J}\nabla H(\phi_t^H(z))\right)$$

leads to

$$\frac{d}{dt}\frac{\partial \phi_t(z)}{\partial z} = \mathbb{J}\nabla^2 H(\phi_t^H(z))\frac{\partial \phi_t(z)}{\partial z}$$

Using the symmetry of the Hessian $\nabla^2 H$ and the properties $\mathbb{J} = -\mathbb{J}^T = \mathbb{J}^{-1}$ one can determine the change in time of the symplectic condition

$$\begin{aligned}
\frac{d}{dt}\left(\frac{\partial \phi_t^H}{\partial z}^T \mathbb{J}\frac{\partial \phi_t^H}{\partial z}\right) &= \frac{d}{dt}\left(\frac{\partial \phi_t^H}{\partial z}\right)^T \mathbb{J}\frac{\partial \phi_t^H}{\partial z} + \frac{\partial \phi_t^H}{\partial z}^T \mathbb{J}\frac{d}{dt}\left(\frac{\partial \phi_t^H}{\partial z}\right) \\
&= \frac{\partial \phi_t^H}{\partial z}^T \left(\nabla^2 H(\phi_t^H(z))\right)^T \mathbb{J}^T \mathbb{J}\frac{\partial \phi_t^H}{\partial z} \\
&\quad + \frac{\partial \phi_t^H}{\partial z}^T \mathbb{J}\mathbb{J}\left(\nabla^2 H(\phi_t^H(z))\right)\frac{\partial \phi_t^H}{\partial z} \\
&= 0
\end{aligned}$$

Since $\phi_0^H(z) = z$ one has

$$\frac{\partial \phi_0^H}{\partial z}^T \mathbb{J} \frac{\partial \phi_0^H}{\partial z} = \mathbb{J}$$

i.e. $\phi_0^H$ is a canonical transformation and so are all $\phi_t^H$. □

### 2.4.6 Liouville's theorem

**Theorem 14.** *Given a set of initial configurations at time $s$ which describe a certain volume in phase space, one may find the same configurations in another phase space at a later time $t$ which has the same volume as the original.*

*Proof.* Let the time evolution of a phase space variable $z$ be a symplectic transformation $z(t) \to Z(t + \delta t)$ with transformation matrix $\mathbb{F}$. Since $\det \mathbb{F} \cdot \det \mathbb{J} \cdot \det(\mathbb{F}^T) = \det \mathbb{J}$, one has

$$\det \mathbb{F} = 1$$

Then, the phase space volume (in three dimensions) is

$$\int d^6 z = \int \det\left(\frac{\partial z}{\partial Z}\right) d^6 Z = \int \det(\mathbb{F}) d^6 Z = \int d^6 Z$$

□

## 2.5 Example: a linear system with a single degree of freedom

The Euler-Lagrange equation is used to solve the functional of the generalized coordinates $q(t)$. Assuming a quadratic potential function being expanded at $q(0)$

$$V(q) = \frac{1}{2}(q - q(0))^T K (q - q(0)) + F^T (q - q(0)) \tag{2.37}$$

and kinetic energy

$$K(\dot{q}) = \frac{1}{2}\dot{q}^T M \dot{q} \tag{2.38}$$

one obtains the equation of motion (2.13)

$$0 = -(K(q - q(0)) + F) - M\ddot{q} \tag{2.39}$$

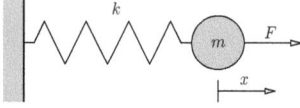

Figure 2.2: A spring and mass with single degree of freedom

## 2.5. Example: linear SDOF system

For a SDOF system ($K \to k$, $M \to m$, figure 2.2) the solution is straight forward. First, the equation is transformed into a homogenious differential equation using the coordinate transformation

$$q = x - K^{-1}F + q(0) \tag{2.40}$$

This leads to

$$0 = m\ddot{x} + kx \tag{2.41}$$

For the solution of this homogenious differential equation one may use the ansatz

$$x = A\exp(\lambda t) \tag{2.42}$$

For $k > 0$ one obtains

$$\lambda_{1,2} = \pm i\omega, \quad \omega = \sqrt{\frac{k}{m}} \tag{2.43}$$

with eigenfrequency $\omega$ and, thus, $x = A\cos(\omega t) + B\sin(\omega t)$. The coefficients $A$, $B$ are determined from the initial conditions, i.e.

$$A = k^{-1}F, \quad B = (m\omega)^{-1}j_0 \tag{2.44}$$

The initial conditions of the next element are computed from

$$q_1 = q_0 + A\cos(\omega h) + B\sin(\omega h) - k^{-1}F \tag{2.45}$$
$$j_1 = m\omega\left(-A\sin(\omega h) + B\cos(\omega h)\right) \tag{2.46}$$

The derived integrator is exact for quadratic potentials. One can verify this by inserting the final result for $q$, $q(t) = A\cos(\omega t) + B\sin(\omega t) + C$, into the elemental action $S^k$. The derivatives of $S^k$ for the parameters $\omega$, $A$, $B$ and $C$ should vanish for the determined values. The derivative of $S^k$ with respect to the step length $h$ is shown to be constant in time, i.e. the discrete energy is independent of the time step length. Due to the theorem of Ge and Marsden [281], an energy conserving symplectic integrator with constant step size describes the exact solution, which is true in this case.

Notice, if the stiffness $k$ is not constant in $q$ or $t$, then the resulting algorithm is neither symplectic nor energy conserving and may even lead to less accurate results than numerical first order methods.

# Chapter 3

# Variational integrators

## 3.1 Introduction

This section presents variational integrators, their derivation, properties, example methods and discretizations of unilateral and bilateral constraints. Variational integrators can be systematically derived by direct discretization of the action integral in Hamilton's principle. They belong to the class of geometric and mechanical integrators. That is, they provide a numerical solution to the discrete equations of motion and at the same time preserve one or more invariants of the time-continuous system. The aims of this chapter are:

1. Section 3.3 presents a methodology to derive variational integrators from Hamilton's principle. This includes the expression in Lagrangian and Hamiltonian form. The reader should understand that both are equivalent and how one transforms into the other. The transformation can be taken out in various ways, either by the Legrende transformation or by the Hamilton-Pontryagin procedure, leading to different formulas of the momentum. In the subsequent sections, the Hamiltonian point of view is preferred. The author believes that it expresses the dynamics in a more natural way. Compare, for example, central differences in Lagrangian notation with Velocity Verlet in Hamiltonian notation. For the first, the initial conditions must be expressed by $(q_{-1}, q_0)$ where $q_{-1}$ is computed by a fictitious backward step at the beginning of the simulation, while Velocity Verlet needs $(q_0, j_0)$ which are naturally given. Furthermore, additional hidden constraints based on constraint rates can be incorporated more easily.
The notation follows the scheme of Generalized Galerkin Variational Integrators [153] and discrete variational mechanics in [137, 267].

2. Variational integrators preserve all invariants of the continuous system except energy which oscillates around the initial value. The discrete counterparts of continuous invariants from section 2 are presented and proved in section 3.4. Understanding the properties of variational integrators helps to accept the constraints on the construction of numerical schemes. A useful

tool to understand the impacts of symplecticity is backward error analysis which explains the good long-term stability of variational integrators: Symplectic integrators can be shown to provide the exact solution to a Hamiltonian system which is very close to the original, see section 3.5. Furthermore, symplecticity simplifies linear stability analysis, see section 3.6.
Proofs and definitions mainly follow the books and lecture notes [84, 86, 97, 242, 267].

3. Example integrators are provided in section 3.7. The list of examples is restricted to schemes which require at most one force evaluation per time step. That is usual in structural dynamics. Schemes of higher than second order in accuracy are uncommon. Beside standard methods (Euler, Velocity Verlet, implicit midpoint), the list provides an almost complete overview on the variety of possible schemes. Therefore, existing symplectic-momentum schemes are reinterpreted as variational integrators and the discrete action is presented.

Multistep, exponential and IMEX integrators are approaches to allow longer time steps in explicit dynamics. They assume an additive split of the potential energy at least into two parts, whereby one of them is small compared with the other. This is an assumption often met in structural dynamics, in particular in problems of model order reduction, where structures are dominated by a linear response. Then only small nonlinear perturbations are of interest or are allowed. Linear forces may be computed very efficiently by a "constant-matrix times vector product" as used in multistep methods. If the stiffness matrix is available, IMEX and exponential integrators are possible. The choice among methods depends on the size of the problem, the desired accuracy and robustness, and the numerical cost required for the computation of the linear parts of the solution. In some cases, these sophisticated methods may be a real alternative to standard algorithms.

The aim of this chapter is two-fold: First, one should understand IMEX integrators as a numerically cheap implementation of the matrix exponential. From the same point of view, multi time step methods are a similar approximation to the exact solution of the linear forces in exponentially fitted schemes. Therefore, all modifications to a specific integrator can be applied to other methods in the same manner, for example MOLLY. A new integrator is presented which is a combination of IMEX and the mollified impulse method.

Very often engineers are more interested in efficiency, i.e. stability, in favour of accuracy. The question, therefore, arises how IMEX, multi time stepping and exponential integration behave if the basic assumption is not satisfied, i.e. if the perturbation is not very small. An experimental analysis is given by model problems from structural dynamics, see section 3.9. An algebraic treatment of stability of nonlinear systems is not easy to derive. Therefore, stability is measured in terms of Lyapunov, see section 3.6.

4. A specific problem is the discretization of constraint equations. Treatment of nonlinear holonomic constraints will be presented. A special class are unilateral constraints which may arise in contact/impact situations. The presented methodology follows the idea of non-smooth dynamics and algorithms given in [36, 58, 142]. The methods are presented modularly, i.e. the derivation allows the combination of presented constraint discretizations with different base integrators. Therefore, individual algorithmic realizations are given together with examples, see section 3.9.

## 3.2 Geometric integrators

A numerical method for solving ordinary differential equations is a mapping $\Phi_h$ defined on the phase space that approximates the time-$h$ flow $\phi_h$. The approximation at time $t = kh$ is obtained by

$$z_k = \Phi_h(z_{k-1}) \tag{3.1}$$

The numerical method $\Phi_h$ may satisfy some of the properties listed in table 3.1 [83]. A numerical method which satisfies at least one of these properties is called a geometric integrator.

Mechanical integrators in the sense of [266] are numerical methods which are well suited for the solution of dynamic problems in classical mechanics. Mechanical integrators are either energy-momentum or symplectic-momentum preserving algorithms. It is generally not possible to construct a numerical constant time step method which is symplectic, momentum and energy preserving [281]. Therefore, the third non-preserved quantity may be selected as a measure of accuracy. The error in energy as a scalar quantity can be easily computed while the error in symplecticity is difficult to measure. Based on these considerations, symplectic-momentum methods can be preferred over energy-momentum methods.

| property | condition |
|---|---|
| $\Phi_h$ is of order $r$ | $\Phi_h(z) = \phi_h(z) + \mathcal{O}(h^{r+1})$ |
| symmetric | $\Phi_h \circ \Phi_{-h} = identity$ |
| energy-preserving | $H(q_n, p_n) = const.$ |
| symplectic | $\nabla_z \Phi_h(z)^T \mathbb{J} \nabla_z \Phi_h(z) = \mathbb{J}$ |
| $\rho$-reversible | $(\rho \circ \Phi_h)(z) = (\Phi_h^{-1} \circ \rho)(z)$ |
| | for all $h$ and all $z$ and if $H(q,p) = H(q,-p)$ |

Table 3.1: Properties of numerical integrators $\Phi_h$

## 3.3 Discrete Euler-Lagrange equation

Equations (2.8) and (2.9) define the map

$$\Phi : (q_k, j_k) \to (q_{k+1}, j_{k+1}) \qquad (3.2)$$

which is defined by the structure of the discrete Lagrangian, see equation (2.7). If an exact solution is not possible or desired, one may assume an interpolation for the coordinates $q$ within each time element being dependent on a set of parameters $u_i^k$ and of an artificial time variable $\alpha$,

$$q = q(\alpha, u_i^k), \quad i = 0, 1, 2, \ldots, m, \quad \alpha = (t - t_0^k)/(t_1^k - t_0^k) \qquad (3.3)$$

where $t_0^k$ and $t_1^k$ define the left and the right boundary of the time element. Stationarity of the elemental action yields the system of equations

$$0 = q(0, u_0^k, \ldots, u_m^k) - q_0^k \qquad (3.4)$$

$$0 = q_1^k - q(1, u_0^k, \ldots, u_m^k) \qquad (3.5)$$

$$0 = j_0^k \frac{\partial}{\partial u_i^k} q(0, u_0^k, \ldots, u_m^k) - j_1^k \frac{\partial}{\partial u_i^k} q(1, u_0^k, \ldots, u_m^k) +$$

$$+ \frac{\partial}{\partial u_i^k} L_d(h, u_0^k, \ldots, u_m^k) \qquad (3.6)$$

which are the discrete equivalent of the Euler-Lagrange equation. These equations implicitly determine $q_1^k$, $j_1^k$, and $u_i^k$. The integrator is obtained from the global action sum as $(q_0^k, j_0^k) := (q_k, j_k)$ and $(q_{k+1}, j_{k+1}) := (q_1^k, j_1^k)$.

The map given by equation (3.2) denotes a variational integrator. It inherits some of the properties of the original system, such as symplecticity and conservation of momentum, which will be shown in the subsequent section.

Rewriting equation (2.9) gives

$$S = \sum_{k=1}^{N-1} \left[ S_d(q_0^k, q_1^k) + j_k(q_0^k - q_1^{k-1}) \right] + \left[ S_d(q_0^0, q_1^0) + j_0(q_0^0 - q_0) \right] + j_N(q_N - q_1^{N-1}) \qquad (3.7)$$

where the elemental action becomes a function of the bounding coordinates $S^k =: S_d(q_0^k, q_1^k)$ and $j_k := j_{k-} = j_{k+}$. Variation of the coordinates yields

$$0 = D_1 S_d(q_0^k, q_1^k) + j_k \qquad (3.8)$$

$$0 = D_2 S_d(q_0^k, q_1^k) - j_{k+1} \qquad (3.9)$$

This form of the discrete Euler-Lagrange equations is also known as discrete Legrende transformation. It does not reveal all parameters describing the motion $q(t)$ within a time element, but it is closer to the formulation known from continuous mechanics compared with equations (3.4)-(3.6).

## 3.4 Preserved discrete quantities

### 3.4.1 Discrete time Noether's theorem

**Theorem 15.** *Let a one parametric symmetric transformation $q' = q'(q, \epsilon)$ with $q'|_{\epsilon=0} = q$ and let the elemental actions be stationary, i.e. $S_d(q_0'^k, q_1'^k) = S_d(q_0^k, q_1^k) \, \forall \epsilon$. Then the discrete time Noether's theorem states that the quantity*

$$\mathcal{I} = \frac{\partial S_d(q_0^k, q_1^k)}{\partial q_1^k} \left. \frac{\partial q_1'^k}{\partial \epsilon} \right|_{\epsilon=0} \tag{3.10}$$

*is preserved [5].*

---

*Proof.* Define

$$\xi_\alpha^k = \left. \frac{\partial q_\alpha'^k}{\partial \epsilon} \right|_{\epsilon=0}$$

Invariance of the elemental action implies invariance of the action sum. Assuming that $q_1^k$, $j_k$ describe a solution trajectory, one obtains for the elemental action,

$$0 = \left. \frac{\partial S_d(q_0'^k, q_1'^k)}{\partial \epsilon} \right|_{\epsilon=0} = D_1 S_d(q_0^k, q_1^k) \xi_0^k + D_2 S_d(q_0^k, q_1^k) \xi_1^k$$

Inserting this into the $\epsilon$-derivative of equation (3.7) and using equations (3.8)-(3.9) one obtains for the action sum

$$0 = \left. \frac{\partial S}{\partial \epsilon} \right|_{\epsilon=0} = -j_0 \xi_0 + j_N \xi_N$$

For all $\xi_k = const.$, one can show that the discrete momentum in direction of $\xi$ is preserved and given by

$$\mathcal{I} := j_0 \xi = j_N \xi = D_2 S_d(q_0^{N-1}, q_1^{N-1}) \left. \frac{\partial q_1'^{N-1}}{\partial \epsilon} \right|_{\epsilon=0}$$

□

---

### 3.4.2 Conservation of linear momentum

**Theorem 16.** *Given an arbitrary translation $q_k' = q_k + \epsilon r$ under which the discrete Lagrangian is invariant, the linear momentum is preserved by the variational integrator.*

*Proof.* Inserting the transformation into Noether's theorem, equation (3.10), one obtains the invariant

$$\mathcal{I} = j_k \left. \frac{\partial q_1'^{k-1}}{\partial \epsilon} \right|_{\epsilon=0} = j_k \cdot I$$

□

### 3.4.3 Conservation of angular momentum

**Theorem 17.** *Given an arbitrary rotation $q'_k = \exp(\epsilon\bar{\omega}_\epsilon)q_k$ under which the discrete Lagrangian is invariant, the angular momentum $L_k$ is preserved by the variational integrator.*

> *Proof.* Inserting the transformation into Noether's theorem, equation (3.4.1), one obtains the invariant
>
> $$\mathcal{I} = j_k \left.\frac{\partial q_1'^{k-1}}{\partial \epsilon}\right|_{\epsilon=0} = j_k \cdot (\bar{\omega}_\epsilon q_1^{k-1}) = j_k \cdot (\omega_\epsilon \times q_1^{k-1}) = \omega_\epsilon \cdot (q_0^k \times j_k) = \omega_\epsilon \cdot L_k$$
>
> □

### 3.4.4 Preservation of the symplectic structure

**Theorem 18.** *A mapping $\phi : (q,p) \to (Q,P)$ is symplectic if and only if there exists locally a function $S(q,p)$ such that*

$$P^T dQ - p^T dq = dS \tag{3.11}$$

This means that $P^T dQ - p^T dq$ is a total differential. $S$ is called a generating function.

> *Proof.* Define the phase space vectors $z = (q,p)$ and $Z = (Q,P)$. The Jacobian can be computed by
>
> $$\frac{\partial Z}{\partial z} = \begin{bmatrix} \frac{\partial Q}{\partial q} & \frac{\partial Q}{\partial p} \\ \frac{\partial P}{\partial q} & \frac{\partial P}{\partial p} \end{bmatrix}$$
>
> This is inserted into the symplecticity condition (12). One obtains that the conditions
>
> $$P_p^T Q_p = Q_p^T P_p, \quad P_p^T Q_q - I = Q_p^T P_p, \quad Q_q^T P_q = P_q^T Q_q$$
>
> are equivalent to (12). Insert $dQ = Q_q^T dq + Q_p^T dp$ into (3.11) to obtain
>
> $$dS = (P^T Q_q - p^T, P^T Q_p)\begin{pmatrix} dq \\ dp \end{pmatrix} = \begin{pmatrix} Q_q^T P - p \\ Q_p^T P \end{pmatrix}^T \begin{pmatrix} dq \\ dp \end{pmatrix}$$
>
> In order to prove that this equation is a total differential, one has to show that the second variation is symmetric (Schwartz' theorem). Derivation for $dz$ leads to
>
> $$\frac{d^2 S}{dz^2} = \begin{pmatrix} Q_q^T P_q & Q_q^T P_p - I \\ Q_p^T P_q & Q_p^T P_p \end{pmatrix} + \sum_i P_i \frac{\partial^2 Q_i}{\partial(q,p)^2}$$
>
> Since the Hessians of $Q$ are symmetric, symmetry of the last equation becomes equivalent to the symplectic condition in equation (3.4.4). □

## 3.5. Error analysis

Generating functions are scalar functions which describe a symplectic mapping. Relation (3.11) suggests to use $(q, Q)$ as independent variables of the mapping $S$. For near-identity mappings as numerical integrators, however, mixed variables such as $(Q, p)$ or $(q, P)$ may be more convenient.

**Theorem 19.** *Let* $(q, p) \to (Q, P)$ *be a smooth transformation, close to the identity. It is symplectic if and only if one of the following conditions holds locally:*

- $Q^T dP + p^T dq = d(P^T q + S_1)$ *for some function* $S_1(q, P)$
- $P^T dQ + q^T dp = d(p^T Q - S_2)$ *for some function* $S_2(Q, p)$

*Proof.* The first characterization follows from $d(Q^T P) = Q^T P + P^T dQ$ and (3.11) if one puts $S_1$ such that $P^T q + S_1 = Q^T P - S_1$. For $S_2$ one uses $d(p^T q) = q^T p + p^T dq$. □

**Theorem 20.** *The numerical method* $(q_n, j_n) \to (q_{n+1}, j_{n+1})$ *defined by equations* (3.4), (3.5) *and* (3.6) *is a symplectic integrator.*

*Proof.* The differential of $L_d^k = L_d^k(q_n, q_{n+1})$ satisfies

$$dL_d^k = j_{n+1}^T dq_{n+1} - j_n^T dq_n$$

which proves symplecticity due to theorem 18. □

### 3.4.5 Discrete energy

**Theorem 21.** *The discrete energy is defined by*

$$E_d^k = -\frac{\partial}{\partial h_k} L_d^k \qquad (3.12)$$

*where $h_k$ is the length of the $k$-th time element.*

Using constant time steps, the discrete energy is generally not conserved.

## 3.5 Error analysis

At least three types of error expansions are possible:

1. **Global error expansions** For example, time-reversible methods like Verlet's algorithm have the expansion

$$z_n = z(nh) + h^2 e_1(nh) + h^4 e_2(nh) + \ldots \qquad (3.13)$$

The coefficients $e_k(t)$ are determined by substituting the expansion into the difference equation, expanding about $t = nh$ and equating like powers of $h^2$. This expansion is used to determine terms of the error that grow in time [161]. One can show that the global error grows linearly for symplectic methods while the error grows quadratically for non-symplectic schemes [32].

2. **Local error expansions**  The local error is the deviation of the result given by a single step of the method $\Phi$ from the exact solution

$$\delta_k^h = \|\Phi(q_{k-1}, j_{k-1}, h) - z(t_k)\| \qquad (3.14)$$

where it is assumed that $q_{k-1}, j_{k-1}$ are exact.

3. **Backward error expansions**  The numerical solution is expressed formally as the exact solution to the Hamiltonian equations (2.36) with a perturbed right hand side. The perturbation is expressed as an expansion in powers of $h$. Using a truncated asymptotic backward expansion yields a more accurate approximation to the numerical solution than a global error expansion [85].

Let us outline the procedure of backward error analysis. For details, see [86, 225]. Consider an ODE given by Hamilton's equations $\dot{z} = f(z)$ and an integrator $z_{k+1} = \Phi(z_k, h)$. For a given initial condition $z_0$, let $z(t)$ be the true and $\{z_k\}$ the numerical solution of the trajectory. Define a modified ODE $\dot{\bar{z}} = \bar{f}(\bar{z})$ in such a way that the numerical method $\Phi$ is the exact solution to it, i.e. $z_k = \bar{z}(t_k)$. The backward error norm is then defined as $\|\bar{f} - f\|$.

One reason why symplectic integrators are preferable over standard methods is their good long-term stability. Although they do not preserve the energy, the error in energy is bounded. Usually the discrete energy oscillates around the true value. Since the numerical method is symplectic, its modified ODE $\bar{f}$ describes a symplectic continuous system. Since symplecticity is equivalent to the existence of a generating function, the modified system $\bar{f}$ is associated with a Lagrangian $\bar{L}$. The corresponding energy function denotes the "shadow Hamiltonian" $\bar{H}$. As a result, the numerical solution of variational integrators has all properties of a conservative mechanical system, such as energy preservation. This is the origin of the good long-term behaviour.

Generally, it is not possible to derive an explicit expression for shadow Hamiltonians. In case of Verlet's method, the first expansion is given by (see [55, 161, 243])

$$\bar{H}(q,p) = H + h^2 \frac{1}{12} T_p V_{qq} T_p - h^2 \frac{1}{24} V_q T_{pp} V_q + \mathcal{O}(h^4) \qquad (3.15)$$

where $T_p$ denotes the derivative of the kinetic energy $T(p)$ with respect to the momentum $p$ and $V_q$ the derivative of the potential energy $V(q)$. It is also possible to derive a closed form for a linear system with a single degree of freedom, see [244].

## 3.6 Linear stability analysis

### 3.6.1 Fundamental matrix

The numerical trajectory can be expressed in the form

$$z_{k+1} = \Phi_k(z_k) = \Phi_k(\Phi_{k-1}(\Phi_{k-2}(\ldots \Phi_0(z_0)\ldots))), \quad z = \begin{pmatrix} q \\ j \end{pmatrix} \tag{3.16}$$

Depending on the considered integrator and given a linear system, this can be simplified to

$$z_{k+1} = F_k z_k = F_k F_{k-1} F_{k-2} \cdots F_0 z_0, \quad F_k = \frac{\partial z_{k+1}}{\partial z_k} \tag{3.17}$$

If the fundamental matrix $F_k$ does not depend on the current state $z_k$, i.e. $F_k = F$ one can write

$$z_{k+n} = F^n z_k \tag{3.18}$$

This assumption simplifies proofs of stability conditions. It holds for most fixed step size methods if the potential function is quadratic. Therefore, it is termed linear stability analysis.

Furthermore, any linear system may be diagonalized. The potential function $V(u) = \frac{1}{2} u^T K u + F^T u$ can be transformed into normal form by the transformation $u \to x(u) = u + K^{-1} F$. The resulting system can further be decomposed using modal analyses, i.e. $x(q) = \Phi q$, such that $\Phi^T K \Phi = \mathrm{diag}(\lambda_i)$ and $\Phi^T M \Phi = I$. Herein, $\lambda_i$ denote the eigenvalues. The motions of each degree of freedom are decoupled from any other coordinate. Therefore, each degree of freedom $q$ may be evaluated separately.

For stability analysis one may use the spectral decomposition of $F$ given by $F = PJP^{-1}$, where $P$ is the matrix of eigenvectors of $F$. $J$ is the Jordan canonical form of $F$ with eigenvalues $\mu_i$ of $F$ on its diagonal. Notice that $F$ may be unsymmetric. Since $P^{-1}P = I$,

$$F^n = PJ^n P^{-1} \tag{3.19}$$

The time stepping scheme is assumed being stable if $\|z_k\|$ does not grow beyond all bounds for $k \to \infty$.

Let $\rho(F)$ be the spectral radius defined as the largest absolute value of the eigenvalues

$$\rho = \max_i |\mu_i| \tag{3.20}$$

If

$$\rho \leq 1 \tag{3.21}$$

then $J^n$ and $F^n$ are bounded for $n \to \infty$. For $\rho < 1$, $F^n \to 0$.

The considered method is symplectic. Therefore, Liouville's theorem can be applied (section 2.4.6). That is $\det(F) = \prod_i \mu_i = 1$. Using $\text{Tr}(F) = \sum_i \mu_i$ a symplectic integrator is stable if

$$|\text{Tr}(F)| \leq 2 \tag{3.22}$$

$F$ being a $2 \times 2$ matrix.

For example, in case of the synchronous Euler method, the propagation matrix of a linear single degree of freedom system is

$$F = \begin{pmatrix} 1 - h^2 k/m & hk/m \\ -hk & 1 \end{pmatrix} \tag{3.23}$$

from which one can conclude that stability is given if $k > 0$ and

$$h\sqrt{k/m} < 2 \tag{3.24}$$

### 3.6.2 Combining different integrators

Consider, for example, the synchronous Verlet scheme. It is obtained from the discrete potential action

$$V_d^V = 0.5h(V(q_0^k) + V(q_1^k)) \tag{3.25}$$

with time step length $h$ leading to the map $\Phi^V(q_k, j_k, h_k)$. One can easily verify the critical time step being $h_{crit}^V = 2(k/m)^{-0.5}$. Further, consider the adjoint integrators $\Phi^1(q_k, j_k, h_k)$ and $\Phi^2(q_k, j_k, h_k)$ which are obtained from the discrete potential actions

$$V_d^1 = hV(q_k^+), \quad \text{and} \quad V_d^2 = hV(q_{k+1}^-) \tag{3.26}$$

For each, the individual critical time step is $h_{crit}^{1,2} = 2(k/m)^{-0.5}$. One could now consider the combination

$$\Phi^3(q_k, j_k, 2h) = \Phi^2\left(\Phi^1(q_k, j_k, h), h\right) \tag{3.27}$$

If the combination would not affect the stability then the critical time step would be $h_{crit}^{3,pred} = h_{crit}^{1,2}$. In fact, it turns out that $\Phi^3(q_k, j_k, 2h)$ is identical to the Verlet scheme with time step length $2h$ and stability limit $h_{crit}^3 = h_{crit}^V/2$. That means, by considering the two base integrators one overestimated the critical time step by a factor 2.

The reason why ensuring stability of the individual methods fails, lies in the properties of the eigenvectors of their propagation matrices. Consider the sequence

$$z_n = F_{n-1} \cdot F_{n-2} \cdots F_1 F_0 z_0 \tag{3.28}$$

$$= P_{n-1} J_{n-1} P_{n-1}^{-1} P_{n-2} J_{n-2} P_{n-2}^{-1} \cdots P_1 J_1 P_1^{-1} P_0 J_0 P_0^{-1} z_0 \tag{3.29}$$

## 3.6. Linear stability analysis

Only if $P_k^{-1} P_{k-1} = I$ then stability can be estimated by the product of the individual eigenvalue matrices $\prod_{k=0}^{n-1} J_k$. One possibility to measure stability is, therefore, to find (or to construct) periodic subsequences.

This strategy is used by [65, 232] to predict the stability of asynchronous and multiple time stepping algorithms. Time intervals are selected with synchronous start and end time. All individual base integrators $\Phi_k$ acting within this interval are collected into a single macro integrator $\hat{\Phi}$. Then, the total scheme can be seen as a sequence of identical maps $\hat{\Phi}$. Stability may be accurately determined by considering the map $\hat{\Phi}$.

### 3.6.3 Lyapunov stability

**Theorem 22.** *An equilibrium point is stable in the sense of Lyapunov if for all $\epsilon > 0$, there exists a $\delta > 0$ such that*

$$\|x(0) - x_e\| < \delta \quad \rightarrow \quad \|x(t) - x_e\| < \epsilon \, \forall t > 0 \tag{3.30}$$

This implies that the solution $x(t)$ stays nearby $x_e$ given a perturbation $x(0)$ whereby "nearby" is defined by $\epsilon$. Simplify the equations by defining the error $y(t) = x(t) - x_e$. Now assume, that the described process is an iteration $x_{k+1} = cx_k$, $c > 0$. Then the error increases by each step

$$\frac{\|y_{k+1}\|}{\|y_k\|} = \frac{\|c(x_k + y_k) - cx_k\|}{\|y_k\|} = c \tag{3.31}$$

where $y_k$ denotes the initial error. After $n$ iterations the error grows to $c^n$. Given this motivation one can define the Lyapunov exponent

$$\lambda(y(0)) = \lim_{t \to \infty} \frac{1}{t} \ln \frac{\|y(t)\|}{\|y(0)\|} \tag{3.32}$$

An equilibrium point $x_e$ is

- asymptotically stable if $\lambda < 0$ or if it is Lyapunov stable and approaches $x(t) \to x_e$ as $t \to \infty$. Then the trajectory approaches an attractor (fixed point; for example damped systems).

- neutrally stable if $\lambda = 0$ or if it is Lyapunov stable, but not asymptotically stable (conservative systems, i.e. Hamiltonian).

- unstable.

Applied to stability analysis, the integrator is considered being stable if the numerical trajectory $z_k$ starting at $z_0$ ($x_e = 0$) is bounded. Then

$$\lambda = \lim_{n \to \infty} \frac{1}{n} \ln \frac{\|z_n\|}{\|z_0\|} = \lim_{n \to \infty} \frac{1}{n} \ln |\mu^n| \tag{3.33}$$

where $\mu$ denotes the largest eigenvalue of the fundamental matrix from equation (3.20). The Lyapunov stability condition is identical with equation (3.21).

Linear stability analysis may not be easily applied to certain integrators, for example to general asynchronous integrators for large problems and variable time step integrators ($\mu$ is unknown or does not exist). Then stability may be at least measured in terms of Lyapunov stability. To this end one tracks the phase space norms

$$\lambda_k = \frac{1}{k} \ln \frac{\|z_k\|}{\|z_0\|} \tag{3.34}$$

for large $k$. Notice, when evaluating the exponent in practice, small positive values may be obtained even if the method is stable. This happens because $\lim_{k \to \infty}$ cannot be evaluated in practice.

## 3.7 Example integrators

This section presents some example variational integrators. The selected methods are restricted to schemes where only one force evaluation is required per time step. Therefore, higher-order methods (Runge-Kutta types, composition methods, etc.) are not mentioned. The aim is to illustrate the great variety of schemes which may be derived from the variational principle. Beside general purpose methods (Euler, Verlet) there exist approaches which make use of second order information or are limited to problems of a special structure, for example where the linear dynamics is dominant.

### 3.7.1 Symplectic Euler

The simplest numerical scheme is obtained by applying a piecewise linear interpolation of the coordinates $q$ in time

$$q(\alpha) = q_0 + \alpha \Delta q, \quad \alpha = 0 \ldots 1 \tag{3.35}$$

The potential energy is integrated by a single integration point per time element, either at the left or right element boundary. The discrete Lagrangian, see equation (2.7), writes

$$L_d = \frac{1}{2h} \Delta q^T M \Delta q - h V(q_0) \tag{3.36}$$

From equations (3.4)-(3.6) one obtains the scheme

$$q_1 = q_0 + h M^{-1} j_1 \tag{3.37}$$
$$j_1 = j_0 - h \nabla V(q_0) \tag{3.38}$$

Compare this with standard Euler

$$q_1 = q_0 + h M^{-1} j_0 \tag{3.39}$$
$$j_1 = j_0 - h \nabla V(q_0) \tag{3.40}$$

## 3.7. Example integrators

An example phase space diagram of both methods is given in figure 3.1 and clearly emphasizes the importance of symplecticity.

The method is explicit and first order accurate. The critical time step is given by $h_{crit} = 2/\omega_{max}$ where $\omega_{max}$ is the largest natural frequency of the linear system.

### 3.7.2 Velocity Verlet

Verlet's method is equivalent to the schemes known as leapfrog, Störmer, explicit Newmark, central difference method. When expressed in phase space variables $(q, j)$ it is known as Velocity Verlet. The coordinates are piecewise linear in time and the potential function is numerically integrated at both time element boundaries, see figure 3.2. The discrete Lagrangian is given by

$$L_d = \frac{1}{2h}\Delta q^T M \Delta q - h\frac{1}{2}\left(V(q_0) + V(q_0 + \Delta q)\right) \quad (3.41)$$

leading to

$$q_1 = q_0 + hM^{-1}j_0 - h^2\frac{1}{2}M^{-1}\nabla V(q_0) \quad (3.42)$$

$$j_1 = j_0 - h\frac{1}{2}\nabla V(q_0) - h\frac{1}{2}\nabla V(q_1) \quad (3.43)$$

The method is explicit and second order accurate. The critical time step is given by $h_{crit} = 2/\omega_{max}$ where $\omega_{max}$ is the largest natural frequency of the linear system. By a modification of the momentum $\bar{j}_k = j_k + h\frac{1}{2}\nabla V(q_1)$ it can be transformed into symplectic Euler working in $(q_k, \bar{j}_k)$ space. Damping and external forces can be incorporated [8]. Given a constant damping matrix $C$ the corresponding forces

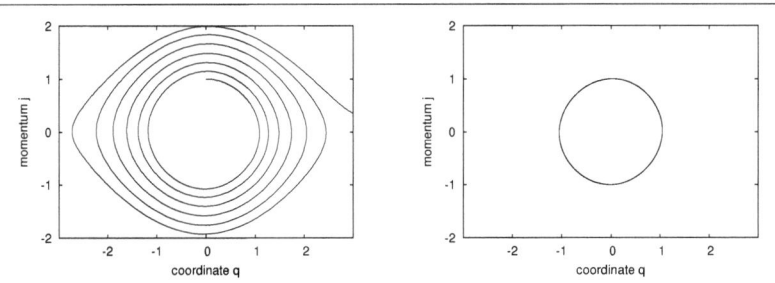

Figure 3.1: Phase space diagram of a harmonic oscillator (nonlinear pendulum). The standard Euler (left) clearly violates energy conservation. It even passes the critical velocity from the domain of oscillations to the domain of turn-overs. The symplectic Euler method (right) exhibits a closed orbit suitable for long term simulation. But the energy is not exactly preserved; the orbit forms an ellipse being sheared horizontally. Both used the same time step $h = 0.05$ (mass $m = 1$, gravity $g = 1$).

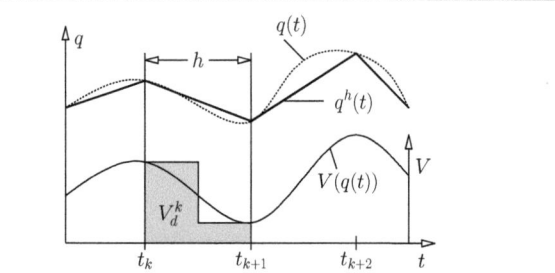

Figure 3.2: Interpolation of $q(t)$ for Verlet. The highlighted area is the approximation to the integral $V_d^k = \int_{t_k}^{t_{k+1}} V(q) dt$.

can be integrated implicitly without influencing the critical time step. With external force vector $F(t)$ one obtains the scheme

$$\Delta q_k = h \left( M + hC \right)^{-1} \left( j_k - \frac{h}{2} \nabla V(q_k) - \frac{h}{2} F(t_k) \right) \quad (3.44)$$

$$q_{k+1} = q_k + \Delta q_k \quad (3.45)$$

$$j_{k+1} = \frac{1}{h} M \Delta q_k - \frac{h}{2} \nabla V(q_{k+1}) - \frac{h}{2} F(t_{k+1}) \quad (3.46)$$

Verlet's method is the standard method in molecular and structural explicit dynamics. It can be easily implemented and is very fast. The mass matrix $M$ is often assumed to be diagonal and constant. The force vector $\nabla V(q_1)$ is temporarily stored and used as $\nabla V(q_0)$ in the subsequent time step. The limitations due to critical time step length are often balanced by the numerical efficiency when compared with implicit methods.

### 3.7.3 Midpoint

The implicit midpoint rule uses a piecewise linear interpolation of the coordinates $q(t)$. Its discrete Lagrangian writes

$$L_d = \frac{1}{2h} \Delta q^T M \Delta q - hV \left( q_0 + \frac{1}{2} \Delta q \right) \quad (3.47)$$

The method is implicit in $q$ and second order accurate. It requires the (generally iterative) enforcement of an equilibrium condition in the center of the time step. The scheme is unconditionally stable for linear systems.

### 3.7.4 Multiple time stepping

Multiple time stepping is often used in molecular dynamics. In structural dynamics it is known as "subcycling". The motivation is to increase the time step which is

## 3.7. Example integrators

limited by a stability barrier in explicit simulation. Therefore, the potential energy is additively split into two parts

$$V(q) = V^{slow}(q) + V^{fast}(q) \qquad (3.48)$$

where $V^{slow}(q)$ is associated with a large and $V^{fast}(q)$ with a small critical time step. Both potentials are evaluated at different frequencies, see figure 3.3. A typical integrator of this family is Verlet-I/r-RESPA (reversible reference system propagator algorithm), an application of Verlet to multiple time scales. More than two levels are possible [258], but not considered at this point.

Now assume that the slow forces are evaluated using a time step length $\Delta t$ and the fast force using a smaller time step $\delta t = \Delta t / \epsilon$ where $\epsilon$ is an integer number. Then the discrete Lagrangian of a "slow" step writes

$$L_d = \sum_{i=0...\epsilon-1} \left( \frac{1}{2\delta t} \left( q^{i+1} - q^i \right)^T M \left( q^{i+1} - q^i \right) - \frac{\delta t}{2} \left( V^{fast}\left(q^i\right) + V^{fast}\left(q^{i+1}\right) \right) \right)$$
$$- \frac{\Delta t}{2} \left( V^{slow}\left(q^0\right) + V^{slow}\left(q^\epsilon\right) \right), \quad q^0 = q_0, \quad q_1 = q^\epsilon \qquad (3.49)$$

The resulting method is summarized in algorithm 3.4.

Stability is limited by the critical time step lengths of the two potentials and by the occurance of resonances. An intuitive explanation for resonances was given in [65], see figure 3.5. Assume that the slow time step $\Delta t$ is exactly the half period of the fast motion. Between two kicks of $V^{slow}$ the vibration phase yields a trajectory with the natural frequency of $V^{fast}$. Now consider a linear oscillator (representing the 'vibration' of the fast forces) with initial conditions $q = 1$ and $j = 0$. The soft kick is then always applied when $q = 1, j \leq 0$ or $q = -1, j \geq 0$. In both cases, the sign of the force is such that the velocity will always increase leading to a monotonous increase of energy.

It can be shown that whenever

$$\Delta t_{resonance} \approx m \frac{T_{eff}}{2} \quad \text{with} \quad T_{eff} = \frac{2\pi}{\omega_{eff}} \qquad (3.56)$$

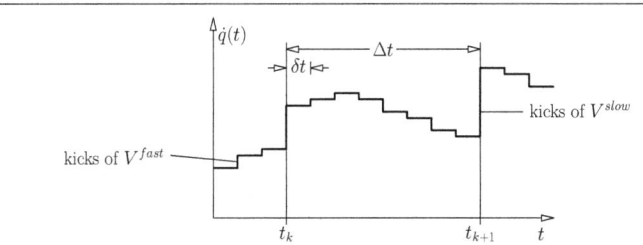

Figure 3.3: Interpolation of velocity using multiple time stepping such as r-RESPA. Each time step of the slow forces $\nabla V^{slow}$ contains a large number of small time steps of the fast forces $\nabla V^{fast}$

> **for** $k = 0$ to $N - 1$ **do**
> half a kick:
> $$q_k^0 = q_k \tag{3.50}$$
> $$j_k^0 = j_k - \frac{\Delta t}{2}\nabla V^{slow}(q_k) \tag{3.51}$$
> $\delta t = \Delta t/\epsilon$
> **for** $i = 0$ to $\epsilon - 1$ **do**
> vibration:
> $$q_k^{i+1} = q_k^i + \delta t M^{-1} j_k^i - M^{-1}\frac{\delta t^2}{2}\nabla V^{fast}(q_k^i) \tag{3.52}$$
> $$j_k^{i+1} = j_k^i - \frac{\delta t}{2}\nabla V^{fast}(q_k^i) - \frac{\delta t}{2}\nabla V^{fast}(q_k^{i+1}) \tag{3.53}$$
> **end for**
> half a kick:
> $$q_{k+1} = q_k^\epsilon \tag{3.54}$$
> $$j_{k+1} = j_k^\epsilon - \frac{\Delta t}{2}\nabla V^{slow}(q_{k+1}) \tag{3.55}$$
> **end for**

Figure 3.4: Verlet-I/r-RESPA

then resonances occur, see [65, 113]. $\omega_{eff}$ denotes the natural frequency of the fast motion and $m$ an integer.

Resonances can be eliminated by application of the mollified impulse method (see [68,113], MOLLY). The idea is to filter the force components which lead to the resonances. This can be done in a symplectic manner by changing the position where the fast potential is evaluated, i.e.

$$V(q) = V^{slow}(q) + V^{fast}(\mathcal{A}(q)) \tag{3.57}$$

leading to a force vector

$$\nabla V(q) = \nabla V^{slow}(q) + \nabla \mathcal{A}^T \nabla V^{fast}(\mathcal{A}(q)) \tag{3.58}$$

The operator $\mathcal{A}$ is an averaging operator of the trajectory's positions, i.e.

$$\mathcal{A}(q) = \frac{1}{\Delta t}\int_0^\infty \phi\left(\frac{t}{\Delta t}\right)\tilde{q}(t)dt \tag{3.59}$$

where $\phi$ is some weighting function and $\tilde{q}(t)$ is the solution to an auxiliary problem

$$\tilde{L} = \frac{1}{2}\dot{\tilde{q}}^T M \dot{\tilde{q}} - V^{fast}(\tilde{q}), \quad \tilde{q}(0) = q, \quad \dot{\tilde{q}}(0) = 0 \tag{3.60}$$

This approach is feasible if the weighting function has compact support in time. Depending on the selected weighting function and the effort spent in solving the

## 3.7. Example integrators

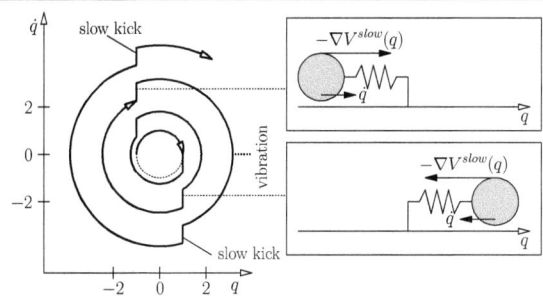

Figure 3.5: Phase space diagram of a harmonic oscillator illustrating r-RESPA resonances. The oscillator is hit by a velocity change every half period. Instead of forming a closed orbit, the kicks lead to monotonous energy growth. In r-RESPA, the fast forces represent the linear oscillator. The slow forces are presented by the velocity changes.

auxiliary problem, the instability is restricted to smaller intervals or may be eliminated (see [113]).

### 3.7.5 Exponential integrators

Accuracy and stability of multiple time step methods can be improved by superimposing the explicit solution of the slow forces with the exact solution of the (assumed linear) fast forces. This approach is feasible if

$$V^{slow}(q) \ll V^{fast}(q)\,\forall q, \quad \nabla^2 V^{fast}(q) = K_0 = const. \tag{3.61}$$

Due to the exact solution of the linear forces the algorithm is based on matrix exponentials replacing the vibration phase in algorithm 3.4. Therefore, this class of algorithms is known as exponential integrators, exponentially fitted integrators and integrating factor method.

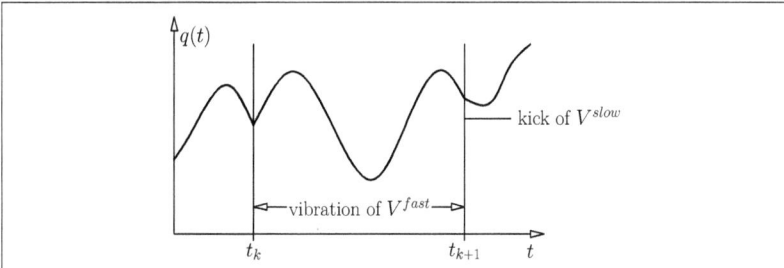

Figure 3.6: Interpolation of coordinates using exponential time stepping. At each time step boundary a kick due to $\nabla V^{slow}$ is applied to the velocities.

For now, confine the discussion to systems with a single degree of freedom. The Lagrange function can be written as

$$L(q, \dot{q}) = \frac{1}{2}m\dot{q}^2 - \omega^2 q^2 - N(q), \quad N(q) := V^{slow}(q) \quad (3.62)$$

The trajectory is interpolated by

$$q(\alpha) = a\cos(\omega h\alpha) + b\sin(\omega h\alpha) + c \quad (3.63)$$

within each time element (figure 3.6). The kinetic energy and the potential of the fast forces are analytically integrated in time. For the integration of the nonlinear forces $N_d = h \int_0^1 N(q(\alpha))d\alpha$ with definitions $\nu = \omega h/2$, $\text{sinc}(x) = \sin(x)/x$ if $x \neq 0$ and $\text{sinc}(0) = 1$, $\text{tanc}(x) = \text{sinc}(x)/\cos(x)$ is chosen:

- Gautschi

$$N_d = \frac{h}{2}\left(N(\mathcal{A}(q_0)) + N(\mathcal{A}(q_1))\right)\text{tanc}(\nu) \quad (3.64)$$

- Deuflhard

$$N_d = \frac{h}{2}\left(N(\mathcal{A}(q_0)) + N(\mathcal{A}(q_1))\right) \quad (3.65)$$

- Hairer-Lubich

$$N_d = \frac{h}{2}\left(N(\mathcal{A}(q_0)) + N(\mathcal{A}(q_1))\right)\text{sinc}(\omega h) \quad (3.66)$$

Therein, $\mathcal{A}(q)$ is the MOLLY averaging operator from equation (3.59). One arrives at the generic scheme (see [41])

$$q_1 = \cos(h\omega)q_0 + \frac{\sin(h\omega)}{m\omega}j_0 - \frac{h^2}{2m}\Psi\nabla N(\Phi q_0) \quad (3.67)$$

$$j_1 = -\omega m\sin(h\omega)q_0 + \cos(\omega h)j_0 - \frac{h}{2}\left(\Psi_0\nabla N(\Phi q_0) + \Psi_1\nabla N(\Phi q_1)\right) \quad (3.68)$$

$$\Phi = \phi(h\omega) \quad (3.69)$$

$$\Psi = \psi(h\omega) \quad (3.70)$$

$$\Psi_0 = \frac{\psi(h\omega)}{\text{tanc}(h\omega)} \quad (3.71)$$

$$\Psi_1 = \frac{\psi(h\omega)}{\text{sinc}(h\omega)} \quad (3.72)$$

$$a = \begin{cases} 0, & \text{if } \cos(h\omega) = 1 \\ \frac{q_1 - q_0 - \sin(h\omega)b}{\cos(h\omega) - 1}, & \text{else} \end{cases} \quad (3.73)$$

$$b = \frac{j_0}{\omega m} \quad (3.74)$$

$$c = q_0 - a \quad (3.75)$$

The scheme stays symplectic if $\omega$ is arbitrarily changed during the simulation (see [41]). Exponential integrators exhibit instabilities due to the same resonances as multiple time stepping schemes. Therefore, MOLLY operators have been introduced as well. The definition of various schemes is given in table 3.2.

## 3.7. Example integrators

| $\psi(x)$ | $\phi(x)$ | name |
|---|---|---|
| $\mathrm{sinc}^2(x/2)$ | 1 | Gautschi [70] |
| $\mathrm{sinc}(x)$ | 1 | Deuflhard [47] |
| $\mathrm{sinc}^2(x)$ | 1 | Hairer-Lubich [87] |
| $\mathrm{sinc}(x)\phi(x)$ | $\mathrm{sinc}(x)$ | mollified Deuflhard, Garcia-Archilla [68] |
| $\mathrm{sinc}^2(x/2)\phi(x)$ | $\mathrm{sinc}(x)$ | mollified Gautschi |
| $\mathrm{sinc}^2(x)\phi(x)$ | $\mathrm{sinc}(x)$ | mollified Hairer-Lubich |

Table 3.2: Definition functions of various exponential integrators

Exponential integrators target at problems where the number of variables is limited and the degree of nonlinearity is very small. Possible candidate problems are those of model order reduction. Using modal analysis, a system with multiple degrees of freedom may be diagonalized. A new space of variables $x$ is introduced with $q = Tx$, $T^T K_0 T = \mathrm{diag}(\omega_i^2)$ and $T^T M T = I$. Then the nonlinear potential is evaluated via

$$N^x(x) = N(T\mathcal{A}(x)) = N(T\Phi x), \quad \Phi = \mathrm{diag}(\Phi_i) \quad (3.76)$$
$$\nabla_x N^x(x) = \Phi^T T^T \nabla_q N(T\Phi x) \quad (3.77)$$

The equations are decoupled and the algorithm defined by equations (3.67)-(3.72) can be applied directly to the modal variables $x_i$.

### 3.7.6 Implicit-explicit integrators

If the fast potential in equation (3.48) is linear one may solve the fast motions implicitly with small effort. The resulting integrator is combined implicit-explicit (IMEX) where the slow potential is explicitly solved. The discrete Lagrangian of such an integrator is given by [252]

$$L_d = \frac{1}{2h}\Delta q^T M \Delta q - \frac{h}{2}\left(V^{slow}(q_0) + V^{slow}(q_0 + \Delta q)\right) - h V^{fast}\left(q_0 + \frac{1}{2}\Delta q\right) \quad (3.78)$$

The vibration phase in algorithm 3.4 is now replaced by the implicit midpoint rule. The instability limits due to resonances are eliminated. The critical time step is defined by $V^{slow}$.

The stability can be easily improved by adding a mollification. Assume the fast forces being linear, i.e.

$$V^{fast}(q) = \frac{1}{2}q^T K q + F^T q \quad (3.79)$$

Furthermore, damping forces with constant damping matrix $C$ can be incorporated. External forces can be defined by a time dependent vector-valued function $F^{ext}(t)$. One obtains the integrator

$$\mathcal{A}(q) = q - \frac{h}{4}\left(\frac{1}{h}M + C + \frac{h}{4}K\right)^{-1}(Kq + F) \tag{3.80}$$

$$\nabla\mathcal{A} = I - \frac{h}{4}\left(\frac{1}{h}M + C + \frac{h}{4}K\right)^{-1}K \tag{3.81}$$

$$\Delta q = \left(\frac{1}{h}M + C + \frac{h}{4}K\right)^{-1}\left(j_k - \frac{h}{2}\nabla\mathcal{A}^T \nabla V^{slow}\left(\mathcal{A}(q_k)\right)\right.$$
$$\left. - \frac{h}{2}(Kq_k + F) - \frac{h}{2}F^{ext}(t_k)\right) \tag{3.82}$$

$$q_{k+1} = q_k + \Delta q \tag{3.83}$$

$$j_{k+1} = \frac{1}{h}M\Delta q - \frac{h}{2}\nabla\mathcal{A}^T \nabla V^{slow}\left(\mathcal{A}(q_{k+1})\right)$$
$$- \frac{h}{2}\left(K\left(q_k + \frac{1}{2}\Delta q\right) - F\right) - \frac{h}{2}F^{ext}(t_{k+1}) \tag{3.84}$$

This integrator is unconditionally stable for linear single degree of freedom systems.

### 3.7.7 Rowlands's method

Symplectic integrators provide an exact solution to a modified mechanical system with a Hamiltonian $\bar{H}$, equation (3.15), which is very close to the Hamiltonian $H$ of the original model. Instead of integrating the original system given by $H$ one could improve the accuracy by integrating a similar system given by the Hamiltonian $G$ such that its shadow Hamiltonian $\bar{G}$ is closer to $H$ than $\bar{H}$. This is the idea of Rowlands's method (see [19, 178, 228, 242]).

A change of variables is introduced

$$Q = q + h^2\lambda T_{pp}V_q + \mathcal{O}(h^4) \tag{3.85}$$

$$P = p - h^2\lambda V_{qq}T_p + \mathcal{O}(h^4) \tag{3.86}$$

which is obtained from an additive modification of the Hamiltonian $H_\chi = h\lambda T_p V_q$, kinetic energy $T$ and momentum $p$. The shadow Hamiltonian of the transformed system becomes

$$\bar{G}(\chi(q,p)) = H + h^2\left(-\lambda + \frac{1}{12}\right)T_p V_{qq} T_p + h^2\left(\lambda - \frac{1}{24}\right)V_q T_{pp} V_q + \mathcal{O}(h^4) \tag{3.87}$$

$$= H + \frac{h^2}{24}V_q M^{-1} V_q + \mathcal{O}(h^4), \quad \lambda = \frac{1}{12} \tag{3.88}$$

The error in this shadow Hamiltonian can be compensated by replacing the potential energy in the original system

$$V_h(q) = V(q) - \frac{h^2}{24}\nabla V(q)^T M^{-1} \nabla V(q) \tag{3.89}$$

## 3.8. Constraints

The processed potential is applied to Verlet's method and one obtains the discrete action

$$S_d = J_0 Q_0 - J_1(Q_0 + \Delta Q) + \frac{1}{2h}\Delta Q^T M \Delta Q - \frac{h}{2}(V_h(Q_0) + V_h(Q_0 + \Delta Q)) \quad (3.90)$$

The original variables are recovered through backtransformation from equations (3.85)-(3.86). The method is integrating the variables $(Q, J)$ because the transformation $Q \to q$ is implicit. The method is of fourth order. In practice, the transformation of the coordinates $Q$ can be neglected on the cost of accuracy. The algorithm is explicit, but one has to perform a matrix-vector product with the tangent Hessian at each time step. The critical time step can be determined as $\sqrt{12}/\omega_{max}$.

A similar method was presented by [178] where not accuracy, but stability is optimized.

## 3.8 Constraints

This section presents approaches which discretize Hamilton's principle such that the constraint equations are satisfied by the solution trajectory while geometric properties are preserved. The considered types of constraints range from linear and nonlinear holonomic constraints to inequality equations.

### 3.8.1 Reduction of degrees of freedom by projection

Certain types of linear constraints may be applied to the system prior discretization of time. Consider for example a constraint associating a set of coordinate variables collected in $x^{(2)}$ with another set of variables collected in $x^{(1)}$ by a projection matrix $C$ and an offset $b^{(2)}$,

$$x^{(2)} = Cx^{(1)} + b^{(2)} \quad (3.91)$$

Since both sets of variables belong to the same set $x$, one may rearrange the ordering of the degrees of freedom such that the vector of coordinates is splitted into blocks of dependent and independent degrees of freedom,

$$x = \begin{pmatrix} x^{(0)} \\ x^{(1)} \\ x^{(2)} \end{pmatrix} = \begin{pmatrix} I & 0 \\ 0 & I \\ 0 & C \end{pmatrix} \begin{pmatrix} x^{(0)} \\ x^{(1)} \end{pmatrix} + \begin{pmatrix} 0 \\ 0 \\ b^{(2)} \end{pmatrix} \quad (3.92)$$

Many interface problems, for example constraints arising in mortar methods, may be expressed in this form. More generally, one may define a new set of generalized coordinates $q$ from which the original set of variables $x$ can be recovered by a linear transformation, i.e. with

$$\begin{pmatrix} x^{(0)} \\ x^{(1)} \end{pmatrix} \to q, \quad \begin{pmatrix} 0 \\ 0 \\ b^{(2)} \end{pmatrix} \to b \text{ and } \begin{pmatrix} I & 0 \\ 0 & I \\ 0 & C \end{pmatrix} \to A \quad (3.93)$$

one may express a constraint from equation (3.91) in the form

$$x = Aq + b, \quad \dot{x} = A\dot{q} \quad (3.94)$$

such that the Lagrangian becomes

$$L(q, \dot{q}) = \frac{1}{2}\dot{q}^T A^T M^x A\dot{q} - V^x(Aq + b) \quad (3.95)$$

where $M^x$ denotes the mass matrix and $V^x$ the potential function with respect to the original variables. The equivalent quantities of the reduced system are given by

$$\begin{aligned} M &= A^T M^x A & (3.96) \\ V(q) &= V^x(Aq + b) & (3.97) \\ \nabla V(q) &= A^T \nabla V^x(Aq + b) & (3.98) \\ \nabla^2 V(q) &= A^T \nabla^2 V^x(Aq + b) A & (3.99) \end{aligned}$$

For constraints given by equation (3.91) one assumes the setting

$$M^x = \begin{pmatrix} M_{00} & M_{01} & M_{02} \\ M_{01} & M_{11} & M_{12} \\ M_{01} & M_{12} & M_{22} \end{pmatrix} \quad (3.100)$$

$$\nabla V^x(x) = \begin{pmatrix} F_0 \\ F_1 \\ F_2 \end{pmatrix} \quad (3.101)$$

$$\nabla^2 V^x(x) = \begin{pmatrix} K_{00} & K_{01} & K_{02} \\ K_{01} & K_{11} & K_{12} \\ K_{01} & K_{12} & K_{22} \end{pmatrix} \quad (3.102)$$

and obtains

$$M = \begin{pmatrix} M_{00} & M_{01} + M_{02}C \\ M_{01} + C^T M_{20} & M_{11} + C^T M_{21} + M_{12}C + C^T M_{22}C \end{pmatrix} \quad (3.103)$$

$$\nabla V(q) = \begin{pmatrix} F_0 \\ F_1 + C^T F_2 \end{pmatrix} \quad (3.104)$$

$$\nabla^2 V(q) = \begin{pmatrix} K_{00} & K_{01} + K_{02}C \\ K_{01} + C^T K_{20} & K_{11} + C^T K_{21} + K_{12}C + C^T K_{22}C \end{pmatrix} \quad (3.105)$$

Such a strategy is related to the nullspace method presented in [159] and is used in various FEM implemenations, for example [57]. The advantages of this approach to constraint handling are that

1. standard integrators can be applied directly to the reduced system.

2. Furthermore, both conditions in equation (3.94) are satisfied at all times.

3. The critical time step can be easily estimated.

## 3.8. Constraints

The implementation in software codes becomes more complicated, though, and numerical efficiency can be reduced:

1. The structure of the system matrices is changed. Elements far off the diagonal can appear. The mass matrix is no longer diagonal.

2. Two variable spaces must be maintained. For example, finite elements and contact algorithms are expressed in unconstrained space. Quantities such as restoring forces, contact forces, gradients must be transformed at each time step.

### 3.8.2 Lagrange-d'Alembert principle

In order to discretize nonlinear constraints, one has to introduce them to Hamilton's principle. This is performed using the d'Alembert-Lagrange procedure:

**Theorem 23.** *Given a mechanical system with Lagrange equation (2.1) and a time dependent external force $F(t)$ the principle of d'Alembert (Lagrange-d'Alembert) states that the solution trajectory is determined by*

$$0 = \delta \int_{t_0}^{t_1} L(q,\dot{q})dt + \int_{t_0}^{t_1} F(t)\delta q\, dt \qquad (3.106)$$

If the forces $F$ can be derived from a potential function they are conservative. Else they can be used to introduce dissipative effects.

Consider a mechanical system with coordinates $q$ that are constrained by a set of smooth holonomic equations summarized in the vector $g$,

$$0 = g(q) = \begin{pmatrix} g_1(q) \\ \vdots \\ g_c(q) \end{pmatrix} \qquad (3.107)$$

which Jacobian is of full rank, i.e.

$$\mathrm{rank}(\nabla g(q)) = c, \; \forall q | g(q) = 0 \qquad (3.108)$$

Associate these constraints with a potential function

$$V^g(q, g(q)) \geq 0 \, \forall q, \quad V^g(q, g(q)) = 0 \Leftrightarrow g(q) = 0 \qquad (3.109)$$

for example

$$V^g(q, g(q)) = \epsilon g(q)^T g(q) \qquad (3.110)$$

which penalizes violations of the constraints. The larger the penalty parameter $\epsilon$, the smaller the violation, i.e.

$$\lim_{\epsilon \to \infty} g(q(t)) = 0 \, \forall t \qquad (3.111)$$

Define Lagrange multipliers [224]

$$\lambda = \lim_{\epsilon \to \infty} \epsilon g(q) \tag{3.112}$$

such that $V^g(q,\lambda) = g(q)^T\lambda$. Therefore, Hamilton's principle extended to holonomic constraints $g$ requires stationarity of the augmented action

$$S^g = \int_{t_0}^{t_1} \left( L(q,\dot{q}) - g(q)^T\lambda \right) dt \tag{3.113}$$

wherein $q = q(t)$ and $\lambda = \lambda(t)$ are vectors of generalized coordinates. The resulting Euler-Lagrange equations are

$$0 = L_q - \frac{d}{dt} L_{\dot{q}} - \nabla g(q)^T \lambda \tag{3.114}$$

$$0 = g(q) \tag{3.115}$$

### 3.8.3 SHAKE

**Theorem 24.** *Given a discrete Lagrangian system $L_d$ with holonomic constraint $g : Q \to \mathbb{R}^c$ the constrained discrete Euler-Lagrange equations (see [185]) are given by*

$$D_2 L_d(q_{k-1}, q_k) + D_1 L_d(q_k, q_{k+1}) = \nabla g(q_k) \lambda_k \tag{3.116}$$

$$g(q_k) = 0 \tag{3.117}$$

A method enforcing these constraints is SHAKE [229]. It arises from an augmented discrete Lagrangian $L_d^A$ wherein the Lagrange multipliers $\lambda$ are considered as a part of the vector of generalized coordinates $q$. $\lambda \cdot g(q)$ is considered as a part of the potential function $V(q, \lambda)$. The potential function is integrated by Verlet, see equation (3.41) and [159], i.e.

$$V_d = \frac{h^k}{2} \left( V(q_0^k) + V(q_1^k) \right) + \frac{h^k}{2} \left( g(q_0^k)^T \lambda_0^k + g(q_1^k)^T \lambda_1^k \right) \tag{3.118}$$

The multipliers are continuous in time. One arrives at the scheme

$$M \left( \frac{q_{k+1} - 2q_k + q_{k-1}}{h} \right) + h \nabla V(q_k) + \nabla g(q_k) \lambda_k = 0 \tag{3.119}$$

$$g(q_{k+1}) = 0 \tag{3.120}$$

The system of equation is nonlinear and its tangential coefficient matrix is not symmetric (if $\nabla g \neq const.$).

When using implicit methods, an enforcement of the constraint at the same time where the potential forces are evaluated is desirable. For the implicit midpoint method this leads to the discrete potential

$$V_d = -\frac{h^k}{2} \left( V \left[ \left( \frac{q_0^k + q_1^k}{2} \right) \right] + g \left[ \left( \frac{q_0^k + q_1^k}{2} \right) \right]^T \lambda^k \right) \tag{3.121}$$

## 3.8. Constraints

The system of equation is nonlinear, its tangential coefficient matrix is symmetric. The scheme is, however, unstable and not suitable for long-term simulation. While the constraints $g$ are satisfied in the time step center they are not at the time step boundaries. The violation at the boundaries grows during the time leading to instabilities in terms of artificial oscillations, see [115].

The SHAKE algorithm requires initial conditions $(q_{-1}, q_0)$ satisfying the constraints. This must be ensured before starting the simulation.

### 3.8.4 RATTLE

Using the map $(q_k, j_k) \to (q_{k+1}, j_{k+1})$ SHAKE may not be applied. SHAKE requires the enforcement of two constraint sets, at $q_0$ and $q_{-1}$. When expressing the integrator in phase space variables, one again has to enforce two constraints on the initial conditions $q_0$ and $j_0$. Those are $g(q_0) = 0$ and the hidden constraint

$$\dot{g}(q_0) = \nabla g(q_0)^T M^{-1} j_0 = 0 \tag{3.122}$$

**Theorem 25.** *Given a discrete Lagrangian system $L_d$ with holonomic constraint $g : Q \to \mathbb{R}^c$ the constrained discrete Hamiltonian map (see [185]) is given by*

$$j_0 = \mathbb{P}_{q_0}(-D_1 L_d(q_0, q_1)) \tag{3.123}$$
$$j_1 = \mathbb{P}_{q_1}(D_2 L_d(q_0, q_1)) \tag{3.124}$$
$$g(q_1) = 0 \tag{3.125}$$

*Here, $\mathbb{P}_q$ is a projection matrix*

$$\mathbb{P}_q = I - \nabla g \left[ (\nabla g)^T M^{-1} \nabla g \right]^{-1} (\nabla g)^T M^{-1} \tag{3.126}$$

An application to Velocity Verlet is RATTLE (see [3]). It is given by

$$q_1^k = q_0^k + h M^{-1} \left( j_0^k - \frac{h}{2} \nabla V(q_0^k) - \frac{h}{2} \nabla g(q_0^k) \lambda_0^k \right) \tag{3.127}$$
$$j_1^k = \frac{1}{h} M(q_1^k - q_0^k) - \frac{h}{2} \nabla V(q_1^k) - \frac{h}{2} \nabla g(q_1^k) \lambda_1^k \tag{3.128}$$
$$0 = g(q_1^k) \tag{3.129}$$
$$0 = (\nabla g(q_1^k))^T M^{-1} j_1^k \tag{3.130}$$

see algorithm 3.7. Given an initial condition $(q_0, j_0)$ satisfying $g(q_0) = 0$ one first projects $j_0^+ = \mathbb{P}_{q_0} j_0$. Then, in each step the coordinates $q_1^k$ and multipliers $\lambda_0^k$ are iteratively computed enforcing $g(q_1^k) = 0$. After that, $j_1^k$ and multipliers $\lambda_1^k$ are computed from a linear system of equation enforcing $\dot{g}_1^k = 0$. The last operation is equivalent to application of the projection $\mathbb{P}_{q_1}$ to $j_1^k$. This illustrates the relation to SHAKE: RATTLE is SHAKE with an additional projection of the momentum $j_{k+1}$ onto the constraint manifold.

---

**Given** $(q_0, j_0)$
**Enforce hidden constraint:** $j_0 = \mathbb{P}_{q_0} j_0$
**for** $k = 0$ to $N-1$ **do**
    half a kick:
$$j_{k+0.5} = j_k - \frac{h}{2}\nabla V(q_k)$$
    constrained drift:
$$j^\lambda_{k+0.5} = j_{k+0.5} - \frac{h}{2}\nabla g(q_k)\lambda_0^k$$
$$q_{k+1} = q_k + hM^{-1}j^\lambda_{k+0.5}$$
$$0 = g(q_{k+1})$$
    half a kick:
$$j^-_{k+1} = j^\lambda_{k+0.5} - \frac{h}{2}\nabla V(q_{k+1})$$
    enforce hidden constraint:
$$j_{k+1} = \mathbb{P}_{q_{k+1}} j^-_{k+1}$$
**end for**

Figure 3.7: RATTLE

---

### 3.8.5 Unilateral constraints

Consider a Hamiltonian system subject to a set of inequations collected in a constraint vector $g$, i.e.

$$g(q) = \begin{pmatrix} g_1(q) \\ \vdots \\ g_c(q) \end{pmatrix} \leq 0 \qquad (3.131)$$

Inequalities require a nonsmooth setting. At any time, only a subset or no constraint may be active. As soon as a constraint $g_j$ is activated, that is $\lim_{h \to 0, h > 0} g_j(q(t_c - h)) < 0 \to g_j(q(t_c)) = 0$, the trajectory of the generalized coordinates must be modified to stay feasible. The involved velocity changes generally are discontinuous (the trajectory is nonsmooth).

Parameterize the time with respect to a fictitious time $\alpha$. Assume that the given set of inequations is active once during the considered time interval and that all constraints are active at the same time $t \in \{t(\alpha_c^-) \ldots t(\alpha_c^+)\}$. The action integral becomes

$$S = \int_{t(\alpha_0)}^{t(\alpha_c^-)} L(q, \dot{q})\frac{\partial t}{\partial \alpha} d\alpha + \int_{t(\alpha_c^-)}^{t(\alpha_c^+)} \left(L(q, \dot{q}) - g(q)^T \lambda\right)\frac{\partial t}{\partial \alpha} d\alpha + \int_{t(\alpha_c^+)}^{t(\alpha_1)} L(q, \dot{q})\frac{\partial t}{\partial \alpha} d\alpha \qquad (3.132)$$

## 3.8. Constraints

where the activation times $\alpha_c$ belong to the unknown variables (see [58]). Variation yields

$$\begin{aligned}
\delta S &= [L_{\dot q}(q,\dot q)]_{t(\alpha_0)}^{t(\alpha_1)} + \\
&+ \int_{t(\alpha_0)}^{t(\alpha_c^-)} \left(L_q - \frac{d}{dt}L_{\dot q}\right)\delta q\, dt + \int_{t(\alpha_c^+)}^{t(\alpha_1)} \left(L_q - \frac{d}{dt}L_{\dot q}\right)\delta q\, dt + \\
&+ \int_{t(\alpha_c^-)}^{t(\alpha_c^+)} \left(\left(L_q - \frac{d}{dt}L_{\dot q} - g_q^T \lambda\right)\delta q - g^T \delta\lambda\right)\frac{\partial t}{\partial \alpha}d\alpha + \\
&+ \left[-g\left[q(t(\alpha))\right]^T \lambda \frac{\partial t}{\partial \alpha}\delta\alpha\right]_{\alpha_c^-}^{\alpha_c^+}
\end{aligned} \quad (3.133)$$

where the variation $\delta \dot q$ is transformed into $\delta q$ using integration by parts and where the variation of the (de)activation times $\alpha_c$ is determined from $\frac{\partial \int_a^b f(x)dx}{\partial b} = f(b)$, $\frac{\partial \int_a^b f(x)dx}{\partial a} = -f(a)$.

For further discussion a few simplifications are assumed: (1) Inequalities are assumed to be active at discrete infinitesimal time steps, i.e. $h = (\alpha_c^+ - \alpha_c^-)$ with $\lim_{h\to 0}$. (2) Each inequation $g_j$ may become active at individual times $\alpha_{c,j}$, but assume that an active set $g^a$ is established at a global activation time $\alpha_c$; these are all constraints $g_j(q(t(\alpha_c))) \geq 0$. (3) Let it be sufficient to check the active sets at the end of each time step. The strategy is illustrated in figure 3.8.

Apply a linear interpolation for $q(t)$ to obtain the discrete action for the infinitesimal element,

$$S^k = \lim_{h \to 0}\left(L_d + j_0(q(0) - q_0) + j_1(q_1 - q(h)) - h\left(g(q(0)) + g(q(h))\right)^T \mu\right) \quad (3.134)$$

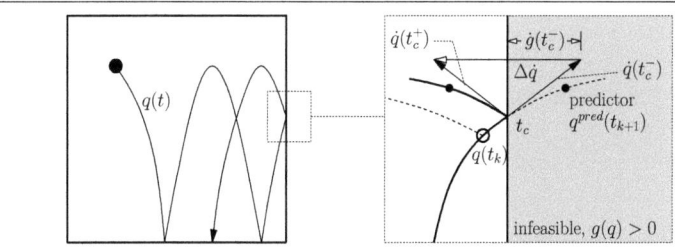

Figure 3.8: Collision of a particle under gravity. The trajectory $q(t)$ is illustrated. The change of momentum at the boundary of the infeasible domain (or velocity, respectively) obviously is a non-smooth process. The constraint can be evaluated at the predictor coordinate $q^{pred}(t_{k+1})$. If $t_c$ is unknown the collision is energy preserving if it is applied at $t_k$.

with discrete Lagrangian $L_d$ from equation (2.7) and multiplier $\mu$. The variation yields

$$0 = q_1 - q_0 - u_1 \tag{3.135}$$

$$0 = \frac{\partial L_d}{\partial u_0} + j_0 - j_1 - h\nabla\left(g(q_0) + g(q_0 + u_1)\right)\mu \tag{3.136}$$

$$0 = \frac{\partial L_d}{\partial u_1} - j_1 - h\nabla g(q_0 + u_1)\mu \tag{3.137}$$

$$0 = h(g(q_0) + g(q_0 + u_1)) \tag{3.138}$$

With $v := \frac{u_1}{h}$ and observing that $V(q)$ and its derivatives are bounded such that their action approaches zero as $h \to 0$, one obtains

$$q_1 = q_0 + hv \tag{3.139}$$

$$j_1 = j_0 - \nabla\left(g(q_0) + g(q_0 + u_1)\right)h\mu \tag{3.140}$$

$$0 = j_0 - Mv - \nabla g(q_0)h\mu \tag{3.141}$$

$$0 = g(q_0 + u_1) \tag{3.142}$$

$$g(q_0 + u_1) = g(q_0) + \sum_{i=1}^{\infty} \nabla^i g(q_0)^T \frac{(vh)^i}{i!} \tag{3.143}$$

Defining $\lambda := h\mu$ and since $g(q_0) = 0$, one obtains

$$q_1 = q_0 + \mathcal{O}(h) \tag{3.144}$$

$$j_1 = j_0 - 2\nabla g(q_0)\lambda + \mathcal{O}(h) \tag{3.145}$$

$$0 = j_0 - Mv - \nabla g(q_0)\lambda \tag{3.146}$$

$$0 = \nabla^i g(q_0)^T v + \mathcal{O}(h) \tag{3.147}$$

Since $h \to 0$ one can neglect all bounded terms multiplied with $h$. First one solves $v = M^{-1}(j_0 - \nabla g(q_0)^T\lambda)$ and inserts this into the last equation to obtain

$$\lambda = \left(\nabla g(q_0)^T M^{-1} \nabla g(q_0)\right)^{-1} \nabla g(q_0)^T M^{-1} j_0 \tag{3.148}$$

The resulting map defines a collision integrator solving inequality constrained dynamics. The determination of the active set of constraints is called collision detection. The collision integrator is given by

$$q_1 = q_0 \tag{3.149}$$

$$j_1 = j_0 - 2\nabla g(q_0)\left(\nabla g(q_0)^T M^{-1} \nabla g(q_0)\right)^{-1}\left(\nabla g(q_0)^T M^{-1} j_0\right) \tag{3.150}$$

This is equivalent to a projection by $\mathbb{P}_q^c$ of the momentum "against" the constraint manifold such that the constraint rate changes its sign, i.e.

$$\mathbb{P}_q^c = I - 2\nabla g\left[(\nabla g)^T M^{-1} \nabla g\right]^{-1}(\nabla g)^T M^{-1} \tag{3.151}$$

It does change the velocities in a discontinuous setting without modifying the coordinates. It can only be used to avoid constraint violations, but generally not to recover a trajectory once it is in the infeasible domain. This is because the only available information are an indicator for the active set, the constraint gradients and the assumption that the last iterate has been feasible.

The collision integrator may be applied as an additional operation between any two time elements. Due to the discrete nature of collision detection certain iterates may be infeasible. In such cases, the map would enforce infeasibility since it is not aware if a constraint is approaching from the feasible or infeasible domain. Define the constraint rate (constraint velocity),

$$\dot{g}_0 := \nabla g(q_0)^T M^{-1} j_0 \qquad (3.152)$$

This quantity may be used to filter slightly violated and active constraints which will be feasible at the next time increment and which would be modified to stay infeasible by the collision integrator. Then the active set $\mathcal{G}^A$ is

$$\mathcal{G}^A = \{i \,:\, g_i(q) \geq 0,\, \dot{g}_i(q, j) > 0\} \qquad (3.153)$$

The collision integrator preserves the energy $E$ (with $\nabla g := \nabla g(q_0)$)

$$\begin{aligned} E_c^+ - E_c^- &= \left(\tfrac{1}{2} j_1^T M^{-1} j_1 + V(q_1)\right) - \left(\tfrac{1}{2} j_0^T M^{-1} j_0 + V(q_0)\right) && (3.154) \\ &= -2\lambda^T \nabla g M^{-1} j_0 - 2 j_0^T M^{-1} \nabla g \lambda + \\ &\quad + 4\lambda^T \nabla g^T M^{-1} \nabla g \lambda && (3.155) \\ &= -4 j_0^T M^{-1} \nabla g \left(\nabla g^T M^{-1} \nabla g\right)^{-1} \nabla g^T M^{-1} j_0 \\ &\quad + 4\lambda^T \left(\nabla g^T M^{-1} \nabla g\right) \left(\nabla g^T M^{-1} \nabla g\right)^{-1} \nabla g^T M^{-1} j_0 && (3.156) \\ &= 0 && (3.157) \end{aligned}$$

The presented approach is known as Decomposition Contact Response (see [36]). It can be combined with plastic unilateral contact laws and friction. It is designed to efficiently solve collisions in explicit dynamics.

## 3.9 Examples

### 3.9.1 A perturbed linear oscillator

Compare the stability of the presented methods when applied to the linear oscillator given in figure 3.9. The material properties are given by $k = 1$, $m = 1$. The

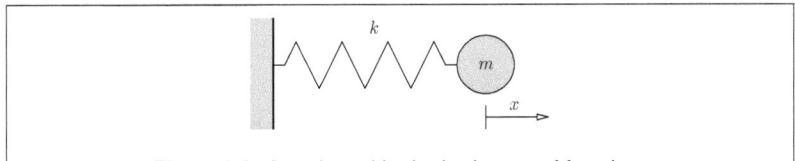

Figure 3.9: A spring with single degree of freedom

stiffness is assumed to be $k_0 = 2/3$ which is used in the implicit/vibrational parts of the methods. Then the potential is given by

$$V(q) = \frac{1}{2}kq^2 = V^{fast}(q) + V^{slow}(q) = \frac{1}{2}k_0 q^2 + \frac{1}{2}(k - k_0)q^2 \qquad (3.158)$$

The initial conditions are $q_0 = 0$ and $\dot{q}_0 = 0.2$ ($x \to q$).

500 time steps are simulated for each configuration. For r-RESPA the "fast" time steps are defined by $\delta t = \Delta t/10$. The methods are analysed regarding their stability for different time step lengths ranging from 0.05 to 15. The Lyapunov exponent is approximated from equation (3.34). The results for different methods are summarized in figure 3.10 with time step length on the abcissa and the Lyapunov exponent as ordinate. Figure 3.11 plots a simple indicator for the error measured in terms of the maximum deviation of the energy $H(t)$ related to the initial energy $H_0$, i.e. $\max_k |H(k,h) - H_0|/H_0$ as a function of the time step length. It can be interpreted as a measure for the overestimation of the maximum displacements. This may be a more important indicator than the absolute error in $q(t)$ since applications in structural dynamics are more interested in the size of stresses and plastic strains as a result of maximum strains. Phase shifts and stretches usually are of less concern.

For the standard methods Verlet/Rowlands and for IMEX, there exist clear stability bounds. This is emphasized by the energy error. The effect of resonances becomes very clear for r-RESPA. The same effects lead to instabilities for Deuflhard and Gautschi, but with different interval bounds. Hairer-Lubich's method is stable, but the energy error is unacceptable in the regions where Gautschi is instable. If the assumption $k_0$ was closer to the real stiffness $k$, the intervals of instability would be much narrower. Remarkably all mollified methods are stable. Most noteworthy, the energy error of mollified Hairer-Lubich and mollified IMEX are smallest.

## 3.9. Examples

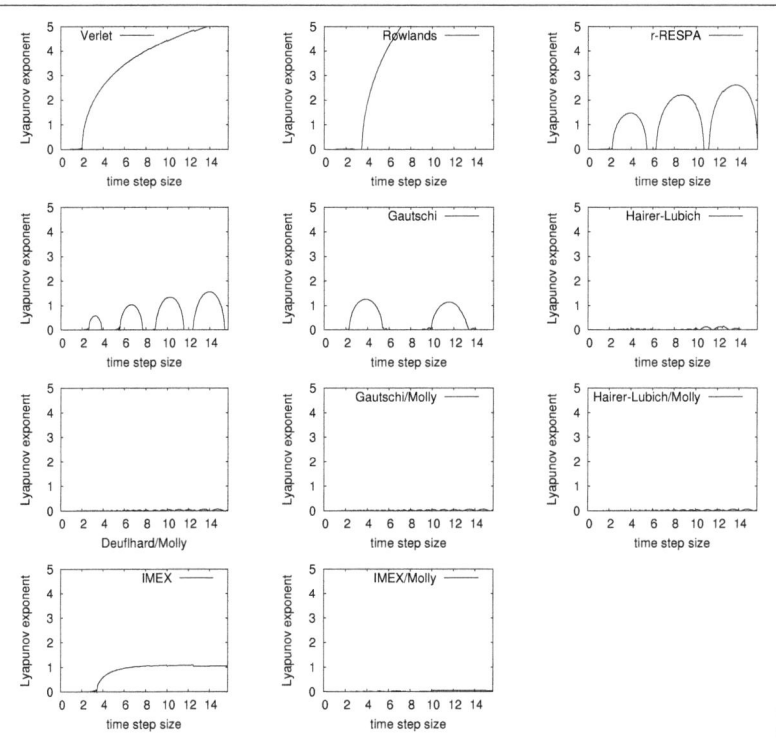

Figure 3.10: Perturbed linear oscillator: Stability of different algorithms measured in terms of the estimated Lyapunov exponent as a function of the time step length $\lambda_k = \lambda_k(h)$.

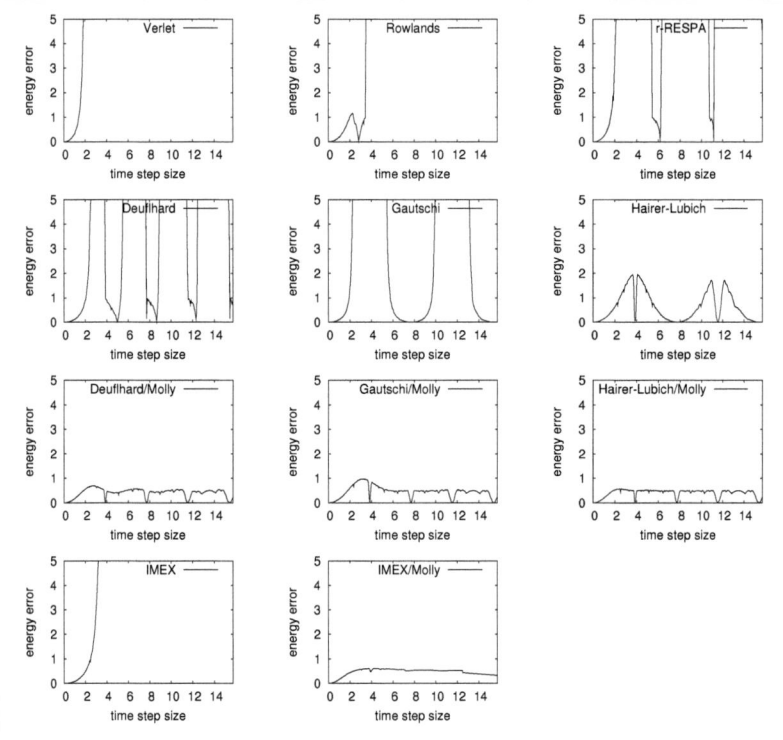

Figure 3.11: Perturbed linear oscillator: Energy error of different algorithms measured in terms of the maximum relative error in energy $\max_k |H(k,h) - H_0|/H_0$ as a function of the time step length.

## 3.9.2 A geometrically nonlinear oscillator

The same setting as in section 3.9.1 will be considered. But this time, a geometrically nonlinear spring is used. With the Green-Lagrangian strain tensor

$$E(q) = 0.5((1 + q/X)^2 - 1) \qquad (3.159)$$

one obtains the potential function

$$V^{fast}(q) = \frac{1}{2}k_0 q^2 \qquad (3.160)$$

$$V(q) = V^{slow}(q) + V^{fast}(q) \qquad (3.161)$$

$$= \left(\frac{1}{2}k_0 E(q)^2 - V^{fast}(q)\right) + V^{fast}(q) \qquad (3.162)$$

The material properties are given by $k_0 = 1$, $m = 1$. The initial conditions are $q_0 = 0$ and $j_0 = 0.2$ ($x \to q$). 500 time steps are simulated for each configuration. For r-RESPA the "fast" time steps are defined by $\delta t = \Delta t/10$.

The methods are analysed regarding their stability for different time step lengths ranging from $0.05$ to $15$. The Lyapunov exponent is approximated from equation (3.34). The results for different methods are summarized in figure 3.12 with time step length on the abcissa and the Lyapunov exponent as ordinate. Figure 3.13 plots a the error in energy $\max_k |H(k,h) - H_0|/H_0$ as a function of the time step length.

For the methods Verlet, Rowlands and r-RESPA, there exist relatively clear stability bounds which are narrower than in the linear case. This is underlined by the energy error. At random points, no instability was measured by the Lyapunov exponent. The instabilities of the base integrators in r-RESPA dominate the resonances. This is different for Deuflhard, Gautschi, Hairer-Lubich and IMEX where almost randomly set intervals are either stable or instable. The vibrational part is implicitly determined and may filter some instable kicks from the nonlinear forces. Best performs the method of Hairer-Lubich in this regard. The intervals of very large energy error correspond to those of the perturbed linear problem. Again, all mollified methods render a stable Lyapunov exponent estimator. For mollified Gautschi and mollified Deuflhard, however, there appear very large peaks in the energy error which indicate instability as $k \to \infty$.

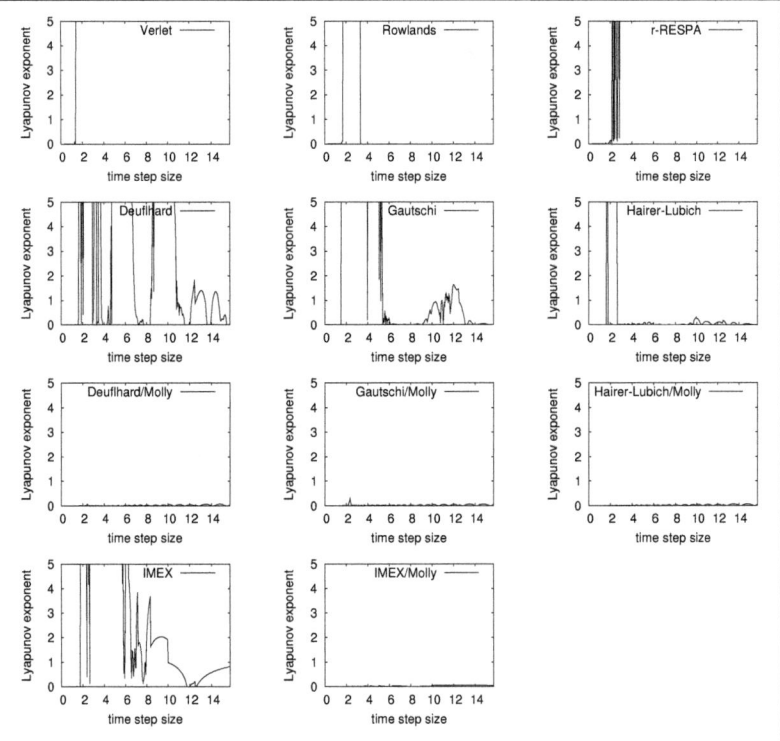

Figure 3.12: Geometrically nonlinear oscillator: Stability of different algorithms measured in terms of the estimated Lyapunov exponent as a function of the time step length $\lambda_k = \lambda_k(h)$.

## 3.9. Examples

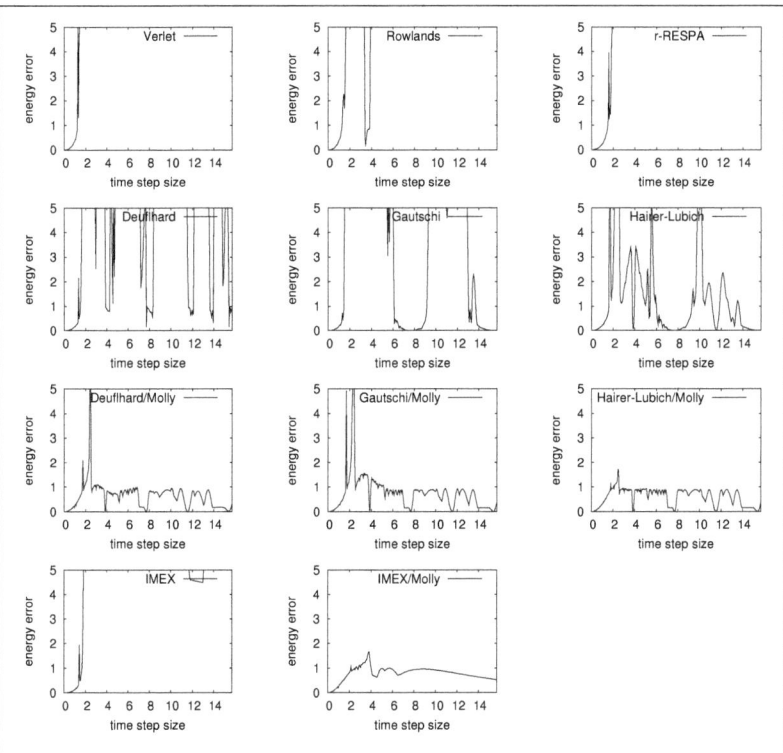

Figure 3.13: Geometrically nonlinear oscillator: Energy error of different algorithms measured in terms of the maximum relative error in energy $\max_k |H(k,h) - H_0|/H_0$ as a function of the time step length.

### 3.9.3 Nonlinear vibration of a cantilever beam

Consider a cantilever beam with square cross section as illustrated in figure 3.14. The geometry is defined by $L = 100$, $B = H = 10$; the linear elastic material is defined by elastic modulus $E = 30 \times 10^9$, Poisson's ratio $\nu = 0$ and mass density $800$ with geometrically nonlinear strains (Green-Lagrange). The beam is discretized by $30 \times 3 \times 3$ 8-noded brick elements. The initial conditions are given by zero displacements $q_0 = 0$ and an 'angular velocity' around the beam's left end, i.e. $v^y_{A,0} = -\omega X_A$ with node $A$ and coordinate horizontal $X_A$. The linear forces are approximated by the initial stiffness matrix of the beam.

This example is used to test the stability of mollified Hairer-Lubich and mollified IMEX when applied to multiple degree of freedom systems. In order to apply the matrix exponential of Hairer-Lubich efficiently to this problem, the number of variables is reduced. This is done by modal reduction. The eigenvectors associated with the 200 smallest natural frequencies are determined and used to diagonalize the system, see section 3.7.5 and equation (3.76). The tested integrators are then individually applied to the decoupled degrees of freedom. The critical time step of Verlet was computed as $h_{crit} = 0.769 \times 10^{-3}$. The reference solution is given by Verlet with $h = h_{crit}/2$. The simulation time is $T = 1$.

**Small nonlinearities**

The initial conditions are given by $\omega = 0.1$. The nonlinearities are small. I.e. the reference simulation obtained an approximate value of $V^{slow}(q) \approx 0.1V(q)$. A critical time step of mollified IMEX could not be found. Up to $h = 4000h_{crit}$ all tested time step lengths were stable. Most interestingly, the mollified Hairer-Lubich method (as all other exponential integrators) did not increase the stability limit compared with Verlet.

Figure 3.15 and 3.16 present the energy $H(q_k, j_k)$ (Notice: not the discrete energy in (3.12) which is nearly constant!) and the maximum displacement vector component $q_{max}(t) = \max_i q_i(t)$ for $h = 20h_{crit}$.

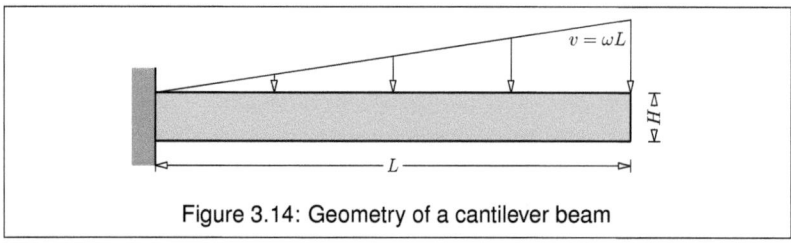

Figure 3.14: Geometry of a cantilever beam

## 3.9. Examples

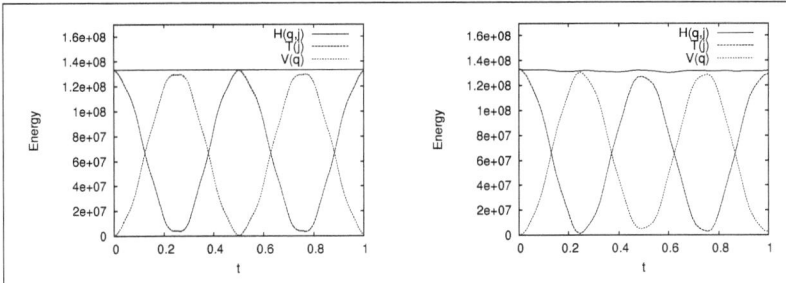

Figure 3.15: Cantilever beam: energy of IMEX/Molly for small nonlinearities.
Left: reference, Right: IMEX/Molly.

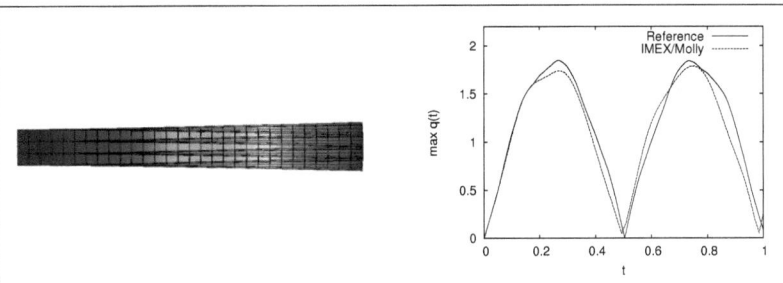

Figure 3.16: Cantilever beam: displacements of IMEX/Molly for small nonlinearities
Left: deformed configurations at different times. Right: $\max_i q_i(t)$.

**Moderate nonlinearities**

The initial conditions are given by $\omega = 0.3$. The nonlinearities are moderate. I.e. the reference simulation obtained an approximate value of $V^{slow}(q) \approx V(q)$. A critical time step of mollified IMEX could not be found. Up to $h = 4000 h_{crit}$ all tested time step lengths were stable.

Figure 3.17 and 3.18 present the energy and the maximum displacement vector component $q_{\max}(t)$ for $h = 20 h_{crit}$.

**Large nonlinearities**

The initial conditions are given by $\omega = 6$. The nonlinearities are large, i.e. the reference simulation obtained an approximate value of $V^{slow}(q) \approx 7V(q)$. A critical time step of mollified IMEX was found at approximately $h_{crit}^{IMEX} \approx 5 h_{crit}^{Verlet}$.

Figure 3.19 and 3.20 present the energy and the maximum displacement vector component $q_{\max}(t)$ for $h = 4 h_{crit}^{Verlet}$.

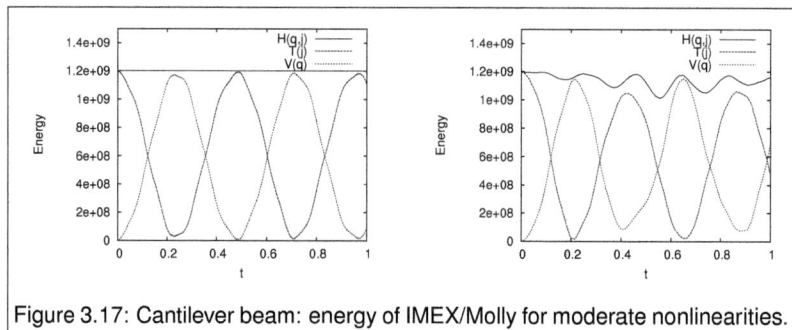

Figure 3.17: Cantilever beam: energy of IMEX/Molly for moderate nonlinearities. Left: reference, Right: IMEX/Molly, $h = 20 h_{crit}$.

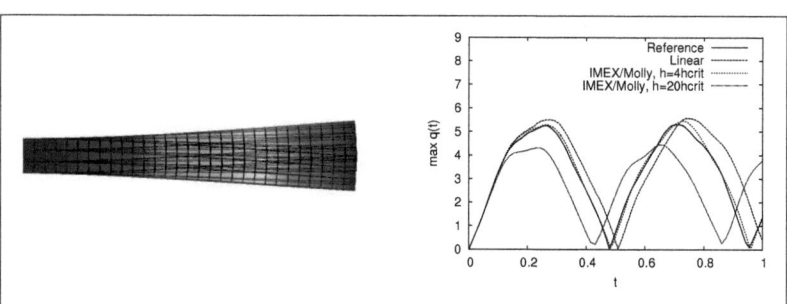

Figure 3.18: Cantilever beam: displacements of IMEX/Molly for moderate nonlinearities
Left: deformed configurations at different times. Right: $\max_i q_i(t)$.

**Large nonlinearities without spectral transformation**

The initial conditions are given by $\omega = 6$. The nonlinearities are large. For this test, all integrators are directly applied to the FEM model, i.e. without modal reduction which may stabilize the kinematics. The critical time step is $h_{crit}^{Verlet} = 0.158 \cdot 10^{-3}$.

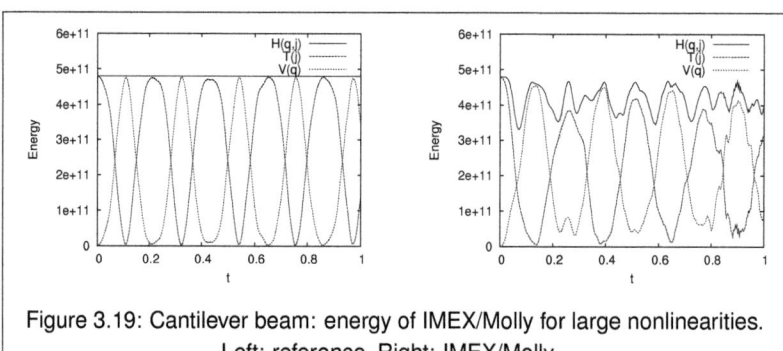

Figure 3.19: Cantilever beam: energy of IMEX/Molly for large nonlinearities. Left: reference, Right: IMEX/Molly.

## 3.9. Examples

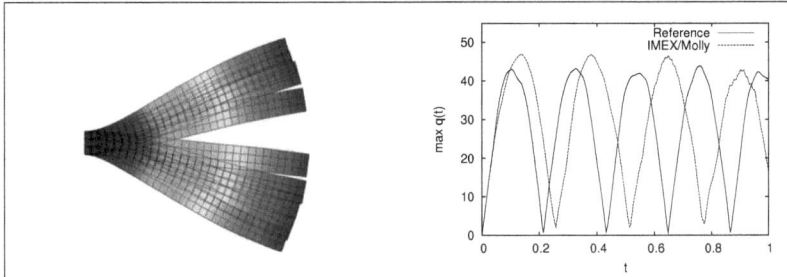

Figure 3.20: Cantilever beam: displacements of IMEX/Molly for large nonlinearities
Left: deformed configurations at different times. Right: $\max_i q_i(t)$.

For IMEX/Molly a critical time step of $h_{crit}^{IMEX} \approx 32 h_{crit}^{Verlet}$ was found for the given initial condition.

Figure 3.21 presents the measured maximum displacement as a function of time. IMEX without MOLLY and Verlet are nearly identical for the same time step. The averaging operator clearly filters important components of the dynamics. Even with identical time step, the frequency is increased and the elongation reduced when compared with the reference solution. The mechanical system is indeed replaced by another one. The perturbation is dependent on the size of the nonlinear forces. Since the averaging operator is a function of the used time step, the response changes with increasing time step.

### 3.9.4 Linear vibration of a beam

In the previous sections the methods IMEX/Molly and Hairer-Lubich/Molly were shown to be unconditionally stable for single degree of freedom systems, but not

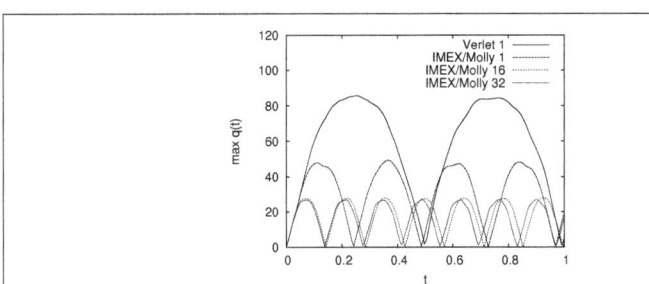

Figure 3.21: Cantilever beam: displacements of IMEX/Molly without modal reduction.
$\max_i q_i(t)$ for various methods: Verlet with $h = h_{crit}^{Verlet}$, IMEX/Molly with $h = h_{crit}^{Verlet}$, $h = 16 h_{crit}^{Verlet}$ and $h = 32 h_{crit}^{Verlet}$.

for more complex systems. Why is that so? The example given in section 3.9.3 is now considered using a geometrically linear strain formulation.

The beam is discretized using standard continuum finite elements (8-noded bricks). The stiffness matrix of the linear forces is assumed to differ from the actual one by a scalar factor $\kappa$, i.e.

$$K_0 = \nabla^2 V^{fast} = \kappa \nabla^2 V(0) \tag{3.163}$$

Using different factors (1.2, 1.5, 2.0) no instabilities were identified for time steps up to $1000 h_{crit}$. The reason is that the system of partial differential equation can be diagonalized. Then the same results apply as for the single degree of freedom system in section 3.9.1.

Let us change the structure of the stiffness matrix. The beam itself is discretized using standard finite elements. For the approximation of the linear forces a different element formulation may be used, such that different couplings between individual degrees of freedoms occur. This can be achived using nodally integrated finite elements, see section 4.6. Although the nodally integrated elements are used without any stabilization, a standard Verlet integration leads to similar results as the standard finite elements. Using IMEX, however, no stability was found with or without mollyfication. Time steps between $10^{-3} h_{crit}$ and $10 h_{crit}$ were tested. The reason is not that the explicit forces (of the perturbation) are too stiff - when considering the critical time steps of the involved potential functions one obtains: for the linear forces $(0.5 q' K_0 q)$: $0.0885 \cdot 10^{-3}$, for the complete potential $(V(q))$: $0.159 \cdot 10^{-3}$ and for the perturbation $(V(q) - 0.5 q' K_0 q)$: $0.106 \cdot 10^{-3}$, which is much larger than the smallest tested time step size.

Therefore, one can assume that the amount of stabilization regarding nonlinearities using mollified impulses in IMEX depends on the structure of the tangential stiffness matrices of the linear and the nonlinear forces. The magnitude of the perturbation itself is less important than the magnitude of (perturbed) couplings between individual degrees of freedom. This is the reason why small geometrical nonlinearities can be handled well: The structure of the stiffness matrix does not change too much.

### 3.9.5 Nonlinear pendulum: Instabilities in implicit midpoint and trapezoidal Newmark

The nonlinear pendulum is used to demonstrate the performance of constrained integrators, see also section 3.9.6. The geometry is illustrated in figure 3.22 with $L = 1$ and $m = 1$. The gravity is ignored. The motion is described in two-dimensional cartesian coordinates. The initial conditions are given $q_0 = (0, -1)$ and $j_0 = (1, 0)$. This is equivalent to the polar coordinates $\theta_0 = 0$ and $\omega_0 = 1$. The motion is subject to the constraint

$$g(q) = q_1^2 + q_2^2 - L^2 \tag{3.164}$$

## 3.9. Examples

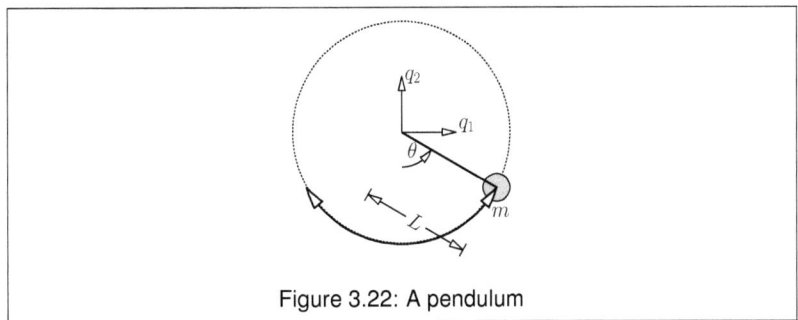

Figure 3.22: A pendulum

Due to the missing external forces the behavior is dependent on the constraint enforcement only.

The instabilities of the implicit midpoint scheme, equation (3.121), can be easily identified. The example is excerpted from [115] who reported these instabilities as well. The time step is chosen $h = 0.04$, the simulation time is $T = 5$. The trajectory is illustrated in figure 3.23 by $x$ and $y$ coordinates. Figure 3.24 presents the corresponding total energy measured in terms of two quantities: The endpoint energy $H(q_k, j_k)$ and the discrete energy $E_d$, see equation (3.12). Both energies grow beyond bounds indicating the instability of the method. The end point energy grows immediately while the discrete energy is approximately constant in the beginning. The trajectory exhibits spurious oscillations in the coordinates around the true solution which grow over time. Indeed, if the considered simulation time is sufficiently small, the instabilities are nearly invisible.

Given linear constraints, the endpoint coordinates would be feasible if the startpoint and midpoint satisfy the constraints (this can be easily proved by inserting the endpoint coordinate of the midpoint scheme into the definition of the linear constraint equation). This is different in case of nonlinear constraints, where the coordinates at the beginning and the end of a time step are generally not feasible if the constraints are enforced in the center. In fact, the given constraint is exactly enforced in the time step center in this example. The start and end point coordinates then start to oscillate around the midpoint. The implicit midpoint scheme

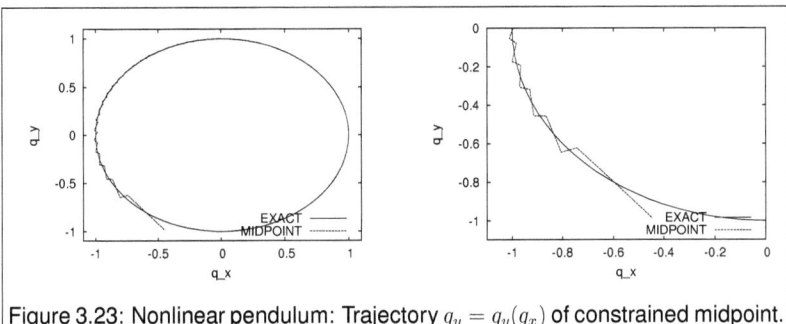

Figure 3.23: Nonlinear pendulum: Trajectory $q_y = q_y(q_x)$ of constrained midpoint.

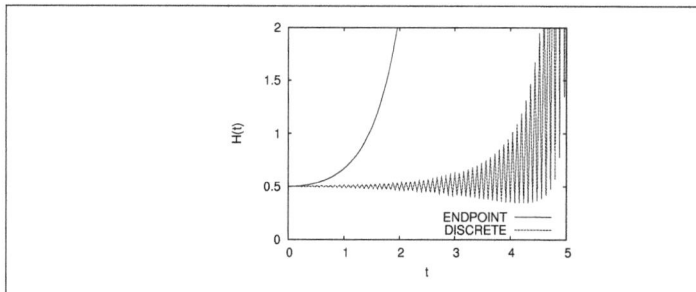

Figure 3.24: Nonlinear pendulum: Total energy of constrained midpoint.

can, therefore, be interpreted in terms of reduced order integration and the instabilities in terms of 'temporal hourglassing'. The choice of the position, where the constraint is satisfied, is not able to enforce a reasonable 'smooth' interpolation of the coordinates $q(t)$ in time.

The observation of instabilities in the implicit midpoint scheme may explain the occurence of spurious oscillations in the velocities in the Newmark method when applied to nonlinear constraints [48]. The trapezoidal Newmark scheme can be interpreted as the implicit midpoint scheme, shifted by a half time step [119]. To illustrate this, express the trapezoidal Newmark scheme without discrete velocities

$$0 = \frac{1}{h^2}(x_{k+2} - 2x_{k+1} + x_k) - \frac{1}{4}A_{k+2} - \frac{1}{2}A_{k+1} - \frac{1}{4}A_k$$
$$A_k = -M^{-1}\nabla V(x_k) \tag{3.165}$$

The implicit midpoint scheme writes

$$0 = \frac{1}{h^2}(q_{k+2} - 2q_{k+1} + q_k) - \frac{1}{2}a_{k+1.5} - \frac{1}{2}a_{k+0.5}$$
$$a_{k+0.5} = -M^{-1}\nabla V(q_{k+0.5})$$
$$q_{k+0.5} = \frac{1}{2}(q_k + q_{k+1}) \tag{3.166}$$

These two schemes are identical with the transformation

$$x_k = \frac{1}{2}(q_k + q_{k+1}) \tag{3.167}$$
$$A_k = a_{k+0.5} \tag{3.168}$$

Introducing the discrete velocities $\dot{x}_k$ to Newmark and $\dot{q}_k = M^{-1}j_k$ to implicit midpoint, both schemes write

$$\dot{x}_k = \frac{x_{k+1} - x_k}{h} + \frac{1}{2}h\left(\frac{A_k + A_{k+1}}{2}\right) \tag{3.169}$$
$$\dot{q}_k = \frac{q_{k+1} - q_k}{h} + \frac{1}{2}ha_{k+0.5} \tag{3.170}$$

Then one obtains for the transformation of the velocities

$$\dot{x}_k = \frac{q_{k+1} - q_k}{h} \tag{3.171}$$

## 3.9. Examples

When introducing nonlinear constraints, then the accelerations $a_k$ are enhanced by the constraint forces

$$a_{k+0.5} = M^{-1}\left(-\nabla V(q_{k+0.5}) + \nabla g(q_{k+0.5})\lambda_{k+0.5}\right) \quad (3.172)$$

This is equivalent to the strategy being applied to the example in this section. When applying the Newmark scheme, the spurious oscillations are expected to appear in the velocities. The endpoint coordinates are enforced being feasible.

The equivalence can be easily verified using this example. Let the initial conditions of Newmark be defined by the first time step of midpoint, i.e. $\lambda_0$=0.24997500, $q_0$=(0,-1), $j_0$=(1,0), $q_1$=(0.03999200,-0.99960012). From the first two midpoint endstep coordinates one can derive the initial coordinate and velocity of Newmark at time $t = h/2$. The initial Lagrange multiplier of Newmark is $\lambda_0$. Figure 3.25 illustrates the trajectory and the total energy. The trajectory seems to be stable, but the energy exhibits growing spurious oscillations. This can be verified by taking a look at the corresponding velocities, see figure 3.26. The Lagrange multipliers which are computed by Newmark and midpoint are identical. The same holds for the discrete energy.

Interestingly, the instabilities can be eliminated by carefully chosing the initial conditions. The exact value of the Lagrange multiplier can be determined being 0.5. If one choses $x_0 = (0,-1)$, $\dot{x}_0 = (1,0)$, $\lambda_0 = 0.5$ for Newmark, then the trajectory is absolutely stable. The Lagrange multiplier is kept constant by the time stepping scheme. Hence, two problems remain: (1) How to chose the initial multiplier? (2) Furthermore, the convergence may appear only in this example, but may not be ensured for other problems. In particular, it may not converge in case of small perturbations in the initial conditions.

A stable discretization of constraints using the Newmark scheme (and the generalized $\alpha$-method) is presented and its stability is proven in [180] where Newmark is applied to the potential energy while RATTLE is adopted for the constraint equations.

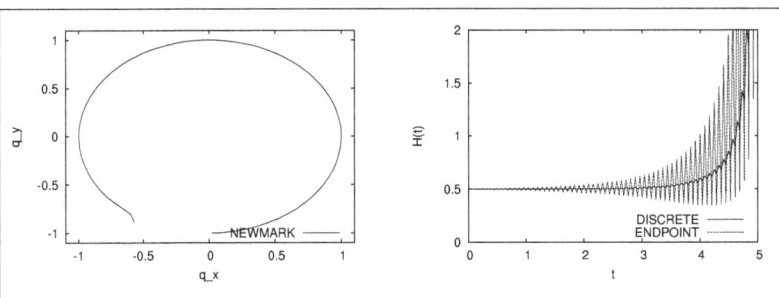

Figure 3.25: Nonlinear pendulum: Trajectory $x_y = x_y(x_x)$ and total energy (as pointwise or discrete energy) of constrained Newmark.

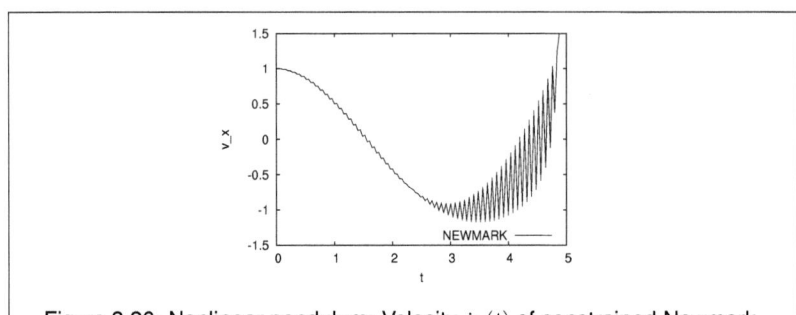

Figure 3.26: Nonlinear pendulum: Velocity $\dot{x}_x(t)$ of constrained Newmark.

The next section presents a stabilizing modification to the constrained midpoint scheme. It combines the midpoint scheme with the collision integrator, equation (3.151), being applied at the end of each step. Unlike RATTLE, this strategy leads to symmetric systems of equations, but requires one additional evaluation of the constraint gradients, i.e. in each time step's midpoint and in each time step's end point. The constrained Newmark scheme is stabilized using the same additional projection. As alternative to the collision integrator, an application of the RATTLE projection is possible, theorem 25, equation (3.126). But it turns out to be unstable in long-term simulation in the example in section 3.9.6.

### 3.9.6 Nonlinear pendulum: Stable discretizations

The example of section 3.9.5 is now considered by numerical methods with stable treatment of nonlinear constraints. The following methods are compared:

- **RATTLE** Algorithm 3.7 on page 66.

- **Collision integrator.** The idea is that infeasible coordinates are allowed. The amount of constraint violation is limited by the time step length. The collision integrator will try to reduce the violation. By application of the collision to the center of each time step the violation at the end of the time step will be reduced further. When combined with Verlet the algorithm, see equations (3.149)-(3.150), writes

$$j_{0.5}^k = j_0^k - \frac{h}{2}\nabla V(q_0^k) \tag{3.173}$$

$$q_{0.5}^k = q_0^k + \frac{h}{2}M^{-1}j_{0.5}^k \tag{3.174}$$

$$j_{0.5+}^k = j_{0.5}^k - 2\nabla g(q_{0.5}^k)\left(\nabla g(q_{0.5}^k)^T M^{-1}\nabla g(q_{0.5}^k)\right)^{-1}\left(\nabla g(q_{0.5}^k)^T M^{-1}j_{0.5}^k\right) \tag{3.175}$$

$$q_1^k = q_{0.5}^k + \frac{h}{2}M^{-1}j_{0.5+}^k \tag{3.176}$$

$$j_1^k = j_{0.5+}^k - \frac{h}{2}\nabla V(q_1^k) \tag{3.177}$$

## 3.9. Examples

- **Modified midpoint.** The constraints are enforced in the time step center, see equation (3.121). The end-step momentum is projected using the collision integrator, i.e. using the projection in equation (3.151) instead of (3.126),

$$q_1^k = q_0^k + hM^{-1}\left(j_0^k - \frac{h}{2}\nabla V\left[\left(\frac{q_0^k + q_1^k}{2}\right)\right] - \frac{h}{2}\nabla g\left[\left(\frac{q_0^k + q_1^k}{2}\right)\right]\lambda_0^k\right) \quad (3.178)$$

$$0 = g(q_1^k) \quad (3.179)$$

$$j_{1-}^k = \frac{1}{h}M(q_1^k - q_0^k) - \frac{h}{2}\nabla V\left[\left(\frac{q_0^k + q_1^k}{2}\right)\right] - \frac{h}{2}\nabla g\left[\left(\frac{q_0^k + q_1^k}{2}\right)\right]\lambda_0^k \quad (3.180)$$

$$j_1^k = \left(I - 2\nabla g(q_1^k)\left(\nabla g(q_1^k)^T M^{-1}\nabla g(q_1^k)\right)^{-1}\left(\nabla g(q_1^k)^T M^{-1}\right)\right)j_{1-}^k \quad (3.181)$$

The time step is chosen to be $h = 0.5$, the simulation time is $T = 500$. The schemes are compared with unconstrained Verlet being applied to polar coordinates and $h = 0.005$. Figure 3.27 illustrates the phase space of the horizontal coordinates $(q_1, j_1)$. Figure 3.28 plots the total energy balance of the three schemes.

Most remarkably, RATTLE as well as the collision integrator conserve the total energy exactly. The modified midpoint scheme does nearly (but not exactly) conserve the energy because it is symplectic. Furthermore, the instabilities arising in midpoint-SHAKE are eliminated. The large errors in the phase space diagram can be regarded to the different points in time where the constraints are enforced ($g = 0$ in the center, $\dot{g} = 0$ at the time step end).

When comparing the actual trajectory of the horizontal coordinate $q_1(t)$ (figure 3.28) one can see that the phase shift is largest for RATTLE and modified midpoint: RATTLE is shortening the phase and MIDPOINT is increasing it. Remarkably, the collision integrator produces the smallest error in phase. The constraint violation of the collision integrator never exceeded $10^{-7}$. Therefore, it is a very efficient alternative to RATTLE.

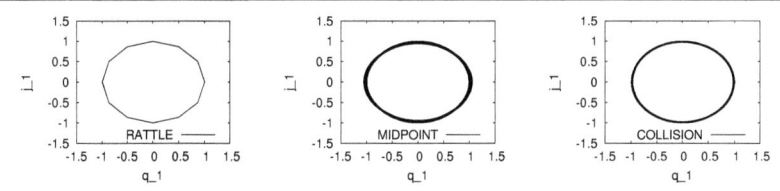

Figure 3.27: Nonlinear pendulum: Phase space diagrams $q_1(t) := q_1(j_1(t))$ for the constrained integrators RATTLE, modified midpoint and collision.

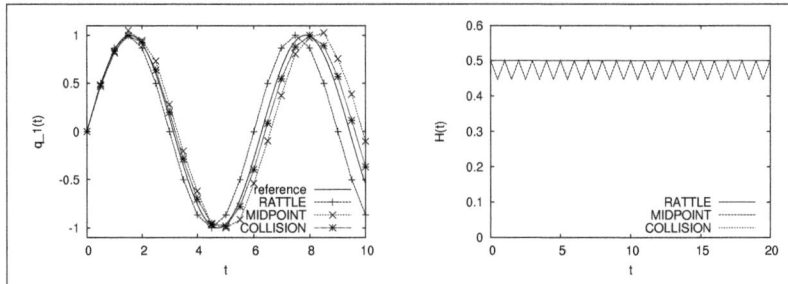

Figure 3.28: Nonlinear pendulum. Left: trajectory of the horizontal coordinate $q_1(t)$ for different schemes. Right: energy at phase space points.

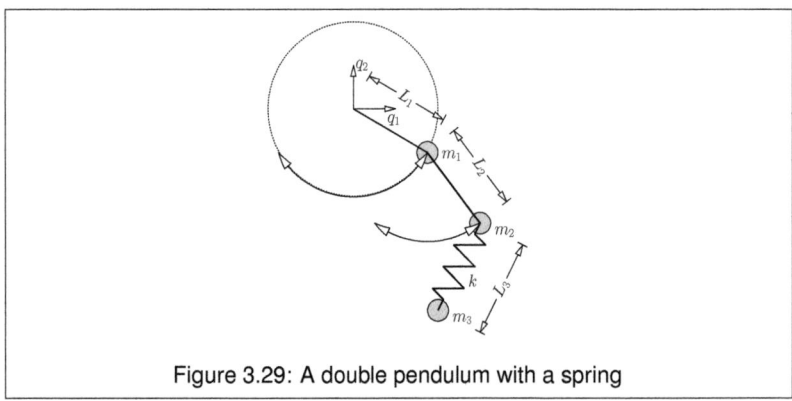

Figure 3.29: A double pendulum with a spring

### 3.9.7 Double pendulum with attached spring and mass

Apply constrained integrators to the double pendulum with an attached spring. The geometry is illustrated in figure 3.29 with $L_1 = L_2 = L_3 = 1$ and $m_1 = m_2 = m_3 = 1$ and $k = 1$. Gravity is considered with $g = 9.81$. The motion of each mass point is described by the cartesian coordinates $(q_1, q_2)$, $(q_3, q_4)$ and $(q_5, q_6)$. The initial conditions are given $q_0 = (0, -1, 0, -2, 0, -3)$ and $j_0 = (1, 0, 0, 0, 0, 0)$. The motion is subject to the constraints

$$g(q) = \begin{pmatrix} q_1^2 + q_2^2 - L_1^2 \\ (q_3 - q_1)^2 + (q_4 - q_2)^2 - L_2^2 \end{pmatrix} \tag{3.182}$$

which enforce the length of the two trusses being constant. The potential function includes gravity and the spring energy, i.e.

$$V(q) = m_1 g q_2 + m_2 g q_4 + m_3 g q_6 + \frac{1}{2} k \left( \sqrt{(q_6 - q_4)^2 + (q_5 - q_3)^2} - L_3 \right)^2 \tag{3.183}$$

The time step is chosen to be $h = 0.1$, the simulation time is $T = 500$. The schemes are compared with the collision integrator using a time step $h = 0.005$. The nonlinear system of equations is solved using Newton's method. A larger

## 3.9. Examples

time step than $h = 0.1$ produced problems in the iterative solution process (no convergence). Using the collision integrator a larger time step is possible, but the violation of the constraint grows inacceptably (for example $\|g\| > 20$ if $h = 0.5$).

Figure 3.30 illustrates the phase space of the horizontal coordinates $(q_1, j_1)$. RATTLE is the only integrator which reproduces the phase space diagram. This behavior is important if a reliable treatment of bifurcations is required. The bad behavior of the other methods may be subject to the different points in time where the momentum and the constraint are measured. This assumption cannot be verified for the collision integrator, which becomes unstable when the collision is applied to the time step's right boundary instead of its center.

Figure 3.31 plots the total energy balance of the compared schemes. RATTLE and the collision integrator exhibit the smallest error in energy.

When comparing the actual trajectory of the horizontal coordinate $q_1(t)$ (figure 3.31) one can see that the phase shift is largest for RATTLE and modified midpoint: RATTLE is shortening the phase and MIDPOINT is increasing it. The collision integrator produces the smallest error in phase of $q_1(t)$. The constraint violation of the collision integrator never exceeded $2 \cdot 10^{-3}$.

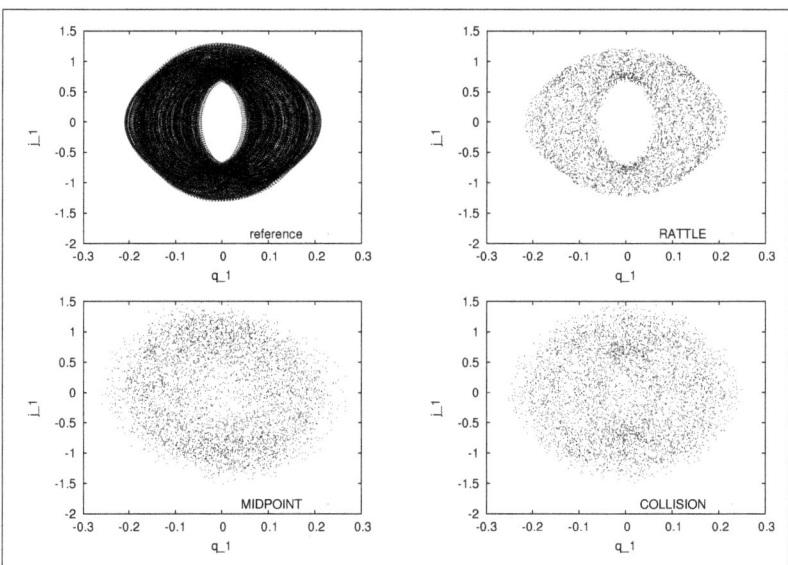

Figure 3.30: Double pendulum: Phase space diagrams $q_1(t) := q_1(j_1(t))$ for the constrained integrators RATTLE, midpoint and collision.

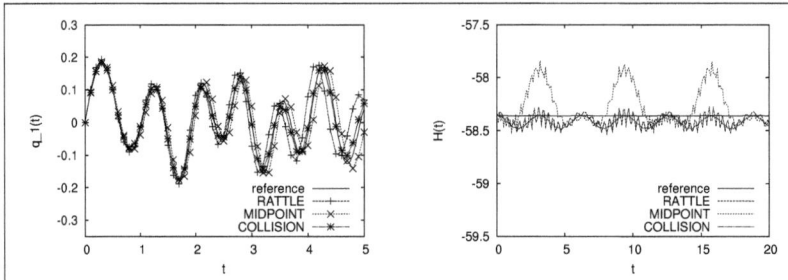

Figure 3.31: Double pendulum. Left: trajectory of the horizontal coordinate $q_1(t)$ for different schemes. Right: energy at phase space points.

### 3.9.8 Collision of two linked linear springs

Compare different collision integrators applied to a small mechanical system as illustrated in figure 3.32. The material data are given by $m_1 = m_2 = 1$ and $k_1 = k_2 = 1$. The system is subject to an inequality constraint

$$g(q) = q_2 - 0.5 \leq 0 \tag{3.184}$$

Application of RATTLE and similar methods is not possible. When combined with an active set strategy which activates only those constraints being violated, these methods lead to numerical damping. Therefore, the following methods are tested:

- **Collision integrator between time steps** A standard Velocity Verlet step is performed. After each step, the constraints are evaluated and the active set is determined at the end-step coordinate. Then, a projection of the momentum $j_k \to \mathbb{P}^c_{q_k} j_k$ takes place using the projection matrix from equation (3.151)

$$\mathbb{P}^c_{q_k} = I - 2\nabla g \left[ (\nabla g)^T M^{-1} \nabla g \right]^{-1} (\nabla g)^T M^{-1} \tag{3.185}$$

This is a simple scheme without predictor step.

- **Collision integrator in time step center** The collision is applied to the center of a time step using the Velocity Verlet algorithm, see equations (3.173)-(3.177). One first computes the mid-step coordinate and momentum. Then a predictor coordinate at the end of the time step is computed (figure 3.33). The constraints are evaluated at the predictor coordinate. Activity is checked using the mid-step momentum. Using the new mid-step

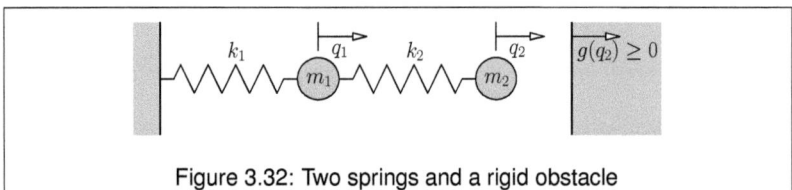

Figure 3.32: Two springs and a rigid obstacle

## 3.9. Examples

momentum a new end-step predictor coordinate can be computed which in turn is used for another constraint evaluation. This is repeated until no constraints are active anymore. A repetition is necessary because the change of momentum due to constraint $g_j$ may lead to a violation of another constraint $g_k$.

Using this methodology, a violation of the inequality constraints at the endpoints is strictly avoided. Only in the time step centers a limited violation may take place. This is irrelevant because no quantities are evaluated at these times.

The choice of the time step center is advantageous in sliding contact. Consider a particle sliding on a surface or even between two surfaces. Regardless which velocity it has or which forces are applied a sliding motion tangential to the adjacent surfaces is possible.

- **Penalty regularization** The standard Velocity Verlet scheme is applied. The constraints are subject to a modified potential function using a quadratic penalty energy function

$$V^c(q) = V(q) + \frac{1}{2}\rho \|\max(0, g(q))\|^2 \qquad (3.186)$$

with $\rho$ being a penalty parameter. The larger $\rho$, the smaller the constraint violations. But the critical time step length can be affected.

Given the stiffness matrix

$$K = \begin{pmatrix} k_1 + k_2 & -k_2 \\ -k_2 & k_2 \end{pmatrix} \qquad (3.187)$$

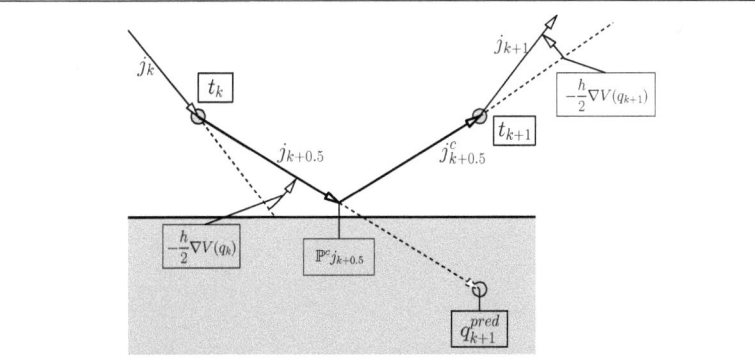

Figure 3.33: Collision integrator at mid-step: Given the initial momentum $j_k$, the mid-step momentum $j_{k+0.5}$ is computed and used to determine the predictor coordinate $q_{k+1}^{pred}$. Active constraints are evaluated at the predictor coordinate. The mid-step momentum is modified and a new end-step coordinate is computed which does not violate the inequalities. The resulting coordinate increment is parallel to the boundary of the infeasible domain.

one obtains the critical time step $h_{crit} = 1.236$. Using a penalty parameter $\rho = 10^5$ the modified stiffness is

$$K^c = \begin{pmatrix} k_1 + k_2 & -k_2 \\ -k_2 & k_2 + \rho \end{pmatrix} \qquad (3.188)$$

and the critical time step is reduced to $h^c_{crit} = 6.32 \cdot 10^{-3}$. The initial conditions are defined by $q_0 = (1,1)$ and $j_0 = (0,0)$. For the collision integrators the time step $h = 0.5$ is used. A larger time step is possible and does not affect stability, but the detection of violated constraints becomes inaccurate. This is because the impact velocity is relatively large compared with the time step length. The penalty method uses the time step $h = 0.001$. This is much less than the critical time step $h^c_{crit}$, but larger values produced a too large energy error during the impact. The mid-step projection is used for a reference solution with $h = 0.001$.

The total energy is plotted in figure 3.34. No drift of the energy can be observed for all methods (there is a very small deviation for the penalty method). Comparing the collision integrators, the mid-step method exhibits a smaller error in energy.

The trajectory of $q_2(t)$ is illustrated in figure 3.34. Given the large time step of the collision integrators, the trajectory agrees well. The reference solution and the penalty solution are almost identical. There exists a phase shift which is largest for the end-step method. The change in phase is related to the accuracy of the predicted contact time. But an accurate computation of the collision time is very inefficient if multiple inequality constraints are present (see [36,58]). Furthermore, the penalty method exhibits a small constraint violation of $\max_t |g(q(t))| = 0.003$, the end-step method $\max_t |g(q(t))| = 0.33$ and the mid-step method did not lead to any violations.

Figure 3.35 shows the active contact force given by

$$F^c(q,j,h) = \frac{2}{h} \nabla g(q) \left[ (\nabla g(q))^T M^{-1} \nabla g(q) \right]^{-1} (\nabla g(q))^T M^{-1} j \qquad (3.189)$$

wherein $q$ and $j$ are the predictor coordinate and the momentum to be modified by the constraint projection. It is obvious that measuring contact forces is not

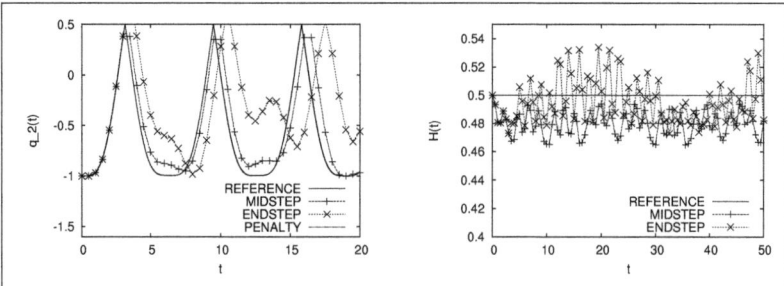

Figure 3.34: Two springs. Left: trajectory of the horizontal coordinate $q_2(t)$ for different schemes. Right: energy at phase space points.

## 3.9. Examples

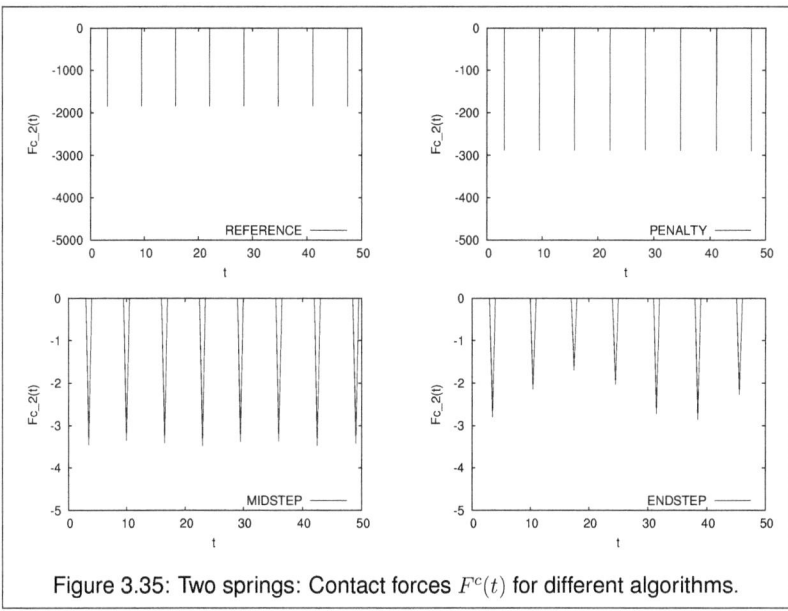

Figure 3.35: Two springs: Contact forces $F^c(t)$ for different algorithms.

recommended although often done in practice. With decreasing time step the contact force grows to infinity for the collision integrators. For penalty methods, the force is dependent on time step and penalty parameter. For an accurate penalty solution with $\rho \to \infty$ the contact forces tend to infinity as well.

## Chapter 4

# Continuous assumed gradient method

## 4.1 Introduction

In recent years a lot of research has been spent in improving the accuracy of low-order continuum finite elements. Numerical efficiency, accuracy and locking behavior are the targets of these approaches, in particular volumetric locking which led to various modifications. Based on assumed strain fields general purpose approaches were also developed, which replace the natural strain field obtained from the finite element shape function derivatives. Such strains can be obtained from assumed fields interpolating the deformation gradient. Methods like nodal integration and the smoothed finite element method (SFEM) belong to this class. There are, however, a few shortcomings: Nodal integration is unstable. It leads to spurious low-energy modes. SFEM provides general numerical integration schemes, but the integration domains have complex geometries in three dimensions and it is not clear how to chose a discretization scheme with desired accuracy and stability. The goal of this chapter is to describe an approach to assumed gradient fields which improves accuracy, is stable and is still simple to implement and numerically efficient.

The chapter is organized as follows: In section 4.2 the fundamentals of continuum mechanics are recalled. Strain and stress measures, constitutive relations and equilibrium conditions are explained. Section 4.3 explains interpolation functions of finite elements. Both sections follow standard text books [8, 183]. Subsequently, the formulation of the assumed gradient field interpolation is presented in section 4.4. The variationally consistent treatment of the assumed gradient is ensured by the principle of Hu-Washizu being added to Hamilton's integral. Choices to solve the three-field functional are discussed. Dual multipliers are used to efficiently reduce the unknowns to the displacements as the only degrees of freedom. A numerical integration scheme of the strain energy is presented which is based on nodal integration. The linearization of the strain energy is derived. The assumed gradient field needs new requirements for mesh generation algorithms which are explained in section 4.5. It also discusses modelling of discontinuities, singular forces and consequences of irregular meshing. Section 4.6 presents nodal integration of finite elements. By interpretation in terms of assumed gradients the occurence of instabilities can be explained. The derivation of stabilizing penalty methods is straightforward. Existing approaches to stabilization are presented and discussed. Section 4.7 presents the smoothed finite element method and puts it into relation with continuous assumed gradient methods. In section 4.8 stable continuous assumed gradient interpolations are presented. Their flexibility with respect to mesh generation and degenerated elements is discussed. A generic scheme is presented how interpolation functions can be derived for arbitrary finite element types. Example interpolation functions are given in appendix **??**. Section 4.9 explains a few aspects of the implementation. Section 4.11 presents a few examples from structural dynamics. They prove the superiority and limitations of the new schemes regarding accuracy, efficiency and stability compared with standard FEM, nodal integration and SFEM. Detailed numerical tests and benchmarks using problems from static structural analysis are given in appendix A.

## 4.2 Fundamentals of continuum mechanics

### 4.2.1 Kinematics

The geometric description of large deformations relies on the consideration of a material body at different configurations. A material body is a physical object, described by properties like stress-strain laws and density, which are distributed as a continuous field in three-dimensional Euclidian space. A motion is a sequence of configurations which are parameterized by the time $t$. Consider two separate configurations. Let the reference configuration $\Omega$ be measured at time $t_0 = 0$ (total

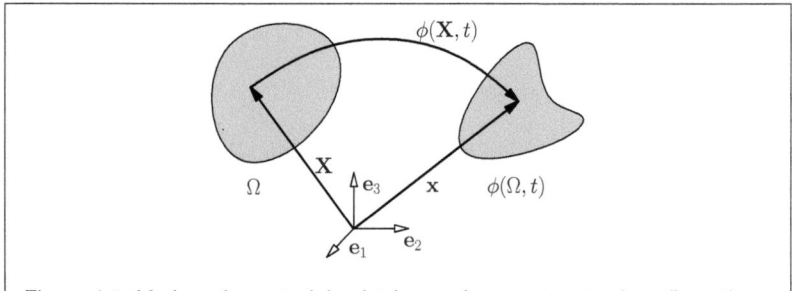

Figure 4.1: Motion of a material point from reference to actual configuration.

## 4.2. Fundamentals of continuum mechanics

Lagrange method). The position vector $\mathbf{X}$ of an arbitrary point is defined by its Lagrangian coordinates $X_i$,

$$\mathbf{X} = X_i \mathbf{e}_i \tag{4.1}$$

The configuration at time $t_1$ is called neighboring configuration. The reference and the neighboring configuration are supposed to be known. The objective is to determine the current configuration $\phi(\Omega, t)$, $t = t_1 + \Delta t$, see figure 4.1, which is a map from the reference configuration through

$$\phi : X \rightarrow x \tag{4.2}$$

The change of the configuration can be discribed by the continuous displacement vector

$$\mathbf{u}(\mathbf{X}, t) = \phi(\mathbf{X}, t) - \mathbf{X} \tag{4.3}$$

The transformation of a differential line element

$$d\mathbf{x} = \mathbf{F} d\mathbf{X} \tag{4.4}$$

is described by the deformation gradient $\mathbf{F}$

$$F_{ij} = \frac{\partial x_i}{\partial X_j} = \left[\frac{\partial \phi(\mathbf{X}, t)}{\partial \mathbf{X}}\right]_{ij} \tag{4.5}$$

which is an asymmetric tensor and represents a linear mapping of tangents of material curves. Since the orientation of bodies during the motion is preserved (no intersection, no reflection), the Jacobian determinant does not change its sign,

$$J = \det(\mathbf{F}) > 0 \tag{4.6}$$

The length of a differential line element reads

$$ds^2 = d\mathbf{x}^T d\mathbf{x} = d\mathbf{X}^T \mathbf{F}^T \mathbf{F} d\mathbf{X} = d\mathbf{X}^T \mathbf{C} d\mathbf{X} \tag{4.7}$$

where $\mathbf{C}$ is the right Cauchy-Green tensor defined as

$$\mathbf{C} = \mathbf{F}^T \mathbf{F} \tag{4.8}$$

The deformation gradient may be decomposed into the rotation matrix $\mathbf{R}$, with $\mathbf{R}^T \mathbf{R} = \mathbf{I}$, and the symmetric dilatation matrix $\mathbf{U}$,

$$\mathbf{F} = \mathbf{R}\mathbf{U} \tag{4.9}$$

such that any given deformation may be interpreted in terms of a dilatation and a subsequent rotation. The Cauchy-Green tensor measures the material strains excluding rigid body kinematics, i.e.

$$\mathbf{C} = \mathbf{U}^T \mathbf{R}^T \mathbf{R} \mathbf{U} = \mathbf{U}^T \mathbf{U} \tag{4.10}$$

The absolute elongation is

$$ds^2 - dS^2 = d\mathbf{X}^T \mathbf{C} d\mathbf{X} - d\mathbf{X}^T d\mathbf{X} = 2 d\mathbf{X}^T \mathbf{E} d\mathbf{X} \tag{4.11}$$

introducing the Green-Lagrangian strain tensor

$$\mathbf{E} = \frac{1}{2}(\mathbf{C} - \mathbf{I}) \tag{4.12}$$

which is quadratic with respect to the deformation gradient,

$$E_{ij} = \frac{1}{2}\left(F_{ki}F_{kj} - \delta_{ij}\right) \tag{4.13}$$

The Green-Lagrangian strain tensor denotes the strain regarding the reference configuration. The variations of the strain tensor and of the deformation gradient denote

$$\delta \mathbf{E} = \frac{1}{2}(\delta \mathbf{F}^T \mathbf{F} + \mathbf{F}^T \delta \mathbf{F}) \tag{4.14}$$

$$\delta F_{ij} = \delta \frac{\partial u_i}{\partial X_j} = \delta \frac{\partial u_i}{\partial x_k}\frac{\partial x_k}{\partial X_j} \tag{4.15}$$

One may map the variation of the strain tensor $E_{ij}$ into the current configuration by using the relation

$$\delta \mathbf{E} = \frac{1}{2}\left(\mathbf{F}^T \delta \frac{\partial \mathbf{u}}{\partial \mathbf{x}}^T \mathbf{F} + \mathbf{F}^T \delta \frac{\partial \mathbf{u}}{\partial \mathbf{x}} \mathbf{F}\right) = \frac{1}{2}\mathbf{F}^T\left(\delta \frac{\partial \mathbf{u}}{\partial \mathbf{x}}^T + \delta \frac{\partial \mathbf{u}}{\partial \mathbf{x}}\right)\mathbf{F} = \mathbf{F}^T \delta \mathbf{e}|_t \mathbf{F} \tag{4.16}$$

or

$$\delta E_{ij} = F_{mi} F_{nj}\, \delta e_{mn}|_t \tag{4.17}$$

Herein, the strain tensor for small strains $e_{ij}|_t$ referred to the current configuration was introduced

$$e_{ij}|_t = \frac{1}{2}\left(\frac{\partial u_i}{\partial x_j} + \frac{\partial u_j}{\partial x_i}\right) \tag{4.18}$$

The small strain tensor $\mathbf{e}$ is obtained by linearizing the Green-Lagrangian strain tensor from the first order Taylor expansion of $E_{ij}$ at $\phi(\mathbf{X}, 0)$ for $\Delta(\frac{\partial u_k}{\partial X_l})$ and is sufficiently accurate for small $\frac{\partial u_j}{\partial X_i}$.

$$e_{ij} = \frac{1}{2}\left(\frac{\partial u_i}{\partial X_j} + \frac{\partial u_j}{\partial X_i}\right) = \frac{1}{2}\left(F_{ij} + F_{ji}\right) - \delta_{ij} \tag{4.19}$$

Due to the symmetry of $e_{ij}$, $E_{ij}$ there exist only 6 independent components. It is common to write the strain tensor in vector form

$$\boldsymbol{\epsilon} = \begin{bmatrix} \epsilon_1 \\ \epsilon_2 \\ \epsilon_3 \\ \epsilon_4 \\ \epsilon_5 \\ \epsilon_6 \end{bmatrix} = \begin{bmatrix} \epsilon_{11} \\ \epsilon_{22} \\ \epsilon_{33} \\ 2\epsilon_{12} \\ 2\epsilon_{23} \\ 2\epsilon_{13} \end{bmatrix} = \begin{bmatrix} \epsilon_{11} \\ \epsilon_{22} \\ \epsilon_{33} \\ \gamma_{12} \\ \gamma_{23} \\ \gamma_{13} \end{bmatrix} \tag{4.20}$$

with $\boldsymbol{\epsilon} \rightarrow \mathbf{e}$ or $\boldsymbol{\epsilon} \rightarrow \mathbf{E}$.

## 4.2. Fundamentals of continuum mechanics

### 4.2.2 Kinetics

Let t be the real stress vector in a material point

$$\mathbf{t} = \frac{\partial \mathbf{f}}{\partial s} \tag{4.21}$$

that is the spatial force vector $\partial \mathbf{f}$ related to the infinitesimal surface $\partial s$, an imaginary cutting plane in the continuum. The orientation of this surface is expressed through its normal vector $\mathbf{n}$. The Cauchy theorem

$$\mathbf{t} = \mathbf{T}\mathbf{n} \tag{4.22}$$

linearly associates each surface normal to a stress vector through the (real) Cauchy stress tensor $\mathbf{T}$. This equation follows from the equilibrium of an infinitesimal tetrahedron element. The Cauchy stress tensor is a function of $\phi(\mathbf{X}, t)$.

The acting body and surface forces for an arbitrary domain satisfy the equilibrium

$$\int_v p_i dv + \int_s t_i ds = 0 \tag{4.23}$$

The second integral can be transformed into a volume integral

$$\int_s t_i ds = \int_s T_{ij} n_j ds = \int_v \frac{\partial T_{ij}}{\partial x_j} dv \tag{4.24}$$

Considering elements of infinitesimal volume the local equilibrium conditions are obtained

$$p_i + \frac{\partial T_{ij}}{\partial x_j} = 0 \tag{4.25}$$

or, respectively,

$$\mathbf{p} + \operatorname{div}\mathbf{T} = \mathbf{0} \tag{4.26}$$

The stress tensor which is energetically conjugated to the deformation gradient is the first Piola-Kirchhoff stress tensor $\mathbf{P}$. The latter is obtained by mapping the spatial force acting on an infinitesimal section surface area $ds$ in the current configuration

$$d\mathbf{f} = \mathbf{T}\mathbf{n}ds \tag{4.27}$$

to the reference configuration. Utilizing the transformation

$$\mathbf{n}ds = \det(\mathbf{F})\mathbf{F}^{-T}\mathbf{N}dS \tag{4.28}$$

yields

$$d\mathbf{f} = \det(\mathbf{F})\mathbf{T}\mathbf{F}^{-T}\mathbf{N}dS = \mathbf{P}\mathbf{N}dS \tag{4.29}$$

where the first Piola-Kirchhoff stress tensor is defined as

$$\mathbf{P} = \det(\mathbf{F})\mathbf{T}\mathbf{F}^{-T} \tag{4.30}$$

Since $\mathbf{P}$ is asymmetric, the second Piola-Kirchhoff stress tensor was introduced

$$\mathbf{S} = \mathbf{F}^{-1}\mathbf{P} = \det(\mathbf{F})\mathbf{F}^{-1}\mathbf{T}\mathbf{F}^{-T} \tag{4.31}$$

which is the conjugate of the Green-Lagrangian strain tensor E. Energetic conjugation is present if the virtual energy defined by the product of the stress with its conjugated virtual deformation quantity is identical within both coordinate systems, i.e.

$$\int_v T_{kl}\, \delta e_{kl}|_t\, dv = \int_V T_{kl}\left(F_{mk}^{-1}F_{nl}^{-1}\delta E_{mn}\right)(\det(\mathbf{F})dV) = \int_V S_{mn}\delta E_{mn}dV \quad (4.32)$$

with

$$S_{mn} = \det(\mathbf{F})F_{mk}^{-1}T_{kl}F_{ln}^{-T} \quad (4.33)$$

### 4.2.3 Constitutive equations

The kinetic relations in Lagrange coordinates denote terms for the components of Green's strain tensor $E_{ij}$. Since it is symmetric, there are only 6 independent variables. They are associated with 6 kinetic equations involving the partial derivatives of the displacements $u$ for the Lagrange coordinates $X$. The 3 equations of motion include the second Piola-Kirchhoff stress tensor $S$, which contains 6 independent components due to its symmetry. Therefore, there exist 9 equations for 15 unknown variables. The missing conditions are defined through constitutive equations. For elastic solids, they generally are of the form

$$E_{ij} = G(S_{kl}) \quad (4.34)$$

In the special case of linear elasticity Hooke's law is denoted by

$$E_{ij} = D_{ijkl}S_{kl} \quad \text{or} \quad S_{ij} = C_{ijkl}E_{kl} \quad (4.35)$$

wherein C denotes the isotropic elastic material stiffness tensor

$$C_{ijkl} = \lambda \delta_{ij}\delta_{kl} + \mu(\delta_{ik}\delta_{jl} + \delta_{il}\delta_{kj}) \quad (4.36)$$

D its inverse and $\lambda$, $\mu$ the Lamé constants. The relations between the Lamé constants and the commonly used Young's modulus $E$ and Poisson's ratio $\nu$ are given as

$$\lambda = \frac{E\nu}{(1+\nu)(1-2\nu)} \quad \text{and} \quad \mu = \frac{E}{2(1+\nu)} \quad (4.37)$$

The elastic constitutive conditions express that the strains depend only on the current stresses, assuming that the bodies are free of stresses if the external loading is removed.

### Geometrically linear analysis

The kinematic relation reads in matrix-vector form

$$\boldsymbol{\epsilon} = \partial_\epsilon \mathbf{u} \quad (4.38)$$

## 4.2. Fundamentals of continuum mechanics

$$\begin{bmatrix} \epsilon_1 \\ \epsilon_2 \\ \epsilon_3 \\ \epsilon_4 \\ \epsilon_5 \\ \epsilon_6 \end{bmatrix} = \begin{bmatrix} \frac{\partial}{\partial x} & & \\ & \frac{\partial}{\partial y} & \\ & & \frac{\partial}{\partial z} \\ \frac{\partial}{\partial y} & \frac{\partial}{\partial x} & \\ & \frac{\partial}{\partial z} & \frac{\partial}{\partial y} \\ \frac{\partial}{\partial z} & & \frac{\partial}{\partial x} \end{bmatrix} \begin{bmatrix} u_x \\ u_y \\ u_z \end{bmatrix} \quad (4.39)$$

The constitutive law can be written as

$$\begin{bmatrix} \sigma_1 \\ \sigma_2 \\ \sigma_3 \\ \sigma_4 \\ \sigma_5 \\ \sigma_6 \end{bmatrix} = \begin{bmatrix} \sigma_{11} \\ \sigma_{22} \\ \sigma_{33} \\ \sigma_{12} \\ \sigma_{23} \\ \sigma_{13} \end{bmatrix} = \mathbf{C} \begin{bmatrix} \epsilon_1 \\ \epsilon_2 \\ \epsilon_3 \\ \epsilon_4 \\ \epsilon_5 \\ \epsilon_6 \end{bmatrix} \quad (4.40)$$

with constitutive matrix C for linear elastic materials

$$C = \frac{E}{(1+\nu)(1-2\nu)} \begin{bmatrix} 1-\nu & \nu & \nu & & & \\ \nu & 1-\nu & \nu & & & \\ \nu & \nu & 1-\nu & & & \\ & & & \frac{1-2\nu}{2} & & \\ & & & & \frac{1-2\nu}{2} & \\ & & & & & \frac{1-2\nu}{2} \end{bmatrix} \quad (4.41)$$

**Material nonlinearity**

The constitutive conditions of elasticity can be divided into three groups:

1. Cauchy material models, which are denoted by a reversible unique relation between stresses and strains, for example linear elasticity. Under certain loading conditions, Cauchy material models may create mechanical energy, leading to the next class, which passes this drawback.

2. Hyperelastic models, eg. rubber materials, which assume the existence of a strain potential, the potential of the energy density $U_0$: $S_{ij} = \frac{\partial U_0}{\partial E_{ij}}$.

3. Hypoelastic models, depending on an incremental relation between stresses and strains: $dE_{ij} = \dot{E}_{ij}dt$ and $dS_{ij} = \dot{S}_{ij}dt$, which are to be distinguished from total differentials.

In elastopolasticity, an incremental model of stresses is used. It writes

$$\mathbf{S}_{t+\Delta t} = \mathbf{S}_t + \int_{\mathbf{S}_t}^{\mathbf{S}_{t+\Delta t}} d\mathbf{S} \quad (4.42)$$

which leads to the nonlinear formulation

$$\mathbf{S} = \tilde{\sigma}(\mathbf{E}, \alpha_{int}, t) \quad (4.43)$$

where $\tilde{\sigma}$ denotes the nonlinear constitutive operator and $\alpha_{int}$ a set of internal variables which describe the state of an inelastic material. In general the internal state variables are path dependent and influenced by the strain history.

### 4.2.4 Equilibrium equations

The system of governing equations and boundary conditions of a continuum $\Omega \in R^3$ (figure 4.2) with boundary $\Gamma$ is

$$\begin{aligned}
\mathbf{f}^B + \partial_\epsilon \sigma &= \mathbf{0} && \text{in } \Omega \\
\epsilon &= \partial_\epsilon \mathbf{u} && \text{in } \Omega \\
\sigma &= \mathbf{C}\epsilon && \text{in } \Omega \\
\mathbf{n}\sigma &= \mathbf{f}^S && \text{on } \Gamma^f \\
\mathbf{u} &= \mathbf{u}^S && \text{on } \Gamma^u \\
\Gamma^u \cup \Gamma^f &= \Gamma \\
\Gamma^u \cap \Gamma^f &= \emptyset
\end{aligned} \qquad (4.44)$$

where $\epsilon$ is the strain vector (see equation (4.39)), $\sigma$ is the stress vector (see equation (4.40)), $\mathbf{f}^S$ is the prescribed traction on the Neumann boundary $\Gamma^f$, the so called natural/static boundary conditions. $\mathbf{u}^S$ is the vector of prescribed displacements on the Dirichlet boundary $\Gamma^u$, known as essential or kinematic boundary conditions. $\mathbf{f}^B$ is the body force vector and $\mathbf{n}$ is the matrix of direction cosine components of a unit normal to the domain boundary (positive outwards)

$$\mathbf{n} = \begin{bmatrix} n_x & 0 & 0 & n_y & 0 & n_z \\ 0 & n_y & 0 & n_x & n_z & 0 \\ 0 & 0 & n_z & 0 & n_y & n_x \end{bmatrix} \qquad (4.45)$$

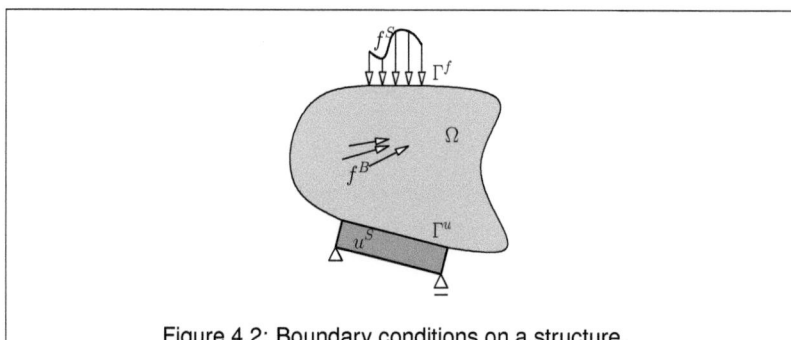

Figure 4.2: Boundary conditions on a structure.

## 4.3 Finite element interpolation of the continuum

Each element $m$ is defined by a predefined number of supporting nodes given by their index $i$. Each node is part of a global finite element structure. Its global index within the structure can be obtained by the map

$$\text{node} : (m, i) \to A \tag{4.46}$$

Continuum elements have three available DOFs $\alpha$ per node $i$, one translation along each coordinate axis. In order to associate a global index to a nodal degree of freedom, one defines some invertible index transformation

$$\text{dof} : (\text{node}(m, i), \alpha) \to A \tag{4.47}$$

that maps the index pair $(A, \alpha)$ to the global space of available DOFs A.

The geometric shape of a continuum element is then defined by

$$\phi(\xi) = N_i(\xi) \mathbf{x}_i \tag{4.48}$$

where $\xi$ is a coordinate within the elemental material coordinate system, $\xi = \xi(\mathbf{X})$, $\mathbf{N}(\xi)$ is the vector of shape functions and $\mathbf{x}_i$ is the vector of nodal deformed coordinates that is the superposition of the nodal referential coordinates $\mathbf{X}_i$ and the nodal displacement vector $\mathbf{u}_i$,

$$\mathbf{x}_i = \mathbf{X}_i + \mathbf{u}_i \tag{4.49}$$

$N_i$ must define the distribution of $\phi$ within the element if $\mathbf{x}_i$ has unit value and all other nodal coordinates are zero. Therefore,

$$\mathbf{N}_i(\xi_j) = \delta_{ij} \tag{4.50}$$

where $\xi_j$ is the local coordinate of node $j$. Consider the case where a rigid body translation is subjected to the element, for example $\mathbf{u}_i = const$. The deformation gradient is zero if the displacement within the element is constant, i.e. $\mathbf{u}(\xi) = \mathbf{u}_i$. Therefore,

$$\sum_i N_i(\xi) = 1 \tag{4.51}$$

Also, from consideration of rigid body motion yielding zero strain one obtains

$$\sum_i \nabla N_i(\xi) = \mathbf{0} \tag{4.52}$$

Standard finite element shape functions are chosen to be Legendre polynomials. Alternatives such as splines and NURBS [104] have been successfully experimented with.

A shape function space is said to have tensor product structure if it can be constructed from products of onedimensional functions, i.e.

$$N_i(\xi_1, \ldots, \xi_n) = \prod_{a=1}^{n} N_{ia}(\xi_a) \tag{4.53}$$

## 4.4 Assumed gradient field

### 4.4.1 Formulation

Given a map $\phi = (\mathbf{X}, t) \to \mathbf{x}$ which transforms the coordinate of a point in the reference configuration $\mathbf{X}$ to the deformed configuration $\mathbf{x}$, the deformation gradient $\mathbf{F}$ at a local coordinate $\xi$ is computed by

$$F_{\alpha\beta}(\xi) = \frac{\partial \phi_\alpha(\xi)}{\partial X_\beta(\xi)} = \frac{\partial \phi_\alpha(\xi)}{\partial \xi_\gamma} \frac{\partial \xi_\gamma}{\partial X_\beta(\xi)}, \quad \alpha, \beta, \gamma = 1 \ldots 3 \quad (4.54)$$

Using isoparametric continuum elements with shape function $N_i$ leads to

$$F^h_{\alpha\beta}(\xi) = \nabla_\gamma N_i(\xi) J^{-1}_{\gamma\beta}(\xi) u_{i\alpha} + \delta_{\alpha\beta} \quad (4.55)$$

where $\mathbf{J}^{-1}$ denotes the inverse of the Jacobian

$$J_{\alpha\beta}(\xi) = \frac{\partial X_\alpha(\xi)}{\partial \xi_\beta} = \nabla_\beta N_i(\xi) X_{i\alpha} \quad (4.56)$$

The resulting field is discontinuous at finite element boundaries. The goal is to replace it by a (for now) arbitrary field with interpolation function $M_i$

$$F^{AN}_{\alpha\beta}(\xi) = M_A(\xi) F_{A\alpha\beta} \quad (4.57)$$

where $\mathbf{F}_A$ are the values at the supporting points $A$ and $\mathbf{F}^{AN}(\xi)$ the assumed natural deformation gradient. The values $\mathbf{F}_A$ are additional degrees of freedom. They are chosen in such a way that the resulting field $F^{AN}_{\alpha\beta}(\xi)$ is a good approximation to the field computed from the finite element shape derivatives $F^h_{\alpha\beta}(\xi)$, i.e.

$$0 = F^{AN}_{\alpha\beta}(\xi) - F^h_{\alpha\beta}(\xi) \quad (4.58)$$

Due to their incompatible function spaces this constraint can not be enforced exactly at all points $\xi$ in general.

### 4.4.2 Principle of Hu-Washizu

A natural approach is to enforce the assumed deformation gradient constraint (4.58) by a least square problem that minimizes the residuum of the constraint equation integrated over the domain, i.e.

$$\int_V \left\| F^{AN}_{\alpha\beta}(\xi) - F_{\alpha\beta}(\xi) \right\|^2 dV \to \min \quad (4.59)$$

Inserting the ansatz functions for the deformation gradient and deriving by $F_{A\alpha\beta}$ one obtains a linear system of equations which can be used to solve $F_{A\alpha\beta}$. Alternatively, one could enforce (4.58) using a continuous field of Lagrange multipliers $\lambda(\xi)$, i.e.

$$\lambda_{\alpha\beta}(\xi) = L_A(\xi) \lambda_{A\alpha\beta} \quad (4.60)$$

## 4.4. Assumed gradient field

with interpolation function $L_A(\xi)$. The multipliers are incorporated by adding the term $\Pi^{HW}$ to the strain energy,

$$\Pi^{HW} = \int_V \lambda_{\alpha\beta}(\xi) \left( F_{\alpha\beta}^{AN}(\xi) - F_{\alpha\beta}^{h}(\xi) \right) dV \qquad (4.61)$$

introducing $\lambda_{A\alpha\beta}$ as additional degrees of freedom. The resulting energy function can be interpreted in terms of the principle of Hu-Washizu and the Lagrange multipliers are identified as 1st Piola-Kirchhoff stress. Deriving the modified strain energy for the discrete multipliers reveals the discretized constraint equations

$$0 = \int_V \left( L_A(\xi) M_B(\xi) F_{B\alpha\beta} - L_A(\xi) F_{\alpha\beta}^{h}(\xi) \right) dV \qquad (4.62)$$

When the same interpolation functions are applied to the multipliers and to the assumed deformation gradient,

$$L_A(\xi) = M_A(\xi) \qquad (4.63)$$

the two formulations (4.59) and (4.61) are equivalent.

### 4.4.3 Solution strategy: Dual multiplier space

In order to solve the discrete values $F_{B\alpha\beta}$ one may chose from the following approaches:

1. One could invert the matrix $\int_V M_A M_B dV$ by assuming a lumping scheme

$$\int_V M_A(\xi) M_B(\xi) dV \to \delta_{AB} \int_V M_A(\xi) dV \qquad (4.64)$$

    Therefore, the support value $F_{B\alpha\beta}$ can be computed internally by

$$F_{A\alpha\beta}^{LM} = \frac{\int_V N_A(\xi) F_{\alpha\beta}(\xi) dV}{\int_V N_A(\xi) dV} \qquad (4.65)$$

2. The discrete values $F_{B\alpha\beta}$ can serve as additional degrees of freedom in order to extend the system of equation. Having $n$ nodes would add $9n$ variables which is undesirable.

3. The values $F_{B\alpha\beta}$ can be eliminated internally using full factorization. One can invert the matrix $\int_V M_A M_B dV$ previous to the simulation, since it is constant. An internal factorization is practically impossible when applied to large systems - the inverse is dense. Therefore, the values $F_{B\alpha\beta}$ would have a global support regarding the displacements $\mathbf{u}$.

4. One could use special multiplier spaces $L_A(\xi) \neq M_A(\xi)$.

A certain choice of $L_A$ in equation (4.62) leads to a lumped matrix structure and, thus, the discrete multipliers can be eliminated internally. The so called dual multiplier space is constructed by the linear combination [272]

$$L_A(\xi) = a_{AB} M_B(\xi) \qquad (4.66)$$

where $a_{AB}$ denotes some coefficient matrix. One can consider each finite element separately. Then, the dual interpolation functions and the coefficients $a_{ij}$ are individually defined for each finite element. Since the dual interpolation is defined by a linear combination of the original interpolation function space, it inherits most of the properties, for example $C^0$ continuity. All volume integrals are restricted to the domain of a single finite element. In order to obtain a lumped matrix scheme, the biorthogonality criterion

$$\int_V L_A(\xi) M_B(\xi) dV = \delta_{AB} \int_V M_A(\xi) dV \qquad (4.67)$$

must be satisfied, written by splitting the integration domain into finite elements

$$\sum_e \int_{V^e} L_{\text{point}(e,i)}(\xi) M_{\text{point}(e,j)}(\xi) dV = \sum_e \delta_{ij} \int_{V^e} M_{\text{point}(e,i)}(\xi) dV \qquad (4.68)$$

Using the definitions $n_{ij} = \text{diag}(\int_{V^e} M_i(\xi) dV)$ denoting the target diagonal matrix, $m_{ij} = \int_{V^e} M_i(\xi) M_j(\xi) dV$ denoting a symmetric matrix, one obtains

$$\int_{V^e} L_i(\xi) M_j(\xi) dV = a_{ik} \int_{V^e} M_k(\xi) M_j(\xi) dV \qquad (4.69)$$

$$n_{ij} = a_{ik} m_{kj} \qquad (4.70)$$

$$a_{ik} = n_{ij} m_{jk}^{-T} = m_{kj}^{-1} n_{ij} \qquad (4.71)$$

Notice, in order to pass the patch test, the coefficient matrix $a_{ij}$ must be individually computed for each finite element and the volume integrals must be accurately evaluated. Else, the assumed gradient operator may become inaccurate in case of distorted elements where the Jacobian determinant is not constant within individual elements.

The discrete constraint yields

$$0 = \delta_{AB} \int_V M_B(\xi) F_{B\alpha\beta} dV - \int_V L_A(\xi) F_{\alpha\beta}^h(\xi) dV \qquad (4.72)$$

and one obtains for the discrete deformation gradient

$$F_{A\alpha\beta} = \frac{\int_V L_A(\xi) F_{\alpha\beta}^h(\xi) dV}{\int_V M_A(\xi) dV} = \frac{\int_V L_A(\xi) \nabla_\gamma N_B(\xi) J_{\gamma\beta}^{-1}(\xi) dV}{\int_V M_A(\xi) dV} u_{B\alpha} + \delta_{\alpha\beta} \qquad (4.73)$$

which is linear in the nodal displacements $u_{B\alpha}$. The discrete deformation gradient $F_{A\alpha\beta}$ has compact support, i.e. it depends only on the degrees of freedom of all finite elements that are adjacent to the supporting point $A$.

**Theorem 26.** *The dual Lagrange multipliers $\lambda_{A\alpha\beta}$ and nodal deformation gradients $F_{A\alpha\beta}$ can be eliminated internally.*

This theorem is easily proved by exploiting the structure of Hu-Washizu's three-field variational principle:

## 4.4. Assumed gradient field

*Proof.* Consider a potential function $\Pi$ dependent on a vector of displacements $\mathbf{u}$ and additional DOFs $\mathbf{q}$. The strain measures in $\Pi$ are replaced by quantities that depend on the assumed deformation gradient $F^{AN}$. This adds the discrete values $F_{Aij}$ to the list of generalized coordinates in $\Pi$. By adding the Hu-Washizu potential $\Pi^{HW}$ one obtains the variational problem

$$0 = \nabla_{\mathbf{q}}\Pi\delta\mathbf{q} + (\nabla_{\mathbf{u}}\Pi + \nabla_{\mathbf{u}}\Pi^{HW})\delta\mathbf{u} + (\nabla_{\mathbf{F}}\Pi + \nabla_{\mathbf{F}}\Pi^{HW})\delta\mathbf{F} + \nabla_{\lambda}\Pi^{HW}\delta\lambda$$

Recall

$$\Pi^{HW} = \int_V \lambda_{\alpha\beta}(\xi)\left(F^{AN}_{\alpha\beta}(\xi) - F_{\alpha\beta}(\xi)\right)dV$$

If one can eliminate $\lambda$ and $\mathbf{F}$ internally, then the variational problem can be written in terms of $\delta\mathbf{q}$ and $\delta\mathbf{u}$ only. The proof, therefore, contains the following steps

- Prove that $\nabla_{\lambda}\Pi^{HW}\delta\lambda = 0$ can be solved internally. That means, $\mathbf{F}$ can be presented as an explicit function of $\mathbf{u}$ and $\delta\mathbf{F}$ by $\delta\mathbf{u}$.
- Prove that $\nabla_{\mathbf{u}}\Pi^{HW}\delta\mathbf{u} + \nabla_{\mathbf{F}}\Pi^{HW}\delta\mathbf{F} = 0$.

The condition $\nabla_{\lambda}\Pi^{HW} = 0$ is satisfied by definition of the Lagrange multiplier field. It was also shown that $F_{A\alpha\beta} = F_{A\alpha\beta}(u_{B\alpha})$. For the second statement, one considers

$$\nabla_{\mathbf{F}}\Pi^{HW}\delta\mathbf{F} + \nabla_{\mathbf{u}}\Pi^{HW}\delta\mathbf{u} = \left(\int_V L_C M_D dV \lambda_{C\mu\nu}\right)\delta F_{D\mu\nu}$$

$$- \left(\int_V L_A(\xi)\nabla_{\gamma}N_B(\xi)J^{-1}_{\gamma\beta}(\xi)dV \lambda_{A\alpha\beta}\right)\delta u_{B\alpha}$$

whereby $L_A = M_A$ in the case of the least-square approximation. For the first term we know

$$\left(\int_V L_C M_D dV \lambda_{C\mu\nu}\right)\delta F_{D\mu\nu} = \lambda_{C\mu\nu}\left(\int_V L_C M_D dV\right) \cdot \frac{\int_V L_D(\xi)\nabla_{\gamma}N_E(\xi)J^{-1}_{\gamma\nu}(\xi)dV}{\int_A M_D(\xi)dV}\delta u_{E\mu}$$

$$= \lambda_{D\mu\nu}\left(\int_V L_D(\xi)\nabla_{\gamma}N_E(\xi)J^{-1}_{\gamma\nu}(\xi)dV\right)\delta u_{E\mu}$$

Inserting it yields

$$\nabla_{\mathbf{F}}\Pi^{HW}\delta\mathbf{F} + \nabla_{\mathbf{u}}\Pi^{HW}\delta\mathbf{u} = \left(\lambda_{D\mu\nu}\left(\int_V L_D(\xi)\nabla_{\gamma}N_B(\xi)J^{-1}_{\gamma\nu}(\xi)dV\right)\right.$$

$$\left. - \lambda_{A\alpha\beta}\left(\int_V L_A(\xi)\nabla_{\gamma}N_B(\xi)J^{-1}_{\gamma\beta}(\xi)dV\right)\right)\delta u_{B\alpha}$$

$$= 0$$

Therefore, when using dual Lagrange multipliers or the least square approximation to solve the discrete deformation gradients, the variational equation simplifies to

$$0 = \frac{\partial\Pi}{\partial q_A}\delta q_A + \left(\frac{\partial\Pi}{\partial u_B} + \frac{\partial\Pi}{\partial F_{C\alpha\beta}}\frac{\partial F_{C\alpha\beta}}{\partial u_B}\right)\delta u_B$$

□

### 4.4.4 Strain energy and numerical integration

Let assume that the applied materials are hyperelastic, i.e. a strain energy potential exists. This assumption is made without loss of generality since we can at least provide a numerical approximation or incremental expression of the strain energy for any constitutive law. The strain energy density $U^d$ is a path dependent function of the strain tensor $\epsilon$, the history variables $\alpha_{int}$ and time $t$

$$U^d = U^d(\epsilon, \alpha_{int}, t) \tag{4.74}$$

wherein the parameters are assumed to depend on the material coordinate $\xi$, i.e. $\epsilon = \epsilon(\mathbf{F}(\xi))$ and $\alpha_{int} = \alpha_{int}(\xi)$. The strain energy density may be discontinuous in space depending on the constitutive laws and material parameters applied to the domain.

By application of the assumed deformation gradient, the polynomial degree of the strains may be equal or even higher than the degree of the element shape functions. Therefore, exact numerical integration would require more integration points when being compared to standard isoparametric elements. Alternatively, one could use a generalization of Simpson's rule, i.e. taking the supporting points as the only integration points.

Let the strain energy density be interpolated using the interpolation function $M_A$

$$U^d(\xi) = \sum_A \sum_{m \in A} M_A^m(\xi) U_A^m \tag{4.75}$$

The supporting points are the integration points $A$ storing one discrete value of the strain energy density for each material definition $m$ adjacent to $A$. If two or more elements share the point $A$ and follow the same constitutive law and material parameters then they share the same support value $U_A^m$. Then one does not need to traverse through all elements, but only through the integration points and materials $m$ adjacent to each point, which greatly reduces the numerical effort when evaluating the material laws. The strain energy becomes

$$U = \int_V U^d(\epsilon, \alpha_{int}, t) dV = \sum_{A,m} W_A^m U_A^m \tag{4.76}$$

where the discrete values are evaluated at the supporting points of the assumed gradient field, i.e.

$$U_A^m = U^m(\epsilon^m(\mathbf{F}_A), \alpha_A^m, t) \tag{4.77}$$

Herein, $U^m$ denotes the constitutive relation defined for material $m$; $\alpha_A^m$ are the history variables at the considered integration point. $\epsilon^m$ defines a function that computes the strain from the deformation gradient $\mathbf{F}_A$. $W_A^m$ is an integration weight denoting the fictive volume of the $m$-th material at integration point $A$, i.e.

$$W_A^m = \int_{V^m} M_A^m(\xi) dV \tag{4.78}$$

### 4.4.5 Linearization

When solving the equation of motion, the strain energy is usually replaced by a quadratic expansion around the current configuration, i.e.

$$U(\mathbf{u} + \Delta\mathbf{u}, t) = U(\mathbf{u}, t) + \nabla U(\mathbf{u}, t)\Delta\mathbf{u} + \frac{1}{2}\Delta\mathbf{u}^T \nabla^2 U(\mathbf{u}, t)\Delta\mathbf{u} \tag{4.79}$$

## 4.5. Regular mesh generation

In structural dynamics, the strain energy gradient $\nabla U$ denotes the restoring force $\mathbf{F}$ and the Hessian $\nabla^2 U$ the stiffness matrix $\mathbf{K}$. They are computed by

$$F_A = \sum_B \sum_{m \in B} W_B^m \sigma_{B\alpha}^m \frac{\partial \epsilon_\alpha^m(\mathbf{F}_B)}{\partial F_{B\beta\gamma}} \frac{\partial F_{B\beta\gamma}}{\partial u_A} \tag{4.80}$$

$$K_{AB} = \sum_C \sum_{m \in C} W_C^m \sigma_{C\alpha}^m \frac{\partial^2 \epsilon_\alpha^m(\mathbf{F}_C)}{\partial F_{C\beta\gamma}\partial F_{C\eta\omega}} \frac{\partial F_{C\beta\gamma}}{\partial u_A} \frac{\partial F_{C\eta\omega}}{\partial u_B} +$$

$$\sum_C \sum_{m \in C} W_C^m C_{C\alpha\delta}^m \frac{\partial \epsilon_\alpha^m(\mathbf{F}_C)}{\partial F_{C\beta\gamma}} \frac{\partial \epsilon_\delta^m(\mathbf{F}_C)}{\partial F_{C\eta\omega}} \frac{\partial F_{C\beta\gamma}}{\partial u_A} \frac{\partial F_{C\eta\omega}}{\partial u_B} \tag{4.81}$$

wherein

$$\sigma_{A\alpha}^m = \frac{\partial}{\partial \epsilon_\alpha^m} U_A^m(\epsilon^m(\mathbf{F}_A), \alpha_A^m, t) \tag{4.82}$$

$$C_{A\alpha\beta}^m = \frac{\partial^2}{\partial \epsilon_\alpha^m \partial \epsilon_\beta^m} U_A^m(\epsilon^m(\mathbf{F}_A), \alpha_A^m, t) \tag{4.83}$$

denote the stress and tangential material tensor.

## 4.5 Regular mesh generation

A finite element mesh is regular, if the interpolation of the deformed geometry is at least $C^0$-continuous at the element interfaces. For example, "hanging nodes" or incompatible shape functions violate this criterion. Hanging nodes are nodes which should be constrained to the motion of the mutual interface of two or more finite elements, but are not due to meshing errors, see figure (4.3).

As a consequence of the assumed gradient field, the regularity requirements must be extended. Not only the deformed geometry must be continuous at the element interfaces, but also the assumed deformation gradient. That is, "hanging nodes" and "hanging integration points" should be prevented, see figure (4.4), unless it is intended to model discontinuous stresses, see figure (4.5). A violation of continuity of the assumed gradient field artificially stiffens the structure.

To illustrate this effect, consider an integration point $A$ which is part of two finite elements. The constitutive law is supposed to be linear with symmetric positive definite material tensor $C$. The discrete deformation gradient is averaged by

$$F_A = (V_1 F_A^1 + V_2 F_A^2)/(V_1 + V_2) \tag{4.84}$$

Assume that the irregular mesh consists of two integration points which were not merged into one. Then the contribution of this integration point to the strain energy in case of the regular mesh is

$$U_r = \frac{1}{2}(V_1 + V_2) F_A^T C F_A \tag{4.85}$$

and for the irregular mesh

$$U_{ir} = \frac{1}{2}\left(V_1 F_A^{1T} C F_A^1 + V_2 F_A^{2T} C F_A^2\right) \tag{4.86}$$

Assuming $U_r \leq U_{ir}$ one obtains (skipping some algebra)

$$0 \leq \left(F_A^1 - F_A^2\right)^T C \left(F_A^1 - F_A^2\right) \tag{4.87}$$

which proves the assumption since $C$ is positive definite. Therefore, the irregular mesh generally measures a strain energy which is too large.

## 4.6 Nodal integration

### 4.6.1 Nodal averages

**Nodal integration**

An obvious choice for the interpolation function of the deformation field is to take the finite element shape functions, i.e. $M_A(\xi) := N_A(\xi)$. This leads to an integration scheme where the integration points coincide with the finite element nodes. For linear tetrahedral elements, this ansatz is equivalent to [22, 133, 169]. For other element types, there are negligible differences in the weighting factors of the nodal averaging operator and in the way how the volume integrals are evaluated.

**Nodal pressure**

A nodally averaged pressure

$$p = \kappa \left(\frac{V}{V_0} - 1\right) = \kappa \left(\det(\mathbf{F}) - 1\right) \tag{4.88}$$

is applied to linear tetrahedra in [116]. It is equivalent to using a nodally averaged determinant of the deformation gradient

$$\det(\mathbf{F})_A = \frac{\sum_{m \in A} \sum_{e \in m} V_A^e \det\left(F_{A\alpha\beta}^e\right)}{V_A} \tag{4.89}$$

This quantity is used to scale the natural deformation gradient used in standard Gaussian quadrature, see also [4]. This variant was implemented in order to benefit from the properties of nodal integration regarding volumetric locking, meanwhile avoiding the instabilities of nodal integration.

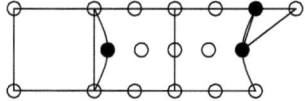

Figure 4.3: Irregular mesh in FEM. Incompatible interfaces (at filled nodes): "hanging node" on the left, incompatible (discontinuous) interpolation on the right

## 4.6. Nodal integration

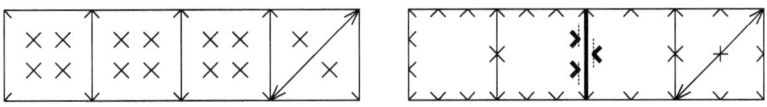

Figure 4.4: Regular mesh with assumed gradient field. Left: Supporting points in nodes and interior - continuous interpolation at interfaces. Right: Supporting points in nodes and on faces - incompatible interpolation at the bold interface.

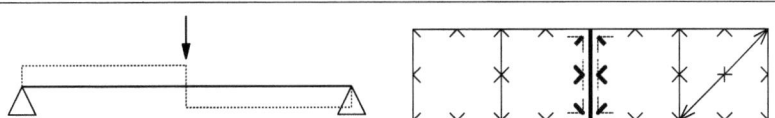

Figure 4.5: Modeling discontinuous stress fields with assumed gradient field. Left: Point load leads to discontinuous shear force distribution. Right: Do not merge integration points of adjacent elements at interface where discontinuous strain is modeled.

### 4.6.2 Analysis of instabilities

Examples show that the nodal averaging operator does not invoke the appearance of spurious zero-energy modes. Unlike in reduced order integration, the kinematic relationship between the assumed gradient and the natural shape function derivatives is established utilizing accurately evaluated integrals over the elemental volumes. Therefore, no rank deficiency occurs.

As being noted by several authors, nodal integration leads to instabilities in terms of spurious low-energy modes. This defect may, however, not appear in many use cases. That is, whenever the respective mode shapes are not excited or when the mesh is fine such that local modes are not visible in the macro scale. The spurious modes become apparent when modal analysis is applied. Therefore, the instabilities in nodal integration are denoted as 'temporal instability' since they are not apparent in many static examples of linear elasticity. Figure 4.18 shows an example of spurious mode shapes.

In order to reduce the degeneracies of the integration scheme, one can test the application of a high order Gaussian integration rule using the assumed strain field. Applying this scheme to 4-noded tetrahedral and 8-noded hexahedral elements leads to the same defects as observed when using NI. On the other hand, if one applies NI utilizing the natural shape derivatives as deformation measure (i.e. individual deformation gradients per local node of each element), one may observe bad accuracy in the sense that the smallest eigenvalues are overestimated, but no spurious modes appear (the behavior could be interpreted as irregular meshing with not merged nodal integration points, see section 4.5). Therefore, one can refer the instabilities to the formulation of the assumed gradient field and not to the integration scheme.

The reason for the instability of NI is the inability of the assumed deformation gradient to capture certain deformation shapes. For example, consider a one dimensional domain with equally sized linear finite elements being deformed as shown in figure 4.6. The interior elements are highly deformed, but the nodal values of the assumed deformation gradient measure zero strain, see figure 4.7. Only the nodal strain values that are located at the boundaries of the structure (where no averaging takes place) describe the deformation accurately. Since the elements at the boundary provide a stiffness, the presented deformation shape is related to an eigenmode whose energy is greater than zero, but greatly underestimated.

In other words, the instabilities arise because the continuous assumed deformation gradient loosely satisfies

$$\int_V F_{\alpha\beta}^{AN} dV \approx \int_V (\nabla_\beta u_\alpha + \delta_{\alpha\beta}) dV \qquad (4.90)$$

that expresses that the deformation gradient obtained from the shape function derivatives and the assumed gradient are equal in the weak sense. But the assumed gradient does not satisfy local equivalence, or, in other words, it does not minimize the local error in the element interior between the assumed field and the field obtained from the shape function derivatives. The latter can be expressed by a global error norm, i.e.

$$\int_V \left\| \nabla_\beta u_\alpha + \delta_{\alpha\beta} - F_{\alpha\beta}^{AN} \right\| dV \to \min \qquad (4.91)$$

Solving equation (4.91) leads to equation (4.59) with $L_A(\xi) = M_A(\xi) = N_A(\xi)$. In that case, both conditions (4.90), (4.91) are satisfied. But this stabilization is, however, not applicable in practice, see section 4.4.3.

### 4.6.3 Penalty regularization

The error in eq. (4.91) can be reduced by application of the penalty method. Herein, the constraint equation (4.59) will be enforced approximately, see figure 4.8. This is done by creating a modified strain energy function $U_{mod}$ through adding a penalizing potential function, such that $U_{mod}(\mathbf{u}, t) = U(\mathbf{u}, t) + P(\mathbf{u})$. $P$ is chosen such that $U_{mod}(\mathbf{u}, t) = U(\mathbf{u}, t)$ if the constraint is satisfied and $P \gg 0$ if the constraint is violated. A possible choice is the quadratic penalty function

$$P(\mathbf{u}) = \rho \int_V \left\| F_{\alpha\beta}^{AN}(\xi) - F_{\alpha\beta}^{h}(\xi) \right\|^2 dV \qquad (4.92)$$

Figure 4.6: One dimensional mesh of 1st order elements. Left: undeformed. Right: deformed.

## 4.6. Nodal integration

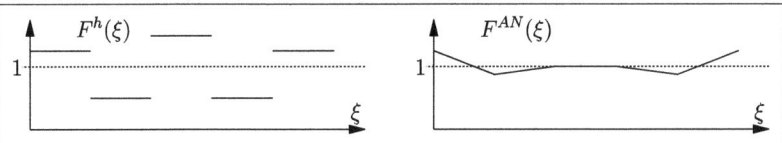

Figure 4.7: Analysis of instabilities in nodally supported assumed gradient field in one dimension. Left: Deformation gradient from shape function derivatives. Right: Assumed deformation gradient based on nodal averages: nearly zero strain!

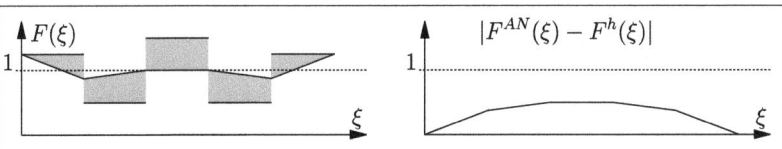

Figure 4.8: Stabilizing nodal integration in one dimension by penalty methods. Left: Difference in deformation gradients from natural and assumed interpolation. Right: Error function to be penalized.

with penalty parameter $\rho$ which adjusts the allowed range of the constraint violation. Typically the penalty parameter is chosen as the minimal value required for stabilization.

Since $\left\| F^{AN}_{\alpha\beta}(\xi) - F_{\alpha\beta}(\xi) \right\|$ may serve as a measure for the deviation of the discrete solution from exact solution, one may chose alternative potential functions. In [9] a penalty potential is presented that assumes that the strain energy is accurately evaluated at the nodes and an error in evaluating the potential appears in the element interior. This error is measured in terms of the strong (point-wise) stress residual, yielding

$$P_{\text{Beissel}}(\mathbf{u}) = \rho \int_V \|\operatorname{div}\sigma(\xi) + \mathbf{b}(\xi)\|^2 \, dV \qquad (4.93)$$

wherein b denotes the vector of body forces and $\sigma$ the Cauchy stress tensor. In order to provide a simple expression for the variations of $P$, the constitutive law is replaced by a linear elastic material. The scaling parameter is determined from $\rho = \frac{\alpha l_c}{E}$ with parameter $\alpha$, characteristic length $l_c$ and elastic modulus $E$. The same penalty has been used to stabilize NS-FEM in [280]. In [26], a penalty of the form

$$P_{\text{Broccardo}}(\mathbf{u}) = \rho \int_V \operatorname{tr}\left[ \left(\mathbf{F}^{AN}(\xi) - \mathbf{F}^h(\xi)\right)^T \left(\mathbf{F}^{AN}(\xi) - \mathbf{F}^h(\xi)\right) \right] dV \qquad (4.94)$$

was implemented which is identical to eq. (4.92).

### 4.6.4 Conforming regularization

In [114,168,215,217] a penalty potential function is presented that acts as a strain energy function which behaves like a Gaussian integrated finite element energy in the limit of infinitesimal strains. Furthermore, the stability and linear exactness could be provided by the stabilization term; material nonlinearities are subject to the nodally integrated parts. The strategy is known as stabilized conforming nodal integration (SCNI) and $\alpha$-FEM. The basic idea is to understand the nodal deformation gradient as a strain measure which is not conforming with the shape function space of the displacements. Then one modifies the strain energy such that

$$U_P(\mathbf{u},t) = U(\mathbf{u},t) - U^{C1}(\mathbf{u}) + U^{C0}(\mathbf{u}) \tag{4.95}$$

Therein, $U(\mathbf{u},t)$ denotes the nodally integrated strain energy. $U^{C1}(\mathbf{u})$ represents a nodally integrated strain energy potential which approximates $\mathbf{U}(u,t)$ using the continuous deformation gradient. $U^{C0}(\mathbf{u})$ denotes a conforming integrated strain energy potential which approximates $\mathbf{U}(u,t)$ using the finite element shape function derivatives. Then

$$U_P(\mathbf{u},t) \approx U^{C0}(\mathbf{u}), \quad \text{if} \quad \epsilon(\xi) \approx 0 \tag{4.96}$$

which does not exhibit any instabilities. For the two additional energy functions a linear elastic material is used in order to provide efficient linearizations, such that the material tensor $\mathbf{C}^s$ of the stabilizing strain energy density $U^s(\xi,\epsilon)$ becomes

$$C^s_{\alpha\beta\gamma\delta}(\xi) = \left.\frac{\partial^2 U^d(\xi,\epsilon,t)}{\partial \epsilon_{\alpha\beta} \partial \epsilon_{\gamma\delta}}\right|_{\epsilon=0,\,t=0} \tag{4.97}$$

In the case of incompressible media this setting would degrade the resistance of nodal integration against volumetric locking. Therefore, the Poisson ratio $\nu$ is decreased in the stabilizing potential such that $\max(\nu) = 0.3$. In the case of nonlinear material laws a material tensor representing the characteristic behavior is recommended in favor of the initial tangent material stiffness.

We now derive a special form in case of infinitesimal strains. The nodally averaged deformation gradient is reexpressed for convenience, $F_{A\alpha\beta} = \frac{1}{V_A}\sum_{e\in A} V_e F_e$ where $F_e$ denotes the deformation gradient at node $A$ obtained from the finite element shape functions at element $e$; $V_A$ is the nodal volume being the sum of the contributions of the surrounding elements $e$, i.e. $V_A = \sum_{e\in A} V_e$. The second variation of the strain energy yields

$$\delta^2 U_P = \sum_A V_A \left(\delta \mathbf{F}_A^T \frac{\partial^2 U_A^d}{\partial \mathbf{F} \partial \mathbf{F}} \delta \mathbf{F}_A - \delta \mathbf{F}_A^T \frac{\partial^2 U_A^s}{\partial \mathbf{F} \partial \mathbf{F}} \delta \mathbf{F}_A + \sum_{e\in A} \frac{V_e}{V_A} \delta \mathbf{F}_e^T \frac{\partial^2 U_A^s}{\partial \mathbf{F} \partial \mathbf{F}} \delta \mathbf{F}_e\right) \tag{4.98}$$

## 4.6. Nodal integration

With eq. (4.96) this can also be expressed as

$$\delta^2 U_P = \delta^2 U^{C0} = \sum_A \sum_{e \in A} V_e \left( \delta \mathbf{F}_e^T \frac{\partial^2 U_A}{\partial \mathbf{F} \partial \mathbf{F}} \delta \mathbf{F}_e \right) \quad (4.99)$$

$$= \sum_A \left[ \sum_{e \in A} V_e \delta \mathbf{F}_e^T \frac{\partial^2 U_A}{\partial \mathbf{F} \partial \mathbf{F}} \delta \mathbf{F}_e + V_A \left( \delta \mathbf{F}_A^T \frac{\partial^2 U_A}{\partial \mathbf{F} \partial \mathbf{F}} \delta \mathbf{F}_A - \delta \mathbf{F}_A^T \frac{\partial^2 U_A}{\partial \mathbf{F} \partial \mathbf{F}} \delta \mathbf{F}_A \right) \right]$$
(4.100)

$$= \sum_A \left[ V_A \left( \delta \mathbf{F}_A^T \frac{\partial^2 U_A}{\partial \mathbf{F} \partial \mathbf{F}} \delta \mathbf{F}_A \right) + \sum_{e \in A} V_e \left( \delta \mathbf{F}_A^T - \delta \mathbf{F}_e^T \right) \frac{\partial^2 U_A}{\partial \mathbf{F} \partial \mathbf{F}} \left( \delta \mathbf{F}_A - \delta \mathbf{F}_e \right) \right]$$
(4.101)

$$= \delta^2 U + \delta^2 \int_V \left( \left\| \mathbf{F}^{AN} - \mathbf{F} \right\|^2_{\frac{\partial^2 U^s}{\partial \mathbf{F} \partial \mathbf{F}}} \right) dV \quad (4.102)$$

Therefore, the conformization can be interpreted in terms of a quadratic penalty term, where the norm of the constraint residual is weighted by a material tensor. It can be combined with a penalty parameter $\alpha$ which may have values between $0$ (not stabilized) and $1$ (fully stabilized). If the infinitesimal strain tensor is used, then the stabilizing stiffness $\mathbf{K}^s = \alpha \nabla_{\mathbf{u}}^2 \left( U^{C0} - U^{C1} \right)$ is constant and can be precomputed previous to the simulation yielding an efficient scheme. It is possible to extend the idea to the case of geometrically nonlinear strains, yielding $U_P^\alpha(\mathbf{u}, t) = U(\mathbf{u}, t) + \alpha \left( U^{C0}(\mathbf{u}) - U^{C1}(\mathbf{u}) \right)$ although this increases the numerical costs.

The presented conforming regularization is a widely-used approach since it stabilizes nodal integration and formulates an artificial energy which is close to the physical model. As any regularization method it exhibits a few limitations, these are

- It adds artificial energy which may become quite large even for relative fine meshes. The artificial energy can be interpreted as a measure of discretization error. It can be reduced by finer remeshing or by decreasing the penalty parameter. The penalty parameter is chosen to be at least large enough to increase the eigenvalues of the spurious modes such that their degree of excitation becomes negligible. Therefore, the optimal choice of the penalty parameter may be dependent on the mesh topology and loading.

- When applied to nonlinear strain measures, the numerical effort increases by a factor three, since three nonlinear strain energy functions must be evaluated at each node. Furthermore, to the author there is no extension of eq. (4.93) or eq. (4.94) known which is invariant to rigid body motion.

- Since the stabilizing energy density is evaluated at the same spatial coordinates as the strain energy density of the given material, both contribute to the modified stress distribution $\sigma_P(\xi) = \sigma(\xi) + \sigma^s(\xi)$. Therefore, the stress arising from the nodally evaluated strain energy density $\sigma$ is smaller compared with unstabilized integration. When nonlinear effects (plastic strains, failure and yield conditions) are of interest, their magnitude is systematically underestimated.

## 4.7 Smoothed Finite Element Method

### 4.7.1 Smoothing operator

In SFEM, the deformation gradient is replaced by a piecewise constant weighted average. To achieve this, the domain $\Omega$ is subdivided into mutually exclusive and collectively exhaustive smoothing cells $\Omega_L$. For each smoothing cell a constant $\mathbf{F}_L$ is computed from

$$\mathbf{F}_L = \int_{\Omega_L} \mathbf{F}^h(\xi) w_L(\xi) dV \qquad (4.103)$$

$\mathbf{F}_L$ is called the smoothed deformation gradient; $w_L$ denotes the smoothing function satisfying the partition of unity condition

$$\int_{\Omega_L} w_L(\xi) dV = 1 \qquad (4.104)$$

and which is typically a Heaviside-type piecewise constant function

$$w_L(\xi) = \begin{cases} 1/V_L, & \xi \in \Omega_L \\ 0 & \xi \notin \Omega_L \end{cases}, \quad V_L = \int_{\Omega_L} dV \qquad (4.105)$$

By application of the divergence theorem, the smoothing operator can be transformed into a surface integral

$$F_{L\alpha\beta} = \delta_{\alpha\beta} + \frac{1}{V_L} \int_{\Gamma_L} u_\alpha(\xi) n_\beta d\Gamma \qquad (4.106)$$

where n denotes the outward normal vector of the boundary surface $\Gamma_L$. In this way, the method is able to provide accurate deformation measures for highly distorted meshes. This is even possible when the Jacobian matrix is singular. By application of the smoothing operator the solution was exemplified to be softer and often more accurate when compared with standard FEM, see [164].

### 4.7.2 Smoothing cells

Different strategies of subdividing the structure into smoothing cells produced a variety of methods, such as SFEM, NS-FEM, ES-FEM and FS-FEM.

Originally, each finite element was individually subdivided into smoothing cells [164]. Hourglassing (appearance of nonphysical zero energy modes) was identified if the number (and location) of the smoothing cells corresponds to the integration points of reduced Gaussian integration in standard FEM, see figure 4.9.

## 4.8. Stable interpolation schemes

SFEM was extended to nodal integration (NS-FEM) [169]. The structure is subdivided into smoothing cells in such a way that each cell prescribes the volume around a single finite element node, see figure 4.9. The properties of the solution are equivalent to other nodal integration approaches. Also, instabilities were reported which manifest in spurious low energy modes [280].

Furthermore, an edge-based method (ES-FEM) has been developed in [167]. The smoothing cells are defined as the surrounding areas of edges in two dimensions, i.e. the areas between the edge and the center point of each adjacent element, see figure 4.9. The method has been extended to tetrahedral meshes in three dimensions, see [203]. The smoothing cells are then defined as the volumes around finite element faces, i.e. the domain between a face and the center point of all adjacent elements.

### 4.7.3 Relation

The Smoothed Finite Element Method creates an assumed gradient field which is piecewise constant and discontinuous at the smoothing cell interfaces. It can be interpreted as a special case of the assumed gradient field in equation (4.57) with interpolation function

$$M_A(\xi) = \begin{cases} 1 & \xi \in \Omega_A \\ 0 & \xi \notin \Omega_A \end{cases} \quad (4.107)$$

and multiplier space $L_A(\xi) = M_A(\xi)$. The biorthogonality criterion, eq. (4.67), is satisfied. The integration points $A$ are located in the centers of the smoothing cells.

## 4.8 Stable interpolation schemes

This section presents interpolation strategies for the assumed gradient field. It is aimed to provide schemes which are stable (with respect to spurious modes), are efficient (accuracy and numerical effort) and lead to regular meshes when applied to heterogeneous finite element types.

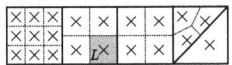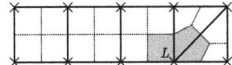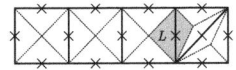

Figure 4.9: Smoothed FEM: two-dimensional smoothing cells schemes for 1st order elements; $L$-th smoothing cell highlighted; smoothing cells marked by crosses. Left: SFEM: subdivided elements. Center: NS-FEM: Regions around nodes. Right: ES-FEM: Areas between edges and adjacent finite element centers.

## 4.8.1 Nodal support with bubble stabilization

Consider the stabilization of nodal integration by increasing the interpolation order of the assumed gradient. Start with the observation that the nodal support values of the deformation gradient are well approximated (if the finite elements would be $C^1$-continuous), but not the values in the element interior.

Therefore, select an integration rule which requires integration points in the nodes and in the element interior. The idea is to use the finite element nodes as supporting points in order to enforce the continuity of the deformation gradient at finite element interfaces. Since the interior of the elements is badly approximated when using the nodal values only, one has to define supporting points in the element interior, see figure 4.10.

The number of required interior points can be determined by considering Gaussian integration schemes in standard FEM. Therein, a minimal order of the numerical integration scheme is required in order to capture all nonzero energy deformation modes. The essence of the reduced Gaussian integration is comparable with the instability of NI. Both are not able to capture certain types of deformation shapes. This is true for NI at least for those finite elements which are not part of the structure's boundary. Therefore, the number and location of interior support points should correspond to the number and location of the integration points of the Gaussian integration scheme with minimal stable integration order. For example, the 4-noded tetrahedron requires one additional supporting point, the 10-noded tetrahedron 4, the 8-nodes hexahedron 8 points, etc., see figure 4.11.

Virtually, the polynomial order of the displacement interpolation is increased by 2. Enriching the interpolation function of the assumed deformation gradient increases its degree by at least 1. Integrating the assumed deformation gradient corresponds to an assumed displacement field with a polynomial order that is increased by at least 2.

The assumed gradient field is continuous. The regularity requirements for finite element meshes using this formulation are identical to standard FEM since the integration points on the boundary coincide with the finite element nodes. The number of integration points is increased by the number of nodes when compared

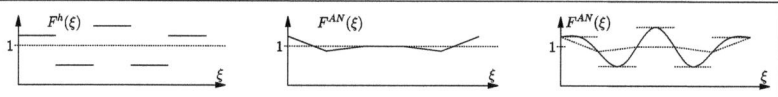

Figure 4.10: Stabilizing assumed gradient field in one dimension by enriched interpolation. Left: Deformation gradient from shape function derivatives. Center: Assumed deformation gradient using shape function for interpolation. Right: Assumed deformation gradient using higher order interpolation.

with standard FEM, but a greater accuracy can be expected due to continuous stresses.

## 4.8.2 Edge support

A stable assumed gradient field is obtained by choosing an interpolation where the support points are located on the finite element edges, see figure 4.11. The number of required integration points per edge is equal to the number of stripes that is required by Simpson's trapezoidal rule, i.e. 1 point for linear edges, 2 points for quadratic edges, etc. The approach is stable; by averaging the deformation gradient on the edges, all nonzero energy deformation shapes are measured as such. This is because the measured gradient along the axis of an edge is independent of the surrounding nodes and, therefore, no spurious modes can be introduced by the actual mesh topology (like in nodal integration).

The assumed gradient field is discontinuous between elements, i.e. it is continuous at its support points on the element interfaces and at all points in the finite elements interior. Discontinuities can be found, for example, at vertices being adjacent to two or more elements. Therefore, it can be expected to be stiffer than the continuous nodal-interior approach.

The regularity requirements for finite element meshes using the edge-based formulation are identical to those of standard finite elements. Because shared edges of adjacent elements always have the same interpolation function, the location of the integration points on all local edges is identical. Therefore, the local points can be merged into a single support point.

The number of edge integration points is large when considering a single finite element. The number of integration points of a finite element assembly may be, however, small when compared with standard FEM. The reason is that multiple elements may share individual edges and, therefore, the same support point is used to define the interpolation of the assumed deformation gradient within all adjacent elements. Since the number of integration points is related to the numerical effort when evaluating the strain energy, the decreased accuracy due to discontinuous stress may be balanced by improved numerical cost.

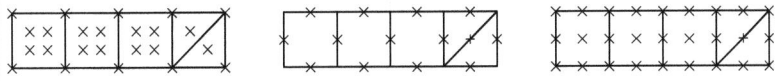

Figure 4.11: Assumed gradient field: Selection of stable interpolation schemes for 1st order finite elements in two dimensions. Left: Support in nodes and in a minimal number of inner-elemental points. Center: Support in points on edges/faces. Right: Support in points of higher order full-tensorial elements.

### 4.8.3 Face support

Face support is equivalent to ES-FEM in two dimensions. The assumed gradient field is discontinuous as in the edge-based approach. For each face the same number of integration points is required as in standard Gaussian integration of the face, i.e. 1 point for linear tetrahedra, 3 points for quadratic tetrahedra, 4 points for bilinear hexahedra. The total number of integration points can be expected to be larger than the edge-based approach, since each face can be part of only one or two elements.

The regularity requirements of finite element meshes using this approach are hard to meet when using heterogeneous element types, see figure 4.12. Although regularity requires a transition element between tetrahedral and hexahedral elements, users typically model faces with a hexahedron on their first side and with two tetrahedra on their other side. In standard FEM, a transition element is at least recommended for first order elements because the curved nature of the bilinear hexahedral face can not be represented by the piecewise linear faces of the tetrahedrons (although often not used in practice). Using node-supported assumed gradient fields, however, leads to a regular mesh regarding the position of the integration points on the interfaces.

Collapsed elements, where individual nodes are merged to create a new element topology, lead to irregular meshes. Following this approach, a hexahedron can be collapsed to a prism, pyramid or tetrahedron, see figure 4.13 for a two-dimensional illustration. No individual implementation of transition elements is necessary in FEM in such cases. With nodal based support this technique is possible. When using edge-based and face-based support, however, degenerated edges/faces lead to support points in nodes and/or edges leading to a discontinuous assumed gradient field. To overcome this problem, one has to ignore the degenerated integration points (the integration weight is automatically zero) and the support values in the interpolation. Therefore, one requires the derivation and implementation of transition elements for edge and face supported elements.

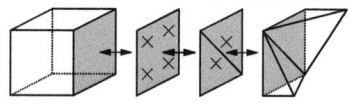

Figure 4.12: Example of irregular mesh for face supported assumed gradient field in three dimensions. Inner-facial support points on 1st order hexahedron do not coincide with facial support points on adjacent 1st order tetrahedra.

## 4.8. Stable interpolation schemes

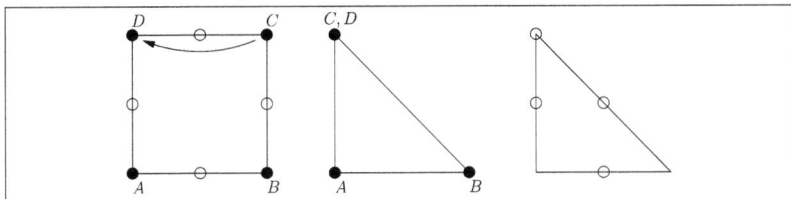

Figure 4.13: Irregular meshes by merging nodes in edge/face based support. Black circles: Nodes; white circles: integration points. Left: hexahedral/quadrilateral element type. Center: Collapsed element by merging 2 nodes. Right: Resulting distribution of support points of the assumed gradient. The support point at node $(C, D)$ must be eliminated due to degenerated edges/faces. Integration weights must be recomputed.

### 4.8.4 Assumed higher order gradient

The edge-based interpolation corresponds to enriching the interpolation of the assumed deformation gradient by polynomials of $2(n-1)$-th order, where $n$ is the number of nodes of the considered edge. Assuming tensor-product finite element shape functions, one could construct an assumed deformation gradient interpolation function of complete tensor-product structure with $2(n-1)$-th order. Hence, stress continuity at element interfaces can be reestablished with minimal additional numerical effort. If a full tensor-product function space is used, one is able to measure the strain energy density at equal distanced points forming a solely homogeneous grid.

Collapsed elements do not lead to irregular meshes because the support points of degenerated edges and faces are merged with the support points of adjacent nodes and edges. There is one limitation: If collapsed elements are allowed, no supporting points must be added to the interior of faces, i.e. the use of tensor-product 2nd order interpolations for 1st order hexahedra is discouraged in favour of a serendipity interpolation. This is because a prism element (obtained from a degenerated brick) would have support points in the center of their faces, but adjacent simplex elements would not have any.

### 4.8.5 Deriving continuum elements

Assuming finite elements with shape functions of tensor-product type, one possible strategy of deriving interpolation functions for the assumed gradient field follows the subsequent scheme:

1. Define the integration scheme and according to that the local coordinates of the integration points.

# Chapter 4. Continuous assumed gradient method

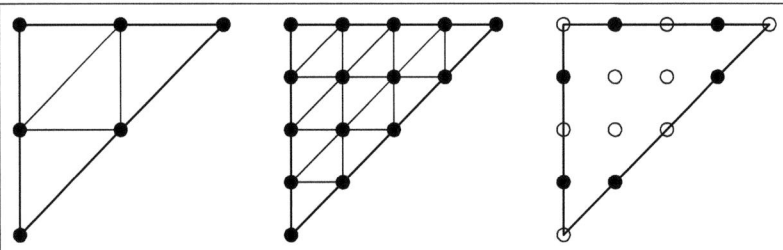

Figure 4.14: Deriving the interpolation function of assumed gradient field. Left: Finite element type: 2nd order triangle. Center: Higher (4th) order full-tensorial element. Right: Reduced element for edge-based interpolation (black circles: support points, white circles: points to be eliminated).

2. Define the "full" element type which corresponds to the aimed integration scheme, but has full tensor-product structure, see figure 4.14 (center). The set of support points of this element is to be reduced to a smaller subset (for example, a 2nd order element contains the support points of a 1st order element with edge integration). Define its shape function vector $\mathbf{N}^f(\xi)$.

3. Assume a scalar field $s$ distributed on the full element. The field values at those nodes of the full element which do not coincide with integration points are dependent on the values at the integration points. A linear transformation is assumed for projecting the function values of the reduced set of support points to the full set, i.e. $s_i^f = T_{ij} s_j^r$. $\mathbf{T}$ is a $n^f \times n^r$ matrix, where $n^f$ denotes the number of nodes in the full element and $n^r$ the number of nodes in the targeted "reduced" element. The row sums of $\mathbf{T}$ must be equal to 1.

4. The reduced shape functions are then computed by $\mathbf{N}^r(\xi) = \mathbf{T}^T \mathbf{N}^f(\xi)$.

The following element types (refering to the interpolation of deformation) are tested as listed in figure 4.15:

- Nodal-interior integrated elements ("xI"= number of inner points)

    - C3D_4N_1I: 1st order 4-noded tetrahedron with integration points in the nodes and one in the center.
    - C3D_8N_8I: The base is is a hexahedron of 3rd order. It has been reduced to an 8-noded element with 8 points in the interior.
    - C3D_8N_1I: The base is is a hexahedron of 2nd order. It has been reduced to an 8-noded element with 1 point in the interior.
    - C3D_10N_4I: The base of this element type is a tetrahedron of 5th order. It is reduced to a 10-noded element with 4 points added in the interior.

## 4.8. Stable interpolation schemes

- Edge integrated elements ("xE"= number of points per edge)

    - C3D_4N_1E: The full element type is a tetrahedron of 2nd order. It is reduced to a tetrahedron with one supporting point in the center of each edge.
    - C3D_8N_1E: The base is a hexahedron of 2nd order. The obtained element type has one supporting point per edge
    - C3D_10N_2E: The base of this element type is a tetrahedron of 5th order. It is reduced to an element with 2 points on each edge.

- Assumed higher order elements: The interpolation functions and integration point coordinates of the element types C3D_4N_10C and C3D_8N_27C are equivalent with the shape functions and nodal coordinates of the 10-noded tetrahedron and, respectively, of the 27-noded hexahedron. "xC" stands for "number of points in a complete interpolation basis"

- Standard finite elements C3D_4N, C3D_8N, C3D_10N.

- Nodally integrated elements without stabilization C3D_4N_NI, C3D_8N_NI and C3D_10N_NI. ("NI" = nodally integrated)

Example shape functions are given in appendix ??, detailed tests are presented in appendix A.

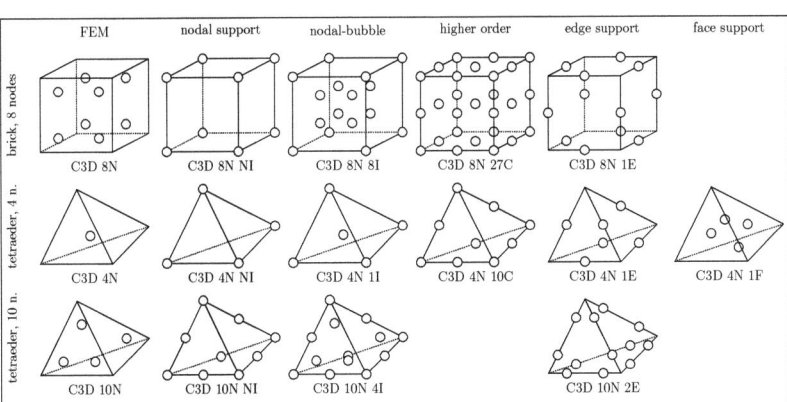

Figure 4.15: Implemented assumed gradient elements. The circles represent the positions of the integration points (= assumed gradient support points).

## 4.9 Implementation

### 4.9.1 Simplified dual mapping

When implementing the assumed gradient fields, one has to compute the coefficients of the dual shape functions used for the Lagrange multipliers. For the purpose of this work, the coefficients of the linear combination defining the discrete deformation gradients are precomputed and stored at the support points. This strategy may take some memory and should be avoided for very large scale problems. Alternatively, one may

- compute the coefficients of the discrete deformation gradients analytically. Since only geometrical quantities are of interest, all integrals are dependent on the a-priori known finite element interpolation functions and can be evaluated analytically.

- or, one can take simple averages of the deformation gradient, i.e.

$$F_{A\alpha\beta} = \frac{\sum_m V_A^m F_{A\alpha\beta}^m}{\sum_m V_A^m} \qquad (4.108)$$

where $V_A^m = \int_{V^m} M_A^m(\xi) dV$ denotes the volume around support point $A$ belonging to element $m$ and $\mathbf{F}_A^m$ is the deformation gradient obtained from the natural shape function derivatives of element $m$ at point $A$. The presented examples were also tested using this approach. The results are nearly identical.

Using the latter simplified approach, the presented method becomes equivalent to the strategy of NICE elements [133]. Then the main difference between NICE and the presented approach is the choice of the positions of the gradient field's support points. There seem to exist, however, a few limitations in the applicability of NICE. For example, the Lamé problem (appendix A.3) can not be solved easily because the stiffness matrix is singular. Responsible are highly distorted elements at the top of the modeled part (at $x = y = 0$). With the full dual multiplier approach, however, the problem is solveable.

### 4.9.2 Integration of volume integrals

**Gauss quadrature**

Consider the integration of the volume integral

$$I = \int_V f(\xi) dV = \int_\Xi f(\xi) \det(\mathbf{J}) d\xi \qquad (4.109)$$

## 4.9. Implementation

with $f$ denoting some function and $\mathbf{J}$ the Jacobian matrix mapping $d\mathbf{X} = \mathbf{J}d\xi$. If the function $f$ is not analytically integrable then the integral is recommended to be solved numerically. This is done using the sum approximation

$$I = \sum_i f(\xi_i) \det(\mathbf{J}(\xi_i)) w_i \tag{4.110}$$

where $\xi_i$ denote the coordinate and $w_i$ the weight of integration point $i$. Depending on the choice of $\xi_i$ and $w_i$ accuracy, stability and efficiency can be adjusted. A commonly used scheme is the Gauss quadrature. This approach obtains the exact solution of a polynomial of degree $2n - 1$ with $n$ supporting points for each dimension. It is obtained by assuming $h(\xi_i) = f(\xi_i)\det(\mathbf{J}(\xi_i))$ being a polynomial of degree $p$. Then the coordinates $\xi_i$ and $w_i$ are computed by minimizing the error

$$\left\| \int_\Xi h(\xi)d\xi - \sum_i h(\xi_i)w_i \right\| \to \min \tag{4.111}$$

between the exact value and the approximation. Since this condition must be satisfied for any polynomial of given order, it is equivalent to enforcing the equality $\int_\Xi h(\xi)d\xi = \sum_i h(\xi_i)w_i$.

One should chose the order of integration according to the polynomial degree of the input function. In practice, one often is interested in reducing the number of integration points due to efficiency penalties. A plain application of this strategy may lead to singular integrals. For example, if the integration order is too small then integrals such as $\int N_i N_j d\xi$ being positive definite matrix functions may have zero eigenvalues. Stabilization schemes may be complex (and often nonphysical), for example hourglass control when integrating the strain energy density by reduced integration orders in explicit dynamic codes.

**Analytical integration**

Theoretically it is possible to analytically evaluate the integrals required by the assumed strain field. Evaluate the numerator of the nodal deformation gradient

$$P_{AB\beta} = a_{AC} \int_\Xi M_C(\xi) \nabla_\gamma N_B(\xi) \text{Adj}(\mathbf{J}(\xi))_{\gamma\beta} d\xi \tag{4.112}$$

with adjoint

$$\text{Adj}(\mathbf{J}) = \begin{bmatrix} J_{22}J_{33} - J_{23}J_{32} & J_{13}J_{32} - J_{12}J_{33} & J_{12}J_{23} - J_{13}J_{32} \\ J_{23}J_{31} - J_{21}J_{33} & J_{11}J_{33} - J_{13}J_{31} & J_{13}J_{21} - J_{11}J_{23} \\ J_{21}J_{32} - J_{22}J_{31} & J_{12}J_{31} - J_{11}J_{32} & J_{11}J_{22} - J_{12}J_{21} \end{bmatrix} \tag{4.113}$$

The Jacobian is

$$J_{\gamma\beta}(\xi) = \frac{\partial}{\partial \xi_\beta} N_i(\xi) X_{i\gamma} \tag{4.114}$$

Given a shape function $N_i$ of order $n$ the integrated polynomial is, therefore, of order $n(n-1)^3$. For example, assuming a quadratic shape function one requires a Gaussian quadrature of 5th order. Using Gaussian quadrature one could precompute the quantity $P_{st\beta\gamma ik\eta\omega\kappa\nu}$ wherein $i$ and $k$ denote the indices of the nodal coordinates which appear as a double sum in $\mathrm{Adj}(\mathbf{J})$. $(\eta,\omega)$ and $(\kappa,\nu)$ denote the indices of each product pair $J_{\eta\omega}J_{\kappa\nu}$ which will be assembled into the Adjoint of $\mathbf{J}$. $\gamma$ is the index of the natural derivative of the $t$-th nodal displacement in $\beta$ direction, which is required to multiply $\partial u_{t\beta}/\partial \xi_\gamma$ with the Jacobian adjoint.

Therefore, in order to compute $P_{st\beta}$ one requires $n^2$ products for each of the 81 product pairs $J_{\eta\omega}J_{\kappa\nu}$, 3 products by $\gamma$ and 1 for the product by $a_{su}N_u$ $(=81n^2+4)$. Compare the sum with the effort of Gaussian integration requiring $9n$ products to compute the Jacobian, 18 products to compute its adjoint, 3 products by $\gamma$ and one by $a_{su}N_u$ $(=9n+22)$. Therefore, one can save significant amounts of computing time using numerical integration.

### 4.9.3 Constitutive laws

The presented methodology can efficiently adopt multiple types of material formulations. Since the structure of $\partial \mathbf{F}/\partial \epsilon$ depends on the mesh geometry there is no further simplification possible regarding special material formulations, except the terms

$$\sigma^m_{B\alpha}\frac{\partial \epsilon^m_\alpha(\mathbf{F}_B)}{\partial F_{B\beta\gamma}}, \quad \sigma^m_{C\alpha}\frac{\partial^2 \epsilon^m_\alpha(\mathbf{F}_C)}{\partial F_{C\beta\gamma}\partial F_{C\eta\omega}}, \quad C^m_{C\alpha\delta}\frac{\partial \epsilon^m_\alpha(\mathbf{F}_C)}{\partial F_{C\beta\gamma}}\frac{\partial \epsilon^m_\delta(\mathbf{F}_C)}{\partial F_{C\eta\omega}} \quad (4.115)$$

which are subjected to the material law implementation. Material implementations of classical FEM provide the quantities $\sigma$ and $C$. Therefore, a wrapper may be implemented supporting various strain formulations. For the following considerations one assumes the notations given in equations (4.20) and (4.40).

**Linearized strain tensor**

The strain vector is computed by

$$\epsilon^T = \begin{bmatrix} F_{11}-1 & F_{22}-1 & F_{33}-1 & F_{12}+F_{21} & F_{23}+F_{32} & F_{13}+F_{31} \end{bmatrix} \quad (4.116)$$

Its energy conjugate stress measure is the Cauchy stress tensor. The strain derivative has sparse structure and can be computed from

$$\delta\epsilon^T = \begin{bmatrix} \delta F_{11} & \delta F_{22} & \delta F_{33} & \delta F_{12}+\delta F_{21} & \delta F_{23}+\delta F_{32} & \delta F_{13}+\delta F_{31} \end{bmatrix} \quad (4.117)$$

Using a compact storage format one can store for each $\epsilon_\alpha$ a list of indices $(\eta\omega)$ of those components in $\mathbf{F}$ for which the strain derivative is nonzero, i.e. $\partial \epsilon_\alpha/\partial F_{\eta\omega} = 1$. The second variation is

$$\delta^2\epsilon = \mathbf{0} \quad (4.118)$$

## 4.10. Error analysis

### Green-Lagrange strain tensor

The strain vector is computed by

$$\epsilon = \begin{bmatrix} -\frac{1}{2} \\ -\frac{1}{2} \\ -\frac{1}{2} \\ 0 \\ 0 \\ 0 \end{bmatrix} + \sum_{\alpha=1}^{3} \begin{bmatrix} \frac{1}{2}F_{\alpha 1}^2 \\ \frac{1}{2}F_{\alpha 2}^2 \\ \frac{1}{2}F_{\alpha 3}^2 \\ F_{\alpha 1}F_{\alpha 2} \\ F_{\alpha 2}F_{\alpha 3} \\ F_{\alpha 1}F_{\alpha 3} \end{bmatrix} \quad (4.119)$$

Its energy conjugate stress measure is the second Piola-Kirchhoff stress tensor. The strain derivative has sparse structure and can be computed from

$$\delta\epsilon = \sum_{\alpha=1}^{3} \begin{bmatrix} F_{\alpha 1}\delta F_{\alpha 1} \\ F_{\alpha 2}\delta F_{\alpha 2} \\ F_{\alpha 3}\delta F_{\alpha 3} \\ F_{\alpha 1}\delta F_{\alpha 2} + \delta F_{\alpha 1}F_{\alpha 2} \\ F_{\alpha 2}\delta F_{\alpha 3} + \delta F_{\alpha 2}F_{\alpha 3} \\ F_{\alpha 1}\delta F_{\alpha 3} + \delta F_{\alpha 1}F_{\alpha 3} \end{bmatrix} \quad (4.120)$$

Herein, for each strain component $\epsilon_\beta$ one stores a list of index pairs $((\eta\omega), (\mu\nu))$. Each pair defines the nonzero components of the strain derivative, i.e. $\partial\epsilon_\beta/\partial F_{\mu\nu} = F_{\eta\omega}$. From the same storage object one can obtain the nonzero values of the second derivatives, i.e. $\partial^2\epsilon_\beta/(\partial F_{\mu\nu}\partial F_{\eta\omega}) = 1$.

## 4.10 Error analysis

Given a numerical model one has the verify and validate the quality of its solution with respect to the physical problem. There basically exist two processes: model validation and verification. Verification [250] is the process of determining if a computational model obtained by discretizing a mathematical model of a physical event represents the mathematical model with sufficient accuracy. Validation, on the other hand, is the process of determining if a mathematical model of a physical event represents the actual physical event with sufficient accuracy [250].

The latter leads to $m$- ('model') and $d$- ('dimension') adaptivity. For example, starting with a linear elastic one-dimensional beam model one may arrive at a material model with Green-Lagrange strains and visco-elastoplacticity, at finite deformation beam theory or even three-dimenional discretization models. Model verification requires lower and upper bounds to the spatial discretization error leading to $h$- ('element size') and $p$- ('polynomial degree') adaptivity. A very important observation can be obtained by the a priori error estimate in the energy norm for linear elasticity

$$\|\mathbf{e}_u\|_a \leq Ch^p |\mathbf{u}|_{H^{p+1}(\Omega)} \quad (4.121)$$

with energy error norm $\|e_u\|_a$, constant $C$ and $H^{p+1}$-seminorm $|u|_{H^{p+1}(\Omega)}$. Hence, the finite element solution converges to the exact solution u of linear order in $h$ and exponentially in $p$.

Some often used error estimators in FEM are (see [250] for an overview and the references therein):

- A priori global error estimators: energy norm, $H^1$-norm, $L_2$-norm.

- A posteriori global error estimators: explicit residual-type error estimators, implicit residual-type error estimators based on local Neumann problems (for example, equilibrated residuals) yielding upper bounds, implicit residual type estimators based on local Dirchlet problems giving a lower bound, hierarchical error estimators, averaging-type estimators (for example, the SPR technique).

- Goal oriented error estimators providing a localized version of global a posteriori estimators suitable for $hp$-adaptivity.

In this work, the displacement error is numerically measured in the example problems in section 4.11 and appendix A. A derivation of error estimates is subject to a future research. It should be noted, however, that lower and upper a priori error bounds were derived for the element-based SFEM in [277]. Furthermore, the CAG method uses stress quantities similar to those of the SPR technique. The difference is that CAG uses these quantities for computation of the displacement field while SPR applies them for postprocessing stresses obtained from FEM. The author believes that this may be a starting point for deriving simple a posteriori error norms.

## 4.11 Examples

### 4.11.1 Vibration analysis of a two-dimensional cantilever

This example is given by a linear elastic cantilever with geometry $L = 100mm$, $h = 10mm$, thickness $b = 1mm$, Young's modulus $E = 210kN/mm^2$, Poisson's ratio $\nu = 0.3$, mass density $8g/cm^3$ (steel), see figure 4.16. The example is excerpted from [174] where a two-dimensional plane stress problem is considered. Therefore, in order to model the problem in three dimensions, the mesh consists

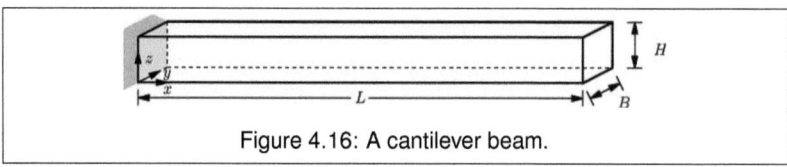

Figure 4.16: A cantilever beam.

## 4.11. Examples

only of 1 row of elements along the $y$ direction. All nodal displacements along the $y$ direction are fixed, i.e. $u_y = 0$. Using the Euler-Bernoulli beam theory, a fundamental frequency of $f_1 = 0.08276 \times 10^4 Hz$ is determined. The mesh used for simulation consists of $20 \times 1 \times 2$ elements. The reference solution is computed using a $100 \times 1 \times 10$ with 27-noded hexahedron elements. The frequencies $f$ are computed from the dynamic eigenvalues $\lambda$ by $f = \sqrt{\lambda}/(2\pi)$.

Table 4.1 lists the first 12 frequencies for different element types. Figure 4.17 plots the relative error $\epsilon = (f - f_{ref})/f_{ref}$.

| reference | C3D_8N | C3D_8N_NI | C3D_8N_8I | C3D_8N_1E | C3D_8N_27C |
|---|---|---|---|---|---|
| 0.08283 | 0.09246 | 0.10630 | 0.09770 | 0.09830 | 0.08781 |
| 0.49558 | 0.55611 | 0.62211 | 0.57930 | 0.58264 | 0.52706 |
| 1.28596 | 1.34849 | 1.0645 | 1.34874 | 1.3489 | 1.34781 |
| 1.30131 | 1.47321 | 1.34929 | 1.50115 | 1.50885 | 1.38955 |
| 2.35724 | 2.70039 | 1.58189 | 2.67576 | 2.68692 | 2.52853 |
| 3.5852 | 4.05023 | 2.55355 | 4.00117 | 4.01279 | 3.86335 |
| 3.85257 | 4.16366 | 2.75567 | 4.03549 | 4.0335 | 4.03821 |
| 4.92538 | 5.80755 | 4.01259 | 5.40281 | 5.4096 | 5.33268 |
| 6.33902 | 6.76582 | 4.02423 | 6.69047 | 6.67921 | 6.71269 |
| 6.40234 | 7.59841 | 4.26421 | 6.83474 | 6.82876 | 6.89852 |
| 7.80019 | 9.50287 | 4.63625 | 8.26831 | 8.23836 | 8.53793 |
| 8.9205 | 9.515 | 5.27116 | 9.29396 | 9.2624 | 9.35967 |
| C3D_4N | C3D_4N_NI | C3D_4N_1I | C3D_4N_1E | C3D_4N_10C | C3D_4N_1F |
| 0.11955 | 0.07495 | 0.09166 | 0.10092 | 0.09813 | 0.11848 |
| 0.70116 | 0.45130 | 0.54797 | 0.60047 | 0.58489 | 0.69569 |
| 1.3512 | 1.19309 | 1.3477 | 1.34906 | 1.3487 | 1.35093 |
| 1.80239 | 1.34599 | 1.43826 | 1.56878 | 1.53094 | 1.79089 |
| 3.20016 | 2.17182 | 2.60151 | 2.82889 | 2.76512 | 3.1844 |
| 4.06039 | 3.30361 | 3.94367 | 4.04407 | 4.04053 | 4.05914 |
| 4.79274 | 4.01747 | 4.03539 | 4.28903 | 4.19721 | 4.77482 |
| 6.50577 | 4.51376 | 5.3954 | 5.87461 | 5.7506 | 6.4873 |
| 6.79651 | 5.74044 | 6.68791 | 6.73615 | 6.72413 | 6.79305 |
| 8.31776 | 6.63062 | 6.89833 | 7.5469 | 7.38568 | 8.29869 |
| 9.55065 | 6.93726 | 8.41526 | 9.26301 | 9.06395 | 9.54405 |
| 10.2019 | 8.06051 | 9.29131 | 9.42524 | 9.38855 | 10.1817 |

Table 4.1: First twelve natural frequencies (in $10kHz$) of a two-dimensional cantilever.

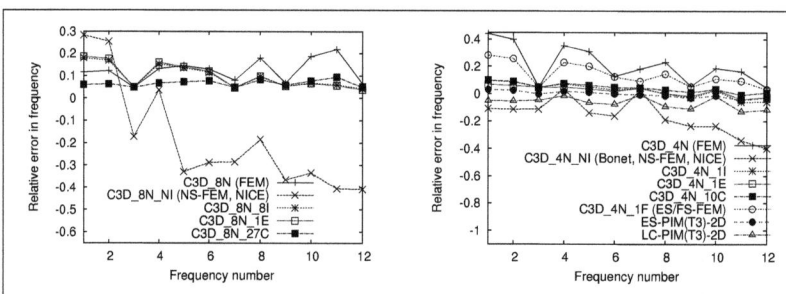

Figure 4.17: Relative error (0...1) of the first twelve frequencies of a slender cantilever for different elements. Left: brick elements. Right: simplex elements.

### 4.11.2 Spurious modes in nodal integration

Modal analysis is a useful tool to inspect instabilities arising in assumed gradient methods such as nodal integration. The method is stable if the mode shapes converge to the true solution. For nodal integration, spurious modes appear even among very low energy modes. Now, apply modal analysis to the example presented in appendix A.7 (a three-dimensional cantilever beam with left-sided support). Figure 4.18 shows the first 10 eigenvectors of the beam discretized by either standard 8-noded hexahedron elements and nodal integrated 8-noded hexahedron elements. Due to symmetry, the bending modes are listed for one direction only. Clearly visible are the spurious modes at mode number 5 and 6.

To be safe, one also has to verify higher mode shapes for column-shaped, thin-walled and compact structures. For example, the 8-noded nodal integrated hexahedron with one interior point C3D_8N_1I converges to the correct eigenvectors up to the 200th eigenmode in this example, but appears to exhibit instabilities if applied to a compact structure as given in section A.5.

### 4.11.3 Forced vibration of a cantilever beam

This example is excerpted from [167]. A cantilever beam is subject to a harmonic vertical tip load. The geometry of the beam is illustrated in figure 4.16 with $L = 4$ and $B = H = 1$. It is discretized by $12 \times 4 \times 4$ elements. The problem is originally considered as plane strain. Therefore, all displacements along the $y$ direction are fixed. The linear elastic material is given through $E = 1$, $\rho = 1$ and $\nu = 0.3$. The motion is subject to Rayleigh damping with damping matrix

$$C = \alpha M + \beta K \tag{4.122}$$

with $\alpha = 0.005$ and $\beta = 0.272$. The loading is given by $F(t) = cos(\omega_f t)$ with $\omega_f = 0.05$. The initial conditions are $(q_0, j_0) = (0, 0)$.

## 4.11. Examples

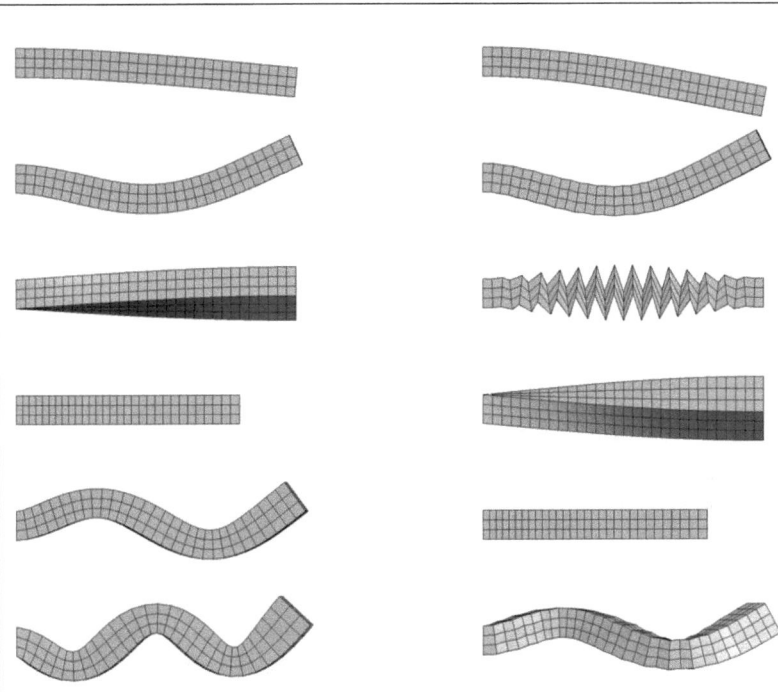

Figure 4.18: Spurious mode shapes in nodal integration (in parantheses: omitted symmetric shapes). Left: The 1st (2nd), 3rd (4th), 5th, 6th, 7th (8th) and 9th (10th) eigenvectors using stable gradient fields. Right: The 1st (2nd), 3rd (4th), 5th (6th), 7th, 8th and 9th (10th) eigenvectors using nodal integration.

The problem is solved using the explicit Velocity Verlet method with implicit damping, see equations (3.44)-(3.46). A time step of $h = 0.9 h_{crit}$ is used in the example.

Figure 4.19 plots the vertical tip displacement for different element types, table 4.2 lists the maximum tip displacements at the third maximum. Standard finite element types with quadratic interpolation are taken as a reference. The CAG hexahedron with complete tensorial interpolation is the only brick element which approaches the reference solution. The nodally integrated element exhibits the smallest elongation (and largest error). The other CAG brick elements are close to the standard FEM solution, but a little stiffer than that. Among the tetrahedron elements, the CAG element with nodal-interior support is very close to the reference solution. The nodally integrated tetrahedra clearly overestimate the maximum displacement. This "overly-soft" behavior is related to the existence of spurious modes in nodal integration. The standard FEM and the FS-FEM elements predict a response which is even $60\%$ too small. The other CAG element types are in between ($\approx 30\%$).

# Chapter 4. Continuous assumed gradient method

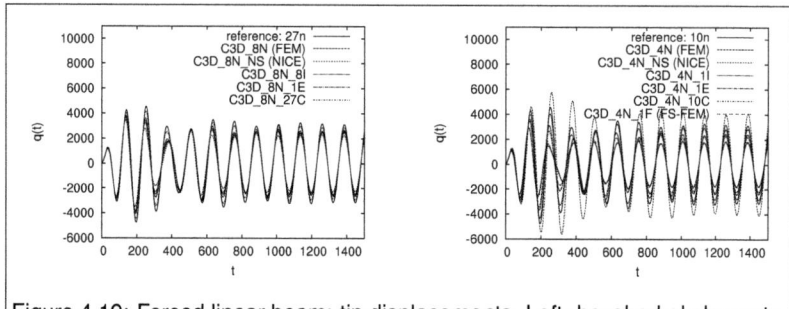

Figure 4.19: Forced linear beam: tip displacements. Left: hexahedral elements, Right: tetrahedra.

Table 4.2 compares the critical time step and the required CPU times for the time step loop. Ideally, the larger critical time step - compared with standard elements - compensates the numerical effort of the CAG field. This is most desirable for the methods which lead to highest accuracy, i.e. C3D_8N_27C and C3D_4N_1I. The C3D_8N_27C requires almost twice as much time as the FEM element. Only the edge-supported brick element is as effective as the FEM counterpart (115% time in the presence of a smaller time step). The pure nodally integrated elements outperform all other hexahedral and tetrahedral element types in numerical efficiency. For the tetrahedral elements, the CAG element with nodal-interior support needs even less time than the standard FEM. This is because the critical time step is almost twice as large and balances the higher numerical effort of C3D_4N_1I.

|  | C3D_8N | C3D_8N_NI | C3D_8N_8I | C3D_8N_1E | C3D_8N_27C |  |
|---|---|---|---|---|---|---|
| $\max_t q$ | 3815 | 3488 | 3728 | 3674 | 4171 |  |
| $h_{crit}$ | 0.0159 | 0.0131 | 0.0140 | 0.0127 | 0.0156 |  |
| time | 260.26 | 70% | 156% | 115% | 176% |  |
|  | C3D_4N | C3D_4N_NI | C3D_4N_1I | C3D_4N_1E | C3D_4N_10C | _4N_1F |
| $\max_t q$ | 2905 | 5754 | 4040 | 3615 | 3451 | 2976 |
| $h_{crit}$ | 0.0138 | 0.0208 | 0.0211 | 0.0126 | 0.0148 | 0.0146 |
| time | 186.36 | 54% | 92% | 203% | 202% | 214 % |

Table 4.2: Forced linear beam: max. tip displacement (at $t \approx 250$, reference: $\max_t q(t) = 4552$ for brick27 and 4561 for tetra10), critical time step $h_{crit}$ and CPU time (percentage relative to FEM).

## 4.11.4 Forced vibration of a geometrically nonlinear cantilever beam

Turn to the example given in section 4.11.3. In this paragraph the model parameters are changed in two ways: (1) A geometrically nonlinear strain formulation is used. (2) The vertical force is given by $F(t) = 10^{-3}\cos(\omega_f t)$.

The results on efficiency are identical to those in section 4.11.3, although the absolute times are larger due to the numerical effort in the nonlinear strain formulation.

Figure 4.20 plots the vertical tip displacement for different element types. Table 4.3 lists the maximum tip displacements of the second maximum. Standard finite element types with quadratic interpolation are taken as a reference. They are used with the same number of elements ($12 \times 4 \times 4$). The relationship in accuracy among the tested first-order elements is similar to the example in section 4.11.3. But the difference of the C3D_8N_27C and C3D_4N_1I elements with respect to their quadratic counterparts C3D_27N and C3D_10N is larger than in the linear case. Therefore, a second reference solution is provided with quadratic elements and same number of degrees of freedom ($6 \times 2 \times 2$ elements). The element types C3D_8N_27C and C3D_4N_1I and the second reference solution give nearly identical results.

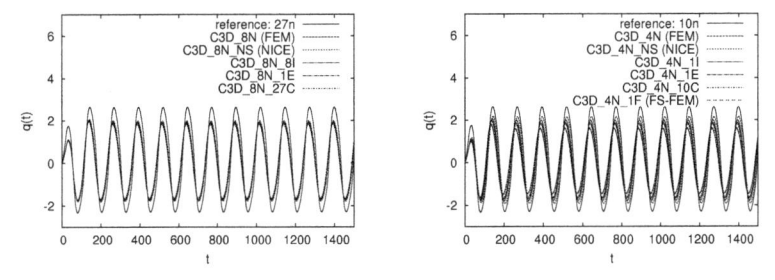

Figure 4.20: Forced nonlinear beam: tip displacements. Left: hexahedral elements, Right: tetrahedra. Reference: $12 \times 4 \times 4$ elements.

|  | C3D_8N | C3D_8N_NI | C3D_8N_8I | C3D_8N_1E | C3D_8N_27C |  |
|---|---|---|---|---|---|---|
| $\max_t q$ | 2.03062 | 1.96195 | 2.01153 | 1.99767 | 2.0715 |  |
|  | C3D_4N | C3D_4N_NI | C3D_4N_1I | C3D_4N_1E | C3D_4N_10C | _4N_1F |
| $\max_t q$ | 1.80577 | 2.18516 | 2.07071 | 1.99139 | 1.95238 | 1.82768 |

Table 4.3: Forced nonlinear beam: max. tip displacement (at $t \approx 265$). Reference ($12 \times 4 \times 4$): $\max_t q(t) = 2.63985$ for brick27 and $2.63563$ for tetra10. Reference2 ($6 \times 2 \times 2$): $\max_t q(t) = 2.08565$ for brick27 and $2.08016$ for tetra10.

# Chapter 5

# Asynchronous variational integration

## 5.1 Introduction

This chapter presents asynchronous variational integrators. They are motivated from the observation that only a small spatial region may be responsible for extraordinary small time steps in explicit integration. For example, finite elements of small geometrical size or elements with very stiff material properties are associated with very large elemental natural frequencies. Those limit the critical time step, although other regions in the structure may be stable at longer time steps if considered separately. The idea of using different time steps for individual spatial domains is not new and known as subcycling, multiple time stepping, multiscale stepping or mixed integration methods. Asynchronous integration is a very general formulation where each domain can be propagated by a basically arbitrarily chosen time step. The derivation from a variational principle guarantees the preservation of mechanical invariants, such as momentum and symplecticity.

The outline of this chapter is as follows: Asynchronous variational integrators (AVIs) are presented in section 5.2. They are illustrated in a general form, i.e. the potential function is seen as a general sum of potential functions. This interpretation is based on [65], but this chapter is limited to the symplectic Euler method.

The general formulation allows asynchronous applications in elasticity, but also molecular dynamics and other fields since it is not tied to the discretization of an elastic space-time continuum. The discretization of the Hamilton's principle in structural dynamics is presented as a special case in section 5.3. The efficient handling of simple constraint equations, i.e. nodal restraints, is discussed in section 5.4. Of particular interest are stability criteria of AVIs which are not simple to derive. Existing approaches use the CFL condition to estimate critical time steps. This strategy is not easily applicable when using continuous assumed gradient elements. The difficulties and existing approaches for time step selection are presented in section 5.5. Two approaches to time step estimation being applicable to continuous assumed gradient methods are presented.

An example is presented in section 5.5 illustrating the efficiency of asynchronous integration.

## 5.2  Asynchronous Euler scheme

Assume that the total potential energy is obtained by some additive composition

$$V(q) = \sum_i V_i(q) \tag{5.1}$$

In case of a synchronous time stepping scheme, the composition $V(q)$ is evaluated at discrete times, i.e. the potentials $V_i(q)$ and their derivatives are computed at synchronous times. The idea of asynchronous integration lies in evaluating the individual potentials at separate times. Asynchronous integration can, therefore, be interpreted as a generalization of multi time stepping. While multiple time stepping schemes create points in time which are synchronous, AVIs can be configured such that synchronous points exist; but their existence is not required.

Assign to each potential $V_i$ a sequence of times $\{0 = t_i^0 < \cdots < t_i^{M_i} = T_i\}$. Another sequence is created by inserting all times $t_i^j$ into a unique and sorted set which then contains all system times $\{\theta^0 < \theta^1 < \cdots < \theta^M\}$, see figure 5.1. The solution trajectory is obtained by interpolating the generalized coordinates along their supports at the system times $\theta_k$, i.e. each $\theta_k$ is associated to a discrete coordinate $q_k$. Furthermore, define the set

$$\mathcal{I}(k) = \left\{i, t_i^j = \theta_k\right\} \tag{5.2}$$

which collects the indices $i$ of all potentials $V_i$ that are evaluated at time $\theta_k$. The function

$$\mathcal{J}(k,i) = j, t_i^j = \theta_k \tag{5.3}$$

determines the index $j$ on the time scale of the $i$-th potential $V_i$ which corresponds to $\theta_k$, see figure 5.2.

Now discretize the solution trajectory $q(t)$ by elements being bounded at the times $\theta_k$ and $\theta_{k+1}$. A linear interpolation of $q(t)$ is assumed within the time element. Define the function

$$\mathcal{A}(k,i) = j, \max_j t_i^j \leq \theta_k \tag{5.4}$$

which determines the index $j$ of time $t_i^j$ where the potential $V_i$ has been evaluated most recently prior system time $\theta_k$. The function

$$\mathcal{K}(i,j) = k, t_i^j = \theta_k \tag{5.5}$$

## 5.2. Asynchronous Euler scheme

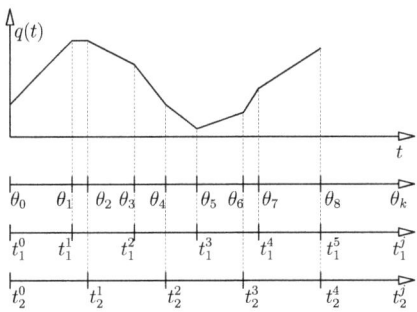

Figure 5.1: Illustration of fixed-step size AVI of a SDOF system with 2 potentials $V_i$ being evaluated either at $t_1^j$ or $t_2^j$. $q(t)$ is interpolated between discrete times $\theta_k$.

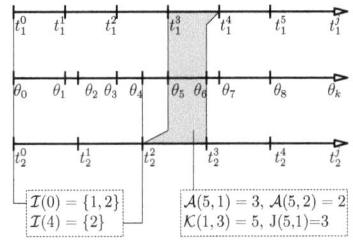

Figure 5.2: Illustration of index maps $\mathcal{A}, \mathcal{I}, \mathcal{J}, \mathcal{K}$ in AVI with 2 potentials $V_i$. The index maps find the indices of the corresponding potentials which are used in the $k$-th time step. For a time element in the interval $[\theta_5, \theta_6]$ the kinetic action is the time integral of the kinetic energy within $[\theta_5, \theta_6]$. There are 2 interpretations for the potential action: (1) The action of potential $V_1$ is the integral of $V_1(q)$ in the interval $[t_1^3, t_1^4]$. The action of potential $V_2$ is the integral of $V_2(q)$ in the interval $[t_2^2, t_2^3]$. (2) When the symplectic Euler method is applied then the discrete energy/Hamiltonian for this time element is the sum of the active kinetic energy and the active potential energies $V_1(q(t_1^{\mathcal{A}(5,1)})) = V_1(q(t_1^3))$ and $V_2(q(t_2^{\mathcal{A}(5,2)})) = V_2(q(t_2^2))$.

returns the index $k$ on the total time scale $\theta$ for a given index pair $(i, j)$ defining the potential $V_i$ and the potential time index $j$. Then the discrete action for a single element writes

$$S_k = j_k^T q_k^+ - j_{k+1}^T q_{k+1}^- + \frac{1}{2(\theta_{k+1} - \theta_k)} \left\| q_{k+1}^- - q_k^+ \right\|_M^2$$
$$- (\theta_{k+1} - \theta_k) \left( \sum_i V_i \left( q_{\mathcal{K}(i, \mathcal{A}(k,i))}^+ \right) \right) \quad (5.6)$$

Herein the kinetic action follows directly from integration of the kinetic energy. For the numerical integration of the potential energy in time, a single integration point is used being located either at the beginning or in the past of the considered time element. If a synchronous time stepping scheme is used, then $q^+_{\mathcal{K}(i,\mathcal{A}(k,i))} = q^+_k$. In such a case, the action of a single element contains only elemental variables (except the multipliers $j_k$, $j_{k+1}$) and each elemental action can be considered individually in order to obtain the time stepping scheme. In the asynchronous case, however, past coordinates are used. If the total action sum is derived by a coordinate $q^+_k$, then the subsequent elements may also contribute to the solution.

The discrete Euler-Lagrange equations are obtained by deriving the sum of all actions $S_k$ for individual $q^+_k$, $q^-_{k+1}$ and $j_{k+1}$:

$$0 = j_k - \frac{1}{\theta_{k+1} - \theta_k} M \left( q^-_{k+1} - q^+_k \right) - \sum_{i \in \mathcal{I}(k)} \left( t_i^{\mathcal{A}(k,i)+1} - t_i^{\mathcal{A}(k,i)} \right) \nabla V_i \left( q^+_k \right) \qquad (5.7)$$

$$0 = -j_{k+1} + \frac{1}{\theta_{k+1} - \theta_k} M \left( q^-_{k+1} - q^+_k \right) \qquad (5.8)$$

$$0 = -q^-_{k+1} + q^+_{k+1} \qquad (5.9)$$

leading to the time stepping scheme

$$j_{k+1} = j_k - \sum_{i \in \mathcal{I}(k)} \left( t_i^{\mathcal{A}(k,i)+1} - t_i^{\mathcal{A}(k,i)} \right) \nabla V_i \left( q^+_k \right) \qquad (5.10)$$

$$q^+_{k+1} = q^+_k + (\theta_{k+1} - \theta_k) M^{-1} j_{k+1} \qquad (5.11)$$

In words, at time $\theta_k$ one determines all potentials which are part of the set $\mathcal{I}(k)$, i.e. which are active at this time. The modification of the momentum is identical to symplectic synchronous Euler except that the time step sizes, which scale the contributions of all active potentials $V_i$, are not identical to the size of the time element $(\theta_{k+1} - \theta_k)$, but are the time steps of the potentials $\left( t_i^{\mathcal{A}(k,i)+1} - t_i^{\mathcal{A}(k,i)} \right)$. The trajectory of the generalized coordinates within the time element is characterized by a constant motion using the modified momentum $j_{k+1}$.

## 5.3 Discretization of the space-time integral

Let us turn to Hamilton's principle of structural mechanics. The action is given through the space-time integral

$$S = \int_0^T \int_V \left( \frac{1}{2} \rho(\xi) \left[ \dot{x}(\xi,t) \right]^2 - U^d(\epsilon(\xi,q),t) \right) dV dt \qquad (5.12)$$

with material coordinate $\xi$, deformed coordinate $x$, mass density $\rho$, strain energy density $U^d$ and volume in reference configuration $V$. In synchronous time stepping schemes, the spatial integral is discretized and evaluated first. Subsequently, the temporal integral is discretized. In asynchronous schemes, both integrals are discretized and solved simultaneously.

For the kinetic action, one choses

$$T_d = \int_{t_a}^{t_b} \frac{1}{2} \dot{q}^h(t)^T M \dot{q}^h(t) dt \qquad (5.13)$$

## 5.3. Discretization of the space-time integral

Therein, $q^h(t)$ is a piecewise linear interpolation of the generalized coordinates $q$ with support at times $\theta_k$. The mass matrix is given by

$$M_{AB} = \int_V \rho(\xi) N_A(\xi) N_B(\xi) dV \tag{5.14}$$

with finite element shape function $N_A(\xi)$. AVIs can only be efficient if a diagonal (lumped) mass matrix is assumed, for example through

$$M_{AB}^{lumped} = \delta_A^B \int_V \rho(\xi) N_A(\xi) dV \tag{5.15}$$

where $\delta_A^B$ denotes the Kronecker delta.

The potential energy can be expressed by a linear combination of discrete values of the strain energy density and some integration weights, see equation (4.76)

$$V_d = \int_{t_a}^{t_b} V(q(t)) dt = \int_{t_a}^{t_b} \left( \sum_{A,m} W_A^m U_A^m(q(t), t) \right) dt \tag{5.16}$$

Utilizing the symplectic Euler scheme one obtains

$$V_d = \sum_i \sum_{j=0}^{n_i-1} h_i^j V_i \left( q\left(t_i^j\right), t \right) \tag{5.17}$$

$$V_i(q, t) = \sum_{m \in i} W_i^m U_i^m(q, t) \tag{5.18}$$

$$h_i^j = t_i^{j+1} - t_i^j \tag{5.19}$$

Therein, $h_i$ denotes the time step of the $i$-th potential $V_i$. $m$ denotes all spatial integration points which belong to $V_i$. If $V_i$ coincides with the energy of a support point $A$ in continuous assumed gradient methods, see equation (4.76), then $m$ denotes the materials adjacent to $A$ and $W_i^m$ their spatial weights.

The discretization is shown in figure 5.3. It also illustrates that the equation systems to be solved are very small compared with the synchronous case: Every spatial integration point is only dependent on the deformed coordinates of the adjacent nodes. Therefore, a force influences only the momentum of these nodes. All other nodes perform a constant motion during this time. It is, therefore, not necessary to update the deformed coordinates of all nodes at all system times $\theta_k$, but only those nodes which are part of the influencing domain of the currently active potential $V_i$.

In the original paper [158], each finite element is associated with an individual time step, i.e. the spatial integration points of each finite element are collected to an individual $V_i$. This strategy simplifies a few things: (1) It is easier to implement AVIs in existing finite element codes since the finite elements are independent of the temporal discretization. (2) An estimation of the critical time step is relatively easy when elemental wave speeds are computed.

When using interpolations of strains and/or stresses which are continuous at finite element interfaces then a finite element based subdivision may not be

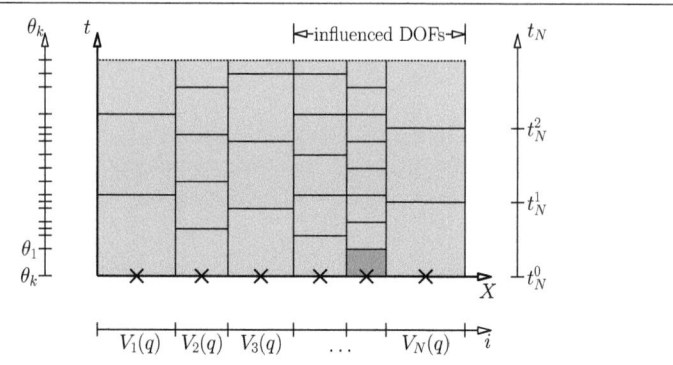

Figure 5.3: Illustration of space-time integration in one spatial dimension. The colored area is the space-time domain to be integrated (material space $X \times$ time $t$). Each cross denotes a single spatial integration point representing a single potential $V_i$. The highlighted cell is a space-time cell associated with a single integration point in space and time.

the best solution. Using classical isoparametric finite elements, the forces of the $i$-th element influence the nodes of the same element only. Using continuous strain interpolations, an integration point located on an element interface may influence the nodes of all surrounding elements. In particular nodal integration points may belong to many finite elements. If the integration points of a single element are collected to build some $V_i(q,t)$ then one needs to update the nodes of all surrounding elements. Furthermore, since the same integration point is part of two or more elements, it may be evaluated more often than necessary. Compare, for example, the number of system times $\theta_k$ with the number of element times $t_i^j$ in figure 5.3, the latter being much smaller. Therefore, it is assumed that every support point $A$ of the continuous gradient interpolation, see equation (4.57), describes an individual potential $V_i(q)$. By doing so, the number of material law evaluations is minimized during the time integration. On the other hand, the number of coordinate updates is larger compared with [158].

Variation of the space-time integral, equation (5.12), leads to the scheme given by equations (5.10)-(5.11). Special care must be spent when updating the minimum required set of coordinates in equation (5.11). Every node $A$ remembers the time $\theta^A$ at which its coordinates and momentum has been updated the last time. Then the next update of the same node refers to the state at time $\theta_{k+1}$, see figure 5.4 for more details. Of course, all nodes may be updated at all system times. But this greatly reduces numerical efficiency.

The resulting scheme is summarized in algorithm 5.5. A main component is the priority queue. It decides which potential is the next one to be evaluated. Each potential appears once in the queue which is sorted according to the next

## 5.3. Discretization of the space-time integral

evaluation times. Since multiple potentials may have identical times, a secondary sort condition is required to make the ordering unique.

As a result, the efficiency of asynchronous integrators is affected by

- Number of material law evaluations. The larger the time step, the better the efficiency. The greater the differences in step size within one model, the larger is the benefit when using AVIs compared with synchronous schemes.

- Number and computing time of drifts. The more drifts are necessary, the longer the computing time. Using standard isoparametric elements, the number of dependent nodes per spatial integration point can be reduced. Furthermore, the number of drifts can be reduced if adjacent spatial integration points can be collected to a single potential $V_i$.

- Priority queue. The complexity of inserting an item into a balanced tree is $\log_2(n)$ where $n$ is the number of potentials (or events). After each event, the information on the next kick must be reinserted into the queue. The smaller the number of events, the better the performance.

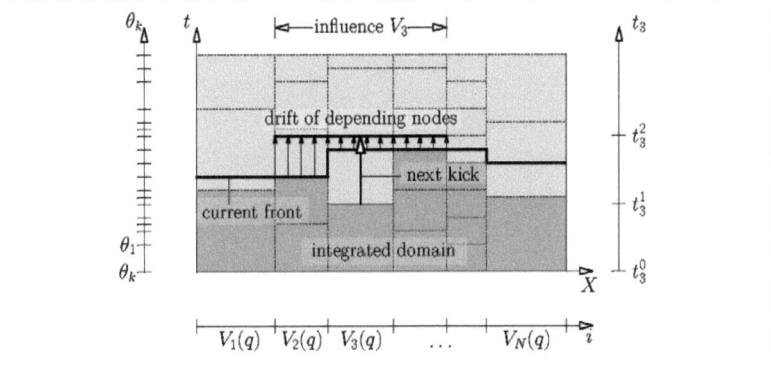

Figure 5.4: Illustration of a space-time front in one spatial dimension: The colored area is the space-time domain to be integrated (material space $X \times$ time $t$). The highlighted area is the domain which is already integrated. The deformed coordinates of all nodes are known at times $\theta^A$ defining the current front. It is not the boundary of the already integrated domain. This is because the space-time cells being integrated at last require the drift of nodes of adjacent finite elements. $V_3$ is "kicked" next at time $t_3^2$ requiring the drift of surrounding nodes.

---

Set initial conditions $q_0, j_0$.
Create global counter $k := 0$
For each potential $V_i$, compute a time step $h_i^0$ and a first kick time $t_i := h_i^0$.
Create a priority queue of indices $i$ which is sorted for $t_i$ in ascending order.
Set global time $\theta_k := 0$
For each node $A$, set a global time $\theta^A := \theta_k$.
**while** $\theta_k < T$ **do**
    Take 1st element from priority queue and remember indices $i$ and time $t_i$.
    $\theta_{k+1} := t_i$
    **for all** nodes $A$ which influence $V_i$ **do**
        *Perform drift:*
        $q_{k+1}^A := q_k^A + h_k j_k^A / M_A, \quad h_k := \theta_{k+1} - \theta^A$
        Set current time of node $A$: $\theta^A := \theta_{k+1}$
        *Perform kick:*
        $j_{k+1}^A = j_k^A - h_i^0 \nabla^A V_i(q_{k+1})$
    **end for**
    *Update state:*
    Set next evaluation time $t_i := t_i + h_i^0$
    $k := k + 1$
    Reinsert $i$ (and $t_i^{j^i}$) into priority queue.
**end while**
*Drift all nodes to the final time:*
**for all** nodes $A$ **do**
    $q_{k+1}^A := q_k^A + h_k j_k^A / M_A, \quad h_k := T - \theta^A$
**end for**

---

Figure 5.5: AVI with constant step sizes

## 5.4 Nodal restraints

Nodal restraints are linear constraint equations which depend only on the deformed coordinates of a single node. A finite element node may be subject to 0 to 3 restraints. Restraints may be used to define simple boundary conditions, for example sliding supports and rigid supports.

A simple way to enforce restraints would be to project the degrees of freedom onto the constraint manifold, see equation (3.94). In this case, every drift phase must check the restraint conditions at runtime. Further algorithms, for example collisions and tie constraints, become more complex. In fact, an application of RATTLE to nodal restraints can be implemented more easily, see section 3.8.4.

Assume that the initial conditions satisfy the restraints, equation (3.107), and the hidden constraints, equation (3.122). If this is the case, linear constraints are automatically satisfied by equation (5.11) in case of a motion with constant velocity. The asynchronous force evaluations lead to modifications of the momentum.

## 5.5. Estimating the time step length

If these kicks are applied such that the hidden constraint is not violated, then the coordinate increments will satisfy the restraints after a drift phase as well.

Given a constraint equation
$$G^T q = 0 \tag{5.20}$$
one projects the momentum after each kick by
$$j_k^+ = \mathbb{P} j_k \tag{5.21}$$
$$\mathbb{P} = I - G \left[ G^T M^{-1} G \right]^{-1} G^T M^{-1} \tag{5.22}$$

Assuming a diagonal mass matrix, the nodal mass can be cancelled out of the fraction. Furthermore, one may alternatively project the increment to the momentum, i.e. the forces $\nabla V_i(q)$ in equation (5.10). The presented approach, however, also corrects initial conditions which do not satisfy the hidden constraint. It is equivalent to RATTLE, i.e. the multipliers $\lambda_0^k$ in equation (3.127) are given by

$$\lambda_k^0 = \left[ G^T M^{-1} G \right]^{-1} G^T \Delta q \tag{5.23}$$

wherein $\Delta q$ is the coordinate increment in a drift phase.

When implementing the projection for nodal restraints, either the partial matrices $\left[ G^T M^{-1} G \right]^{-1}$ and $G$ or the projection matrix $\mathbb{P}$ can be computed prior the simulation. The small size allows the storage of these matrices at the finite element nodes. For the first variant one must save matrices of dimensions $0 \times 0$ to $3 \times 3$ depending on the number of restraints; for the latter case the matrix $\mathbb{P}$ is $3 \times 3$ for all nodes.

## 5.5 Estimating the time step length

The critical time step of synchronous schemes in elastodynamics can be estimated using the CFL condition [38]. Given a one-dimensional space, a rectangular space-time grid is assumed. Then the spatial force of a linear elastic string (or linear strain) at a point $P$ may be approximated by a central difference of the displacements at the neighbouring grid points $O$ and $Q$, see figure 5.6. The discretization in time is performed using a central difference of the point $P$ at the current time step, at the next time step $P'$ and at the previous time step. The cone of dependence is defined, see figure 5.6, with its shape depending on the wave speed. The boundary of the domain of dependence denotes the lines of determination. The state of point $P'$ is influenced by all points which are in the domain of dependence. Stability is ensured if the state of all grid points inside the domain of dependence, i.e. the generalized coordinates $q$ and momentum $j$, is known and used for the computation of the state of $P'$.

For the unstable case on the right of figure 5.6, more points than $O$, $P$ and $Q$ influence $P'$. Consider the grid points at previous times, in particular the grid

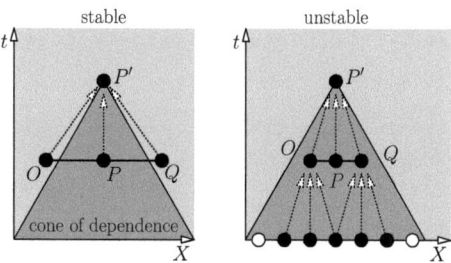

Figure 5.6: Domain of dependence for a rectangular space-time grid in one dimension. At a time $t$ all points in the domain of dependence located on the horizontal line (time $t$) must be known. Left: Stable since $O$, $P$, $Q$ are required to compute the state at $P'$ and all points between them are in the domain of influence. Right: Unstable. The state at $P'$ depends on more information than given by $O$, $P$, $Q$, i.e. the white grid points on the bottom do not contribute to $P'$, but are in the domain of dependence.

points which influence $O$ and $Q$. Then the horizontal gap between the outer grid points and the lines of determination is increasing. It is possible that at a certain time in the past grid points are part of the domain of dependence which are not used to predict the state in $P'$. Therefore, the illustrated case is unstable. The grid can be modified to a stable scheme by increasing the spatial grid size and/or by reducing the temporal grid size.

The difference scheme is stable if the CFL condition

$$c\frac{\Delta t}{\Delta X} \leq C \qquad (5.24)$$

is satisfied where $C$ denotes the Courant number depending on the difference scheme, $\Delta X$ and $\Delta t$ the grid spacing, and $c$ the velocity of wave propagation. For Verlet's method in one-dimensional space, the Courant number is $C = 1$.

This result is equivalent to the critical time step obtained from the largest natural frequency given in equation (3.24). Therefore, one may loosely assume equivalence between estimating the local wave speed and computing the largest natural frequency of single finite elements. In fact, one can show [112] that the largest frequencies of individual elements $\omega_{\max}^{(n)}$ are an upper bound to the largest natural frequency of the entire system $\omega_{\max}$

$$\omega_{\max} \leq \max_n \omega_{\max}^{(n)} \qquad (5.25)$$

as being often used to reduce numerical effort [147].

Some commercial software applications determine the wave speed based on material parameters [89], for example for a linear elastic material with mass density $\rho$, elastic modulus $E$ and Poisson's ratio $\nu$

$$c = \sqrt{\frac{E(1-\nu)}{(1+\nu)(1-2\nu)\rho}} \qquad (5.26)$$

## 5.5. Estimating the time step length

The characteristic spatial grid size is approximated using the finite element volume $V^{(n)}$ and element area $A^{(n)}$

$$\Delta X \approx \frac{V^{(n)}}{A^{(n)}_{\max}} \qquad (5.27)$$

A linear stability analysis of asynchronous integrators is very difficult. A stability analysis was presented for an SDOF system with two asynchronous potential functions in [65]. The analysis was established by assuming the existence of a macro time step with synchronous start and end times. This macro time step always exists - it may cover a time interval which is much larger than the time being simulated. The result is similar to stability conditions of the r-RESPA multiple time stepping algorithm, see section 3.7.4. AVIs and multiple time stepping are subject to resonances, i.e. time steps of one potential being equal to a specific multiple of the natural period of another potential. Resonances may appear at different time step sizes while stable intervals exist in between. Furthermore, a resonance which appears in a small spatial region of a large mesh may not immediately lead to a sudden growth of the phase space variables. This problem may simply lead to permanent energy growth with a small energy jump whenever the resonance appears.

When estimating the critical time step in asynchronous variational integration, one is not interested in the bounds of all stable intervals between resonances. Of more importance is the estimation of the smallest time step leading to resonances. Given isoparametric finite elements the critical time step can be estimated using elemental eigenvalues [158]. This strategy gives a good estimate, but nevertheless authors report that instabilities may sometimes appear for specific problem configurations [65].

Using finite elements with continuous assumed gradients it is not possible to compute an elemental natural frequency. This is because an elemental stiffness matrix does not exist. For general material laws and general finite element types a prediction of the wave speed and of the suitable grid size parameter leads to numerous case studies. A general methodology is required. Two strategies are proposed:

**Local modal analysis** A local eigenvalue problem is solved for each interpolation point of the continuous assumed gradient field, i.e. at each spatial integration point.

Solve the local eigenvalue problem

$$\nabla^2 V_A(q_0)\mathbf{v} = \lambda m_A \mathbf{v} \qquad (5.28)$$

at integration point $A$, wherein $\nabla^2 V_A$ denotes the local initial stiffness, $\mathbf{v}$ the eigenvector, $\lambda$ the eigenvalue and $m_A$ the mass of the integration point

$$m_A = \rho_A W_A \qquad (5.29)$$

determined from the mass density $\rho_A$ and the integration weight $W_A$.

The largest eigenvalue can be easily obtained by the power iteration, see algorithm 5.7. Notice, the stiffness matrix $\nabla^2 V_A$ is very sparse. It is usually stored in a column-wise layout. The matrix-vector product $\mathbf{y} = \nabla^2 V_A \mathbf{x}$ can still be computed efficiently by assuming symmetry, i.e. one multiplies all nonzero components of the $i$-th column with the $j$-th element of $\mathbf{x}$ to obtain the $i$-th element in $\mathbf{y}$. The vectors can be stored in dense format. Although many degrees of freedom may be allocated, the vector norm of $\mathbf{y}$ is computed efficiently since all non-zero components of $\mathbf{y}$ can be determined as the non-empty column indices of the local stiffness matrix.

The critical time step is then estimated from

$$\Delta t_A \leq \frac{2}{\sqrt{\max \lambda_A}} \qquad (5.30)$$

An interpretation in terms of CFL is given in figure 5.8. The current front of all nodes which influence the spatial integration point $A$ must cut the cone of dependence. In order to perform the kick of potential $V_A$ one has to drift all influencing nodes to the desired time $t_A^j$. Then the kick takes place and the momentum and coordinates of all influencing nodes are determined. According to CFL the state of all nodes influencing this operation must have been determined previously. That means that all adjacent integration points which are dependent on at least one of the influencing nodes of $A$ must not be propagated with a larger time step than the time step length being critical for $A$. This cross-check must be done additionally.

**Wave speed** Another strategy is directly estimating the wave speed using material parameters. The wave speed is estimated by

$$c = \sqrt{\frac{\lambda}{\rho}} \qquad (5.31)$$

where $\lambda$ is the maximum eigenvalue of the initial tangential material tensor and $\rho$ the mass density. The characteristic spatial grid size is approximated by the cubic

---

Create a counter $k := 0$
Create a randomly set initial vector $\mathbf{v}_0$
Normalize the initial vector $\mathbf{v}_0 := \mathbf{v}_0 / \|\mathbf{v}_0\|$
$\mathbf{K} = \nabla^2 V_A / m_A$
**while** not converged **do**
$\quad \mathbf{y}_k = \mathbf{K} \mathbf{v}_k$
$\quad \lambda_k = \mathbf{v}_k^T \mathbf{y}_k$
$\quad \mathbf{v}_{k+1} = \mathbf{y}_k / \|\mathbf{y}_k\|$
$\quad k := k + 1$
**end while**

Figure 5.7: Power iteration

## 5.6. Example: Asynchronous integration of a cantilever beam

root of the volume $V_A$ which is associated with the $A$-th spatial integration point, i.e

$$\Delta X = (V_A)^{\frac{1}{3}} \quad (5.32)$$

Then the critical time step is estimated via

$$\Delta t_A \leq \frac{\Delta X}{c} \quad (5.33)$$

Examples show, however, that this estimator is not always conservative.

The presented strategies for estimating the critical time step are necessary, but not always sufficient. This is in agreement with other authors analysing the stability of multiple time stepping schemes in FEM [43, 65]. Therefore, the critical time step length $\Delta t_A$ is associated with a scalar parameter $\beta$ defining a "safety factor"

$$h_A = \beta \Delta t_A, \quad 0 < \beta < 1 \quad (5.34)$$

Numerical experiments show that for linear elasticity values up to $\beta \leq 0.9$ may render a problem stable, while the same problems using a geometrically nonlinear strain formulation often need $\beta \leq 0.6$. Furthermore, the total simulation time influences stability. The appearance of resonances can not be entirely avoided, but their impact is negligable during reasonable small simulation times. For long term simulations, the tested examples were stable and reasonable accurate for $\beta = 0.5$.

## 5.6 Example: Asynchronous integration of a cantilever beam

### 5.6.1 Model problem

Consider a cantilever beam with square cross section as illustrated in figure 5.9. The geometry is defined by $L = 100$, $B = H = 10$. The material is linear elastic

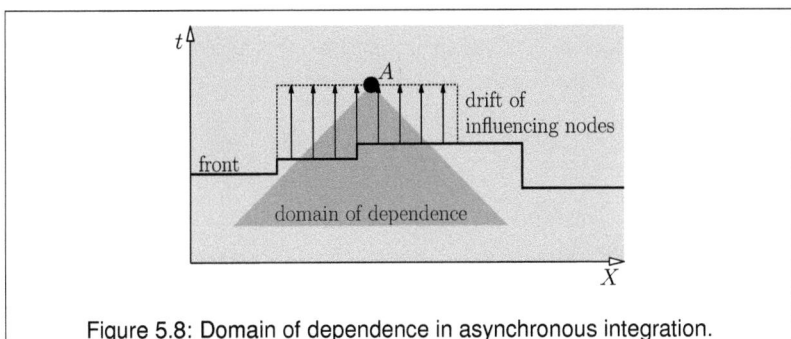

Figure 5.8: Domain of dependence in asynchronous integration.

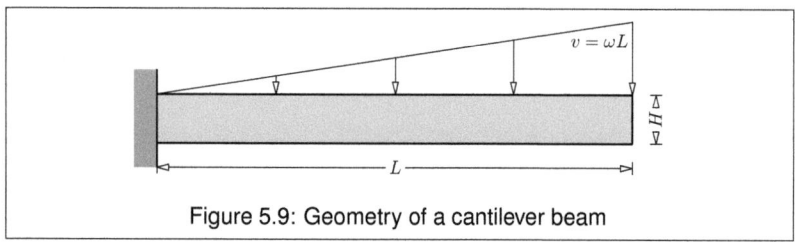

Figure 5.9: Geometry of a cantilever beam

and defined by elastic modulus $E = 30 \times 10^3$, Poisson's ratio $\nu = 0$ and mass density $2400 \times 10^{-9}$. The geometrically nonlinear Green-Lagrange strain tensor is used.

The beam is discretized by $10n \times n \times n$ 8-noded brick elements where $n$ is a mesh parameter. As finite element formulation the continuous assumed gradient element C3D_8N_27C is used, see figure 4.15 in section 4.8.5. The element sizes are chosen to vary along the $x$-axis. The element sizes are small at the support and large at the beam's free side. The $x$-coordinates of the finite element nodes are given by $X_i = L \left( \frac{i}{10n} \right)^2$ where $i$ is a node index $i \in [0, 10n]$. The mesh is illustrated in figure 5.10

The initial conditions are given by zero displacements $q_0 = 0$ and an 'angular velocity' $\omega = 180$ around the beam's left end, i.e. $v^y_{A,0} = -\omega X_A$ with node $A$ and coordinate horizontal $X_A$.

### 5.6.2 Benchmark against synchronous time stepping

This example compares the efficiency of asynchronous integration with standard methods. Velocity Verlet is used as reference solution. Therein, only one restoring force evaluation is implemented (the end step force vector is temporarily stored and used as start step force in the subsequent time step). By doing so, the numerical efficiency is comparable with the asynchronous Euler method with only one force evaluation per step. The critical time step is estimated by modal analysis of the initial stiffness matrix. A safety ratio $\Delta t = 0.5 \Delta t_{crit}$ is chosen. The critical time step of the asynchronous scheme is estimated using local modal analysis, see equation (5.30). The same safety factor is applied. Interestingly, the

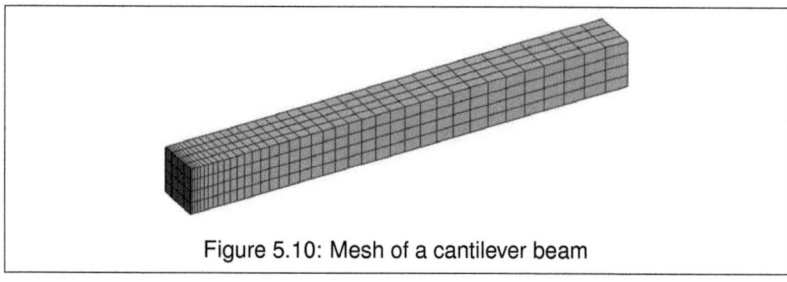

Figure 5.10: Mesh of a cantilever beam

## 5.6. Example: Asynchronous integration of a cantilever beam

local estimation leads to a smallest time step which is approximately half as large as the time step used in the reference solution. Increasing the safety factor such that the smallest time step equals the one used in Verlet, however, may lead to an unstable solution.

Three methods are compared:

1. standard Velocity Verlet,

2. asynchronous Euler with constant step sizes.

3. Furthermore, the asynchronous algorithm is used to emulate the standard Euler by setting all time steps to the smallest time step found in the system. Therefore, the algorithm becomes synchronous. By comparing this strategy with standard Verlet one can measure the computational overhead introduced by the asynchronous procedure (maintaining the queue, numerous drifts, etc.).

A simulation time interval of $T = 0.005$ is used. The total cpu times for the three methods are illustrated in figure 5.11. There is a linear relation between the cpu time and the mesh size for all tested methods. This is a very important observation, since it means that the overhead of the asynchronous scheme does not grow at a different rate as the effort due to a smaller critical time step. For the tested cases, the asynchronous scheme was faster than standard Verlet by a factor of $4 \ldots 6$ with increasing magnitude for larger meshes.

The overhead of the asynchronous procedure is approximately given by a factor of $8$ compared with the standard scheme. This factor can be computed by relating the simulation times of both schemes to the different time step sizes being actually used.

The critical time steps are given in figure 5.12. The critical time step of standard Verlet is shown in the left subfigure. The time step of the synchronous setting of the asynchronous algorithm is computed from the minimum critical time step in the center subfigure. The center and right subfigures compare the step size strategies given by equations (5.30) and (5.33). Equation (5.30) leads to a more efficient scheme being measured by the average time step in the system. The time step sizes from equation (5.33) can be directly related to the actual sizes of finite elements in the structure. One may assume that the latter strategy is more conservative, but this is not true: The rate by which the minimum critical time step in the system decreases with increasing mesh size parameter $n$ is smaller than the rate of the critical time step of standard Verlet. For the case $n = 16$ the minimal estimated critical time step using equation (5.33) is more than twice as large compared with standard Verlet. Therefore, the scheme becomes unstable.

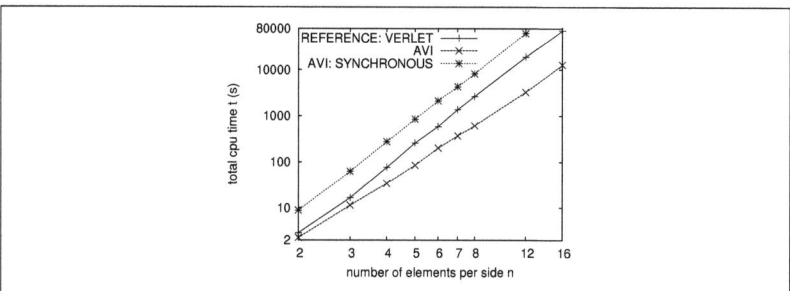

Figure 5.11: Comparison of total cpu time using standard Verlet, AVI and AVI with synchronous time step for different mesh size parameters.

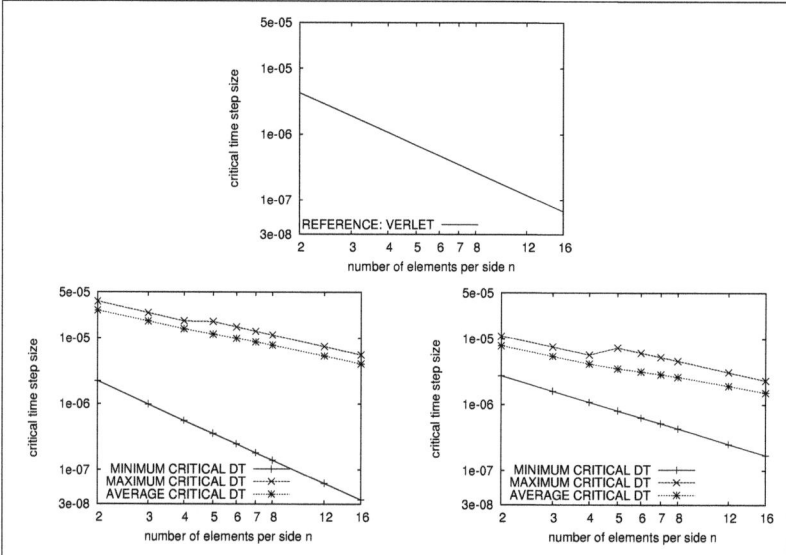

Figure 5.12: Critical time steps computed by different methods for different mesh size parameters. Top: standard Verlet using modal analysis of the complete system. Bottom left: AVI using modal analysis of local properties. Bottom right: AVI using a rough wavespeed estimation.

## 5.6.3 Equally distant nodes

By changing the mesh generation scheme one may create a worst case scenario for the efficiency of asynchronous integrators. Now, each finite element is a cube with uniform edge length $10/n$. The computing times of standard Verlet and the asynchronous case are presented in figure 5.13 for different mesh size parameters. Obviously, the standard method outperforms the asynchronous scheme. This is because there is almost no benefit due to a broad deviation of the critical time step size. For example, given a mesh size parameter $n = 8$, the standard

## 5.6. Example: Asynchronous integration of a cantilever beam

integrator obtains a critical time step of $h_{crit} = 9.238 \times 10^{-6}$. The asynchronous scheme obtains the interval $h_{crit} \in [7.870 \times 10^{-6}, 1.113 \times 10^{-5}]$ with an average of $h_{crit}^{avg} = 1.033 \times 10^{-5}$. Clearly, there is almost no deviation of the critical time step within the spatial domain. Its average value is very close to the one of the standard method. Considering the additional computational effort of the asynchronous algorithm, this results in a less efficient scheme.

This example shows that, although there may be great benefits in numerical efficiency when applying AVIs to certain models, the contrary may happen when applied to other models. Even if some of the numerical overhead can be eliminated, it is unlikely that AVIs may outperform standard Verlet in this example.

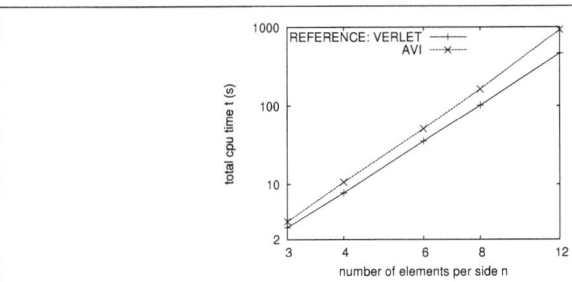

Figure 5.13: Comparison of total cpu time using standard Verlet and AVI for different mesh size parameters with equidistant nodes (worst case).

# Chapter 6

# Variable step size integration

## 6.1 Introduction

Variable time steps may be a useful tool to improve stability and accuracy of time stepping schemes. Target problems are systems where nonlinear effects lead to stiffening, for example penalty regularization of inequality constraints, highly nonlinear potential functions, stiffening due to geometrical nonlinearities in elasticity, etc.

Variable time steps based on variational integration were successfully applied in a symplectic-energy-momentum scheme [158] being presented in the synchronous and asynchronous context. The approach generates, however, at certain points in time equations which are not solvable destabilizing the solution. General time transformations involving a regularizing time step function may be a better choice. The derivation of such time step regularizations within the variational framework is presented in this chapter. Since accurate stability intervals of asynchronous integrators are hard to obtain, the application of variable time steps may be a way to improve stability the method. Therefore, the concept of temporally adaptive time steps is extended to the asynchronous context.

The outline of this chapter is as follows: First, an explicit symplectic energy momentum scheme is derived in section 6.2. The derivation of this variational integrator is simplified by application of the Hamilton-Pontryagin principle. It is extended to arbitrary time transformations in section 6.3. In order to prepare variable time steps in asynchronous integration explicit integrators are reinterpreted as kick and drift operators which are presented in variational form in the subsequent section. A synchronous/asynchronous algorithmic framework is derived and summarized in section 6.5. Some existing approaches to time step regularization are listed in section 6.6. Time step functions which seem suitable for the derived framework are presented in section 6.7. A simple solution procedure of the generated system of equation is presented. Examples are provided in section 6.8 which illustrate the potential advantages, but also problems and limits of variable time step methods.

## 6.2 Explicit symplectic energy momentum integration

A synchronous explicit symplectic energy-momentum integrator is obtained by using an Euler scheme for the integration of the potential action and by introducing the boundary times of the time elements as individual variables. Then, the bounding times $t_k^+$ and $t_{k+1}^-$ are considered as additional "coordinates" and their continuity among adjacent time elements is ensured by a constraint equation with Lagrange multiplier $w_k$.

The elemental action writes

$$S_k^d = j_k^T q_k^+ - j_{k+1}^T q_{k+1}^- + w_k^T t_k^+ - w_{k+1}^T t_{k+1}^-$$
$$+ \frac{1}{2\left(t_{k+1}^- - t_k^+\right)} \left\| q_{k+1}^- - q_k^+ \right\|_M^2 - \left(t_{k+1}^- - t_k^+\right) V\left(q_k^+, t_k^+\right) \quad (6.1)$$

For notational simplicity, the time step length can be eliminated from the kinetic action. The Hamilton-Pontryagin principle is incorporated with momentum $p(t)$ and velocity $v(t)$ which are defined as functions being constant within the time element and outside 0. Then one arrives at

$$S_k^d = j_k^T q_k^+ - j_{k+1}^T q_{k+1}^- + w_k^T t_k^+ - w_{k+1}^T t_{k+1}^-$$
$$+ \left(-H_k^d\right)\left(t_{k+1}^- - t_k^+\right) + p_k^T \left(q_{k+1}^- - q_k^+\right) \quad (6.2)$$
$$H_k^d = p_k^T v_k - \frac{1}{2} v_k^T M v_k + V\left(q_k^+, t_k^+\right) \quad (6.3)$$

with discrete Hamiltonian $H_k^d$. Using this notation, the multipliers $w_k$ can be identified as negative energy. They represent the momentum associated with the time coordinates $t_k$.

Variation leads to

$$\delta q_k^+ : \quad p_k = j_k - \left(t_{k+1}^- - t_k^+\right) \nabla V\left(q_k^+, t_k^+\right)$$
$$\delta q_{k+1}^- : \quad j_{k+1} = p_k$$
$$\delta v_k : \quad v_k = M^{-1} p_k$$
$$\delta p_k : \quad q_{k+1}^- = q_k^+ + \left(t_{k+1}^- - t_k^+\right) v_k$$
$$\delta t_k^+ : \quad 0 = w_k + H_k^d - \left(t_{k+1}^- - t_k^+\right) \frac{\partial}{\partial t_k^+} V\left(q_k^+, t_k^+\right)$$
$$\delta t_{k+1}^- : \quad w_{k+1} = -H_k^d \quad (6.4)$$

The second last equation from (6.4) implicitly defines the time step length. All other variables can be determined explicitly and their determination is identical to

## 6.2. Explicit symplectic energy momentum integration

fixed-step size Euler. The time step length is obtained from a quadratic equation, i.e.

$$t_{k+1}^- - t_k^+ = -\frac{p}{2} \pm \sqrt{\frac{p^2}{4} - q}$$

$$p = 2 \frac{-j_k^T M^{-1} \nabla V \left(q_k^+, t_k^+\right) - \frac{\partial}{\partial t_k^+} V \left(q_k^+, t_k^+\right)}{\nabla V \left(q_k^+, t_k^+\right)^T M^{-1} \nabla V \left(q_k^+, t_k^+\right)}$$

$$q = 2 \frac{w_k + \frac{1}{2} j_k^T M^{-1} j_k + V \left(q_k^+, t_k^+\right)}{\nabla V \left(q_k^+, t_k^+\right)^T M^{-1} \nabla V \left(q_k^+, t_k^+\right)} \tag{6.5}$$

The following cases can be distinguished:

- $\nabla V == 0$ and $\frac{\partial}{\partial t_k^+} V == 0$
  Any time step can be chosen without affecting energy conservation and symplecticity. If the selected time step is too large, however, the stability limits induce a large error and subsequent time steps may be very small or not solveable. If the selected time step is too small, the time step length of subsequent time steps may be very small as well which reduces numerical efficiency of the scheme.

- $\nabla V == 0$ and $\frac{\partial}{\partial t_k^+} V \neq 0$
  The SEM condition becomes linear in the time step length. Depending on the energy rate, a time step can be chosen which is beyond stability limits.

- $\frac{p^2}{4} - q < 0$
  There exists no solution.

- $t_{k+1}^- - t_k^+ \leq 0$
  A nonpositive time step must be avoided. There exists no feasible solution.

- else
  No special considerations.

If no feasible solution exists, one has to chose some time step length. In such a case, the integrator is not symplectic. Furthermore, the stability of the scheme is not guaranteed for the current step, since a heuristic selection of the time step length is identical to a change to another integrator.

The unsolvability of SEM integration is not restricted to the Euler method. Authors reported this problem quite often, for example [118, 158, 238]. Shibberu applied the SEM scheme to the implicit midpoint rule and found some stabilizing inequality constraints [238]. In many cases, the time step length selection in case of infeasible solutions is considered as not crucial, though the considered examples were integrated with very small time step lengths [158].

## 6.3 Time transformations

Synchronous SEM integration can be interpreted by a time transformation. Introduce a fictitious time $\alpha$, such that the discrete Lagrangian becomes

$$L_k^d = \int_{\alpha_k^+}^{\alpha_{k+1}^-} L\left(q^k(\alpha), \frac{\partial q^k(\alpha)}{\partial \alpha}\frac{\partial \alpha}{\partial t}, t^k(\alpha)\right) \frac{\partial t}{\partial \alpha} d\alpha \tag{6.6}$$

A common choice is $\alpha_k^+ = 0$, $\alpha_{k+1}^- = 1$, $\partial t^k/\partial \alpha^k = h_k$ with time step length $h_k$. The time $t^k$ becomes an individual function of fictitious time $\alpha$. Numerical methods integrating the transformed system will compute a solution not only for $q$, but also $t$. The fictive time step length is now $\alpha_{k+1}^- - \alpha_k^+ = 1$. Therefore, the phase space is extended by the time variable, such that

$$Q_k = \begin{pmatrix} q_k \\ t_k \end{pmatrix}, \quad J_k = \begin{pmatrix} j_k \\ w_k \end{pmatrix} \tag{6.7}$$

The discrete elemental action writes

$$S_k^d = L_k^d + J_k^T Q_k^+ - J_{k+1}^T Q_{k+1}^- \tag{6.8}$$

Since exact determination of a time step is not always possible in SEM methods, one may chose a regularization for the time step such that it becomes solvable for any configuration. Apply a Sundman-Poincaré transformation

$$\frac{\partial t^k}{\partial \alpha} = s_k(\alpha, q^k(\alpha), \dot{q}^k(\alpha)) \tag{6.9}$$

For simplicity, the scaling function $s_k$ is usually assumed to be constant during a time step and replaces the time step length $h_k$.

In the discrete setting, let the scaling function be a function of the momentum $p_k$ and the left boundary coordinate $q_k^+$. The left boundary coordinate is chosen in order to stay explicit with respect to the potential energy. In the context of the SEM-Euler method utilizing the Hamilton-Pontryagin principle, one has to add the following term to the right hand side of equation (6.2)

$$\mu_k \left(t_{k+1}^- - \left(t_k^+ + s_k\left(q_k^+, p_k\right)\right)\right) \tag{6.10}$$

## 6.3. Time transformations

That means, the elemental action of the SEM method is subject to a constraint (with multiplier $\mu_k$) which enforces the time step length to be equal to the scaling function $s_k$. Variation leads to

$$
\begin{aligned}
\delta q_k^+ : &\quad p_k = j_k - \left(t_{k+1}^- - t_k^+\right)\nabla V\left(q_k^+, t_k^+\right) - \mu_k \frac{\partial}{\partial q} s_k\left(q_k^+, p_k\right)\\
\delta q_{k+1}^- : &\quad j_{k+1} = p_k\\
\delta v_k : &\quad v_k = M^{-1} p_k\\
\delta p_k : &\quad q_{k+1}^- = q_k^+ + \left(t_{k+1}^- - t_k^+\right) v_k + \mu_k \frac{\partial}{\partial p} s_k\left(q_k^+, p_k\right)\\
\delta t_k^+ : &\quad \mu_k = -w_k - H_k^d + \left(t_{k+1}^- - t_k^+\right)\frac{\partial}{\partial t_k^+} V\left(q_k^+, t_k^+\right)\\
\delta t_{k+1}^- : &\quad w_{k+1} = -H_k^d - \mu_k\\
\delta \mu_k : &\quad t_{k+1}^- = t_k^+ + s_k\left(q_k^+, p_k\right)
\end{aligned} \tag{6.11}
$$

The equations are similar to those of SEM Euler, but the steps to solve the system of equation change significantly. Given the momentum $p_k$, one can explicitly solve for $j_{k+1}$, then $v_k$, $\mu_k$, $q_{k+1}^-$, $w_{k+1}$ and $t_{k+1}^-$. In order to solve $p_k$ one has to eliminate the time step length and the multiplier $\mu_k$ from the conditional equation, i.e.

$$
\begin{aligned}
p_k = j_k - s_k\left(q_k^+, p_k\right)\nabla V\left(q_k^+, t_k^+\right) - \frac{\partial}{\partial q} s_k\left(q_k^+, p_k\right) \cdot \\
\left(-w_k - \frac{1}{2} p_k^T M^{-1} p_k - V\left(q_k^+, t_k^+\right) + s_k\left(q_k^+, p_k\right)\frac{\partial}{\partial t_k^+} V\left(q_k^+, t_k^+\right)\right)
\end{aligned} \tag{6.12}
$$

This equation is usually nonlinear in $p_k$ and must be solved iteratively.

When compared with the standard Euler method, the derivation shows why an adhoc adaptation of time steps may lead to unstable time stepping. In order to stay symplectic when varying the time step length, one must add the terms $-\mu_k \frac{\partial}{\partial q} s_k\left(q_k^+, p_k\right)$ and $\mu_k \frac{\partial}{\partial p} s_k\left(q_k^+, p_k\right)$ to the increments of momentum and coordinates, respectively. This is equivalent to integrating an alternative system with modified Hamiltonian $\hat{H} = s(H + w)$ with the fixed-step size Euler and step size 1. In other words, the symplectic variable time step Euler is equivalent to the discrete Hamiltonian equations

$$
\begin{aligned}
Q_{k+1} &= Q_k + \frac{\partial \hat{H}_k^d(Q_k, J_{k+1})}{\partial J_{k+1}}\\
J_{k+1} &= J_k - \frac{\partial \hat{H}_k^d(Q_k, J_{k+1})}{\partial Q_k}\\
H_k^d(q, j) &= \frac{1}{2} j^T M^{-1} j + V(q)\\
\hat{H}_k^d(Q, J) &= s_k(q, j)\left(H_k^d(q, j) + w\right)
\end{aligned} \tag{6.13}
$$

## 6.4 Variational kick and drift operators

When combining variable time steps and asynchronous integrators, an interpretation of explicit schemes in terms of kick and drift operators is helpful.

Define the kick operator which changes the momentum as

$$\Phi_h^K : (q, p) \to (q, p - h\nabla_q H(q,p)) \quad (6.14)$$

and the drift operator

$$\Phi_h^D : (q, p) \to (q + h\nabla_p H(q,p), p) \quad (6.15)$$

which solves the motion with assumed constant velocity. Using these notations, the Velocity Verlet scheme can be written as $\Phi_h^{VV} = \Phi_{h/2}^K \circ \Phi_h^D \circ \Phi_{h/2}^K$. Its leapfrog representation becomes $\Phi_h^{LF} = \Phi_{h/2}^D \circ \Phi_h^K \circ \Phi_{h/2}^D$. The symplectic Euler can be expressed as $\Phi_h^E = \Phi_h^D \circ \Phi_h^K$.

In order to derive the kick and drift operators from a variational principle, two subsequent time elements are used, see figure 6.1. The length of the first element is infinitesimal small. For both elements, the potential energy is integrated using a single integration point located at the left boundary of the first element. The time between both elements is denoted by the index $k + \tau$. Then the elemental actions using the Hamilton-Pontryagin principle are defined by

$$S_k^d = j_k^T q_k^+ - j_{k+\tau}^T q_{k+\tau}^- - H_k^d \left(t_{k+\tau}^- - t_k^+\right) + p_k^T \left(q_{k+\tau}^- - q_k^+\right)$$

$$S_{k+\tau}^d = j_{k+\tau}^T q_{k+\tau}^+ - j_{k+1}^T q_{k+1}^-$$
$$\quad - H_{k+\tau}^d \left(t_{k+1}^- - t_{k+\tau}^+\right) + p_{k+\tau}^T \left(q_{k+1}^- - q_{k+\tau}^+\right)$$

$$H_k^d = p_k^T v_k - \frac{1}{2} v_k^T M v_k + V\left(q_k^+, t_k^+\right)$$

$$H_{k+\tau}^d = p_{k+\tau}^T v_{k+\tau} - \frac{1}{2} v_{k+\tau}^T M v_{k+\tau} + V\left(q_k^+, t_k^+\right) \quad (6.16)$$

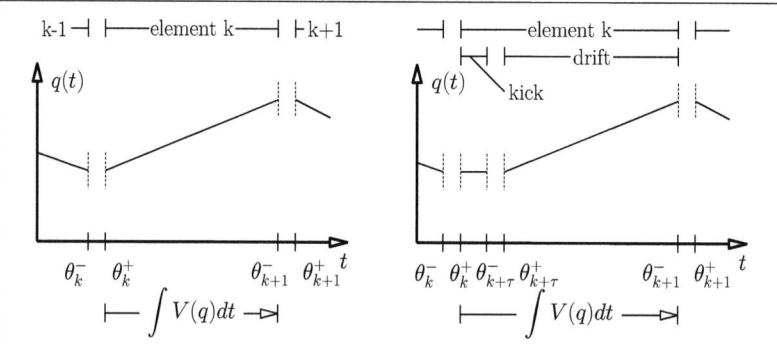

Figure 6.1: Variational kick and drift elements. Element $k$ is split into kick and drift. $V(q)$ is integrated with one integration point per element, both located at $q_k^+$. The kick is of infinitesimal duration.

## 6.5. Asynchronous variable time steps

Variation of the sum of both elemental actions leads to the following equations:

$$\begin{aligned}
\delta q_k^+ : & \quad p_k = j_k - \left(t_{k+1}^- - t_k^+\right) \nabla V\left(q_k^+, t_k^+\right) \\
\delta v_k : & \quad v_k = M^{-1} p_k \\
\delta p_k : & \quad q_{k+\tau}^+ = q_{k+\tau}^- = q_k^+ + \left(t_{k+\tau}^- - t_k^+\right) v_k \\
\delta q_{k+\tau}^- : & \quad j_{k+\tau} = p_k \\
\delta q_{k+\tau}^+ : & \quad p_{k+\tau} = j_{k+\tau} \\
\delta v_{k+\tau} : & \quad v_{k+\tau} = M^{-1} p_{k+\tau} \\
\delta p_{k+\tau} : & \quad q_{k+1}^- = q_{k+\tau}^+ + \left(t_{k+1}^- - t_{k+\tau}^+\right) v_{k+\tau} \\
\delta q_{k+1}^- : & \quad j_{k+1} = p_{k+\tau}
\end{aligned} \quad (6.17)$$

With $\lim_{t_{k+\tau}^- > t_k^+}\left(t_{k+\tau}^- - t_k^+\right) \to 0$ and $\lim \left(t_{k+1}^- - t_{k+\tau}^+\right) \to t_{k+1}^- - t_k^+$ one finds the kick and drift operators $\Phi_h^K : (q_k^+, j_k) \to (q_{k+\tau}^-, j_{k+\tau})$ and $\Phi_h^D : (q_{k+\tau}^+, j_{k+\tau}) \to (q_{k+1}^-, j_{k+1})$ with $h = t_{k+1}^- - t_k^+$.

## 6.5 Asynchronous variable time steps

From the synchronous variable step size methods one can see that right-boundary coordinates and momenta are dependent on the energy error within the considered time element. When applied to asynchronous settings then this dependency may create an implicit relation between $p_k$ and $p_{k+1}$ in the $k$-th time element. This situation, where future variables are required to compute the variables at present, must be avoided when constructing the time stepping scheme. Particular care must be spent regarding the start from a synchronous initial condition. How this can be achieved, is presented in this section.

For each potential $V_i$ one may define individual variable sets for the evaluation times $\{t_i^j, j = 0, \ldots, N_i\}$ and associates an individual time step function $s_i(q,p)$. The time step function defines the time step length $h_i^j$ between the evaluation times $t_i^j$ and $t_i^{j+1}$. In order to stay explicit in $q$ one may chose

$$h_i^j = s_i^j \left(q_{\mathcal{K}(i,j)}^+, p_{\mathcal{K}(i,j)}\right) \quad (6.18)$$

The time stepping scheme is interpreted as a sequence of pairs of kick and drift operators, see figure 6.2. The drift operators propagate the solution assuming constant velocity between all $\theta_k$. Every $t_i^j/\theta_k$ is associated with a kick operator which modifies the momentum and determines the next time $t_i^{j+1}$ where $V_i$ is evaluated. Each kick operator is followed by a drift. If two or more potentials have equal evaluation times, for example at the beginning where all potentials are synchronous, then the drift operators between the evaluation times are of infinitesimal length. This is equivalent to having kick operators being successively applied before the next drift of finite time length takes place. The order in which synchronous kicks take place is predefined and subject to properties of the potential, for example largest wave speed or largest force. That means, even if the numerical value of two $t_i^j$ is equal, an ordering $t_{i^1}^{j^1} < t_{i^2}^{j^2}$ or $t_{i^1}^{j^1} > t_{i^2}^{j^2}$ is assumed if $i^1 \neq i^2$. As a consequence, if $n$ potentials are synchronous at a time $\theta_k$ then one assumes the existence of $n$ individual times $\theta_{k+l} = \theta_k + 2l|\tau|$ with $\lim \tau \to 0$ and $l = 1, \ldots, n$. Before $\theta_k$ and after $\theta_{k+n}$ a drift of finite time takes place. Between the individual $\theta_{k+l}$ kicks and drifts of infinitesimal length are applied according to the respective potential $V_i$.

The action of each element is defined by

$$S_k^d = J_k^T Q_k^+ - J_{k+1}^T Q_{k+1}^- - \left(\theta_{k+1}^- - \theta_k^+\right) H_k^d + p_k^T \left(q_{k+1}^- - q_k^+\right) \quad (6.19)$$

$$H_k^d = p_k v_k - v_k^T M v_k + \sum_i V_i \left(q_{\mathcal{K}(i,\mathcal{A}(k,i))}^+\right) \quad (6.20)$$

$$Q = \begin{pmatrix} q \\ \theta \end{pmatrix}, \quad J = \begin{pmatrix} j \\ w \end{pmatrix} \quad (6.21)$$

The phase space is extended by the times $\theta$ and their conjugate momenta $w$. The discrete Hamiltonian $H_k^d$ is the sum of the current kinetic energy and all potentials $V_i$ being evaluated at the last position $q_{\mathcal{K}(i,\mathcal{A}(k,i))}^+$. The action sum is the sum of all elemental actions enhanced by conditions to predict the evaluation time steps and the length of the kick elements

$$S_d = \sum_{k=0}^{N-1} S_k^d + \sum_{i=1}^{m} \sum_{j=0}^{N_i-1} \mu_i^j \left(\theta_{\mathcal{K}(i,j+1)}^- - \left(\theta_{\mathcal{K}(i,j)}^+ + s_i^j\right)\right)$$

$$+ \lim_{\tau \to 0, \tau > 0} \sum_{i=1}^{m} \sum_{j=0}^{N_i-1} \kappa_i^j \left(\theta_{\mathcal{K}(i,j)+1}^- - \left(\theta_{\mathcal{K}(i,j)}^+ + \tau\right)\right) \quad (6.22)$$

The variables $t_i^j$ are considered as substitutes for the corresponding $\theta_k$. This notion is used for simplicity, because adding extra constraint equations to couple both variables would cause even more complexity in the derivation. Also notice, there is only a single potential $V_i$ active at each time $\theta_k$. Further, the bounding

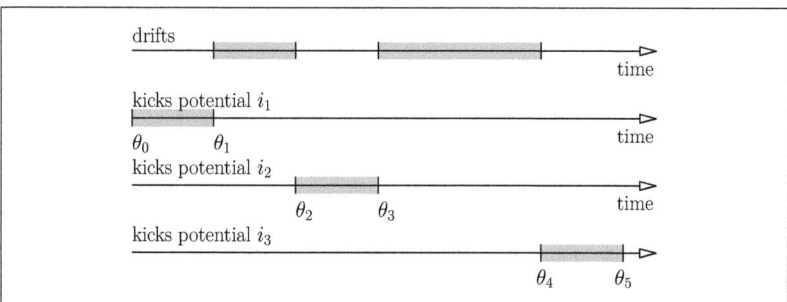

Figure 6.2: AVI: kick and drift elements. If $V_{i_1}$ and $V_{i_2}$ are synchronous at $\theta_0$ then $\theta_2 = \theta_1 + \lim_{\tau \to 0} \tau$.

## 6.5. Asynchronous variable time steps

times of a kick element are known. Initially, all bounding times are set to zero. During the simulation, the kicks compute the right boundary of later drift elements. From that the left boundary time of the adjacent kick element is computed while its right boundary time is always assumed to be infinitesimal larger than its left boundary time.

**Drift** Variation leads to the following equations. If $\theta_k^+$ denotes the begin of a drift phase then no potential is active and the discrete Euler-Lagrange equations are

$$\begin{aligned}
\delta q_k^+ : & & p_k &= j_k \\
\delta v_k : & & v_k &= M^{-1} p_k \\
\delta p_k : & & q_{k+1}^- &= q_k^+ + \left(\theta_{k+1}^- - \theta_k^+\right) v_k \\
\delta q_{k+1}^- : & & j_{k+1} &= p_k
\end{aligned} \quad (6.23)$$

These equations describe a constant velocity as in the fixed step size synchronous case. They are obtained from the derivates for $q_k^+$, $v_k$, $p_k$ and $q_{k+1}^-$ (in the respective order). The times $\theta$ are determined by previous kick operators and are not subject to any drift operator.

**Kick** If $\theta_k^+$ is coincident with the evaluation of the $i$-th potential, then variation leads to the following system of equations

$$\delta q_k^+ : \qquad p_k = j_k - \mu_i^j \frac{\partial s_i^j}{\partial q_k^+} - s_i^j \frac{\partial V_i\left(q_k^+, \theta_k^+\right)}{\partial q_k^+} \qquad (6.24)$$

$$\delta v_k : \qquad v_k = M^{-1} p_k \qquad (6.25)$$

$$\delta p_k : \qquad q_{k+1}^- = q_k^+ + \tau v_k + \mu_i^j \frac{\partial s_i^j}{\partial p_k} \qquad (6.26)$$

$$\delta q_{k+1}^- : \qquad j_{k+1} = p_k \qquad (6.27)$$

$$\delta \kappa_i^j : \qquad \theta_{k+1}^- = \theta_k^+ + \tau \qquad (6.28)$$

$$\delta \theta_k^+ : \qquad \mu_i^j = w_k + H_k^d - s_i^j \frac{\partial V_i\left(q_k^+, \theta_k^+\right)}{\partial \theta_k^+} - \kappa_i^j \qquad (6.29)$$

$$\delta \theta_{k+1}^- : \qquad w_{k+1} = -H_k^d + \kappa_i^j \qquad (6.30)$$

$$\delta \mu_i^j : \qquad \theta_{\mathcal{K}(i,j+1)} = \theta_k + s_i^j \qquad (6.31)$$

$$\delta \theta_{\mathcal{K}(i,j+1)}^- : \qquad w_{\mathcal{K}(i,j+1)} = \mu_i^j - H_{\mathcal{K}(i,j+1)-1}^d \qquad (6.32)$$

$$\delta \theta_{\mathcal{K}(i,j)+1}^+ : \qquad w_{k+1} = -H_{k+1}^d \qquad (6.33)$$

$\lim \tau \to 0$ eliminates $v_k$ in (6.26). There exists a coupling between $\mu_i^j$ and $\kappa_i^j$ which can only be solved from the discrete Hamiltonian of the $(k+1)$-th time element. This is the reason for the existence of drift elements after each kick. Even when two potentials are synchronous, then there is a (fictitious) drift element (of sometimes infinitesimal length) between both kicks. They help to solve $\kappa_i^j$. The right boundary time of the drift element was determined from the latest kick of the subsequent potential. The left boundary time of the drift element is an individual degree of freedom. Its variation leads to condition (6.33) for $w_{k+1}$. A drift operator (6.23) of infinitesimal length is the identity map. Therefore, $H^d_{k+1} = H^d_k$ and $\kappa^j_i = 0$ from equation (6.30). Equation (6.24) implicitely defines $p_k$ which must be solved iteratively. To do this, $\mu^j_i$ from equation (6.29) and $H^d_k$ from equation (6.20) are inserted such that $p_k$ remains as the only unknown. After solving $p_k$, all other variables are determined explicitly.

The resulting time stepping scheme is summarized in algorithm 6.3.

The most problematic issues in the construction of variable step size AVIs are: (1) the definition of a discrete energy which is solved by assuming a discrete Hamiltonian which is active during the interval of each time element, but which may depend on coordinates $q$ which lie outside of the element. (2) a step size function methodology which allows asynchronous time stepping starting from a synchronous setting. If a single step size function would be used, then an implicit relation between subsequent time steps would be the result (and in MDOF systems even spatial coupling can occur). The latter issue was solved by defining individual step size functions for each potential. Therefore, the step size is subsequently determined for each potential. Time elements of infinitesimal length (kick+drift) support the start from a synchronous state.

## 6.6. Example time step functions

---

Set initial conditions $q_0$, $j_0$
Set start times for each potential $V_i$, $t_i^0 = 0$, and initial energy errors $\mu_i^0 = 0$
Set start multiplier to the negative initial energy $w_0 = -H_0$.
Create a priority queue of indices $i$ which is sorted for $t_i^j$ in ascending order. If multiple times are equal, additional conditions must ensure uniqueness of the ordering.
Set global time $\theta_k := 0$
Create global counter $k := 0$
Create counters for each potential $j^i := 0$
**while** $\theta_k < T$ **do**
  Take 1st element from priority queue and remember index $i$ and time $t_i^{j^i}$ and last error $\mu_i^{j^i}$.
  $\theta_{k+1} := t_i$
  **if** $\theta_{k+1} \neq \theta_k$ **then**
    *Perform drift:*
    $h_k := \theta_{k+1} - \theta_k$
    $q_{k+1} := q_k + h_k M^{-1} j_k$
    $j_{k+1} := j_k$
    *Update state:*
    $k := k + 1$
  **end if**
  *Perform kick:*
  Compute discrete Hamiltonian $H_{k-1}^d$ of the last time element (if $k = 0$, take $H_0$).
  $w_k := -H_{k-1}^d + \mu_i^{j^i}$
  Solve equation (6.24) for $p_k$
  $v_k := M^{-1} p_k$
  $q_{k+1}$ from eq. (6.26)
  $j_{k+1} := p_k$
  $\theta_{k+1} := \theta_k$
  $\mu_k$ from eq. (6.29)
  *Update state:*
  Set next evaluation time $t_i^{j^i+1}$ according to eq. (6.31), $t_i^{j^i+1} := t_i^{j^i} + s_i^{j^i}$
  $k := k + 1$
  $j^i := j^i + 1$
  Reinsert $i$ (i.e. $j^i$, $t_i^{j^i}$, $\mu_i^{j^i} = \mu_k$) into priority queue.
**end while**

Figure 6.3: AVI with variable time steps

## 6.6 Example time step functions

In the context of symplectic variable step size integration, a number of scaling functions has been proposed. Some of them are listed below

- **Arc-length parametrization** The basic idea is to establish equidistant steps in phase space, i.e. $\Delta \|(q,p)\| \approx \text{const}$. This leads to a step size function

$$s(q,p) = \epsilon \left( p^T M^{-1} p + \nabla_q V(q)^T M^{-1} \nabla_q V(q) \right)^{-1/2} \quad (6.34)$$

where the inverse mass matrix is introduced in the norm to enforce equal units. This step size function was used in [81, 100]. Hairer suggested to solve the nonlinear equation for $p$ by introducing the variable $\beta = p^T M^{-1} p$. Then, the iterative solver is applied to the solution of the scalar $\beta$ instead of vector $p$.

A definition with lower and upper bounds of the time step is suggested in [100].

- **$q$-dependent arc-length** Since the Hamiltonian is approximately constant if the time stepping scheme does not diverge, the step size function in (6.34) can be replaced by a function which is independent from momentum $p$. Therein, $p$ was eliminated using the kinetic energy, i.e.

$$s(q) = \epsilon \left( 2\left( H_0 - V(q) \right) + \nabla_q V(q)^T M^{-1} \nabla_q V(q) \right)^{-1/2} \quad (6.35)$$

Numerical experiments [81] give nearly identical results for the Verlet scheme, but only as long as the discrete energy is similar to the constant $H_0$. For very large time steps the true arc-length method seems to be more robust, while the simplified version becomes unstable.

- **Logarithmic scaling** The implicit nature of symplectic variable step sizes arises from the coupling of the gradient of the step size function multiplied with the discrete Hamiltonian. If the step size gradient would have a structure where the Hamiltonian is in the denominator and the target variables are separable in the enumerator then the scheme would be completely explicit.

Early approaches in this regard were developed by Mikkola [189]. The modified Hamiltonian $\Gamma_k^d = s_k \left( H_k^d + w_k \right)$ is split into $\Gamma = \Gamma_0 + \Gamma_1$. The choice of $s_k$ then depends on the structure of $H_k^d$ and must lead to a splitting which is separable into momenta $p$ and coordinates $q$. The presented method is, however, tuned to the Keplerian equation.

In N-body problems of astrophysics and molecular dynamics the gravitational potential grows to infinity as relative distances approach zero. This behaviour is equivalent to collisions which are employed by a penalizing energy potential. Then, a scaling function in the form $s = \epsilon \left( V(q) \right)^{-\beta}$ may be used which reduces the time step near singularities where $\beta$ is some parameter [191].

The inverse potential energy approach was combined with the splitting approach and applied to more general systems of the form $H(q,p) = V(q) +$

## 6.6. Example time step functions

$T(p)$ in [190]. It is observed that the modified Hamiltonian $\Gamma = s(H - H_0)$ is zero along the true solution, but generally not separable. The idea is to integrate an alternative system with identical exact solution, that is with modified Hamiltonian $\tilde{\Gamma} = \log(1 + \Gamma)$. Given a scaling $d\alpha = V(q)dt$ with fictive time $\alpha$, one obtains for $s = dt/d\alpha$

$$s(q,p) = \frac{1}{T(p) - H_0}$$
$$\tilde{\Gamma} = \log(T(p) - H_0) - \log(V(q)) \qquad (6.36)$$

This transformation was successfully applied to a Keplerian multi-body system.

- **Generalized logarithmic scaling** In [20, 214] a generalization of the logarithmic transformation was employed. The used time step function is

$$s(q,p) = \frac{f(T(p) - H_0) - f(-V(q))}{H(q,p) - H_0} \qquad (6.37)$$

with $f(u)$ being some function with $f'(u) > 0$. The choice $f(u) = \log(u)$ gives Mikkola's method. Since the scaling function may be ill-conditioned (denominator close to zero), the time step length can be approximated via $s \approx f'(-V(q))$. The Hamiltonian to solve is

$$\tilde{\Gamma} = f(T(p) - H_0) - f(-V(q)) \qquad (6.38)$$

which is separable such that explicit integrators can be applied. The following basis functions are proposed for the Kepler problem [214]

$$f(x) = \begin{cases} \frac{\epsilon}{1-\gamma} x^{1-\gamma} & \text{if } \gamma \neq 1 \\ \epsilon \log x & \text{if } \gamma = 1 \end{cases} \qquad (6.39)$$

- **SEM integration** The asynchronous SEM Euler scheme is obtained if one interprets $s_i^j$ in equations (6.24)-(6.33) in terms of a variable instead of a function. Then, the action sum is derived for $s_i^j$ to obtain an additional conditional equation, that is

$$\mu_i^j = 0 \qquad (6.40)$$

The resulting scheme preserves the discrete energy, but inherits all solution problems of its synchronous counterpart.

Notice that there exist some nonsymplectic step size strategies which also target at long-time stability. Such can be applied to time-reversible systems and symmetric base methods [100, 110, 253].

## 6.7 Time step selection and solution

The presented scaling functions were mostly applied to atomic or gravitational systems. In structural dynamics, however, the requirements of well suited scaling functions may be different:

- $p$-**dependent scaling functions** must be avoided in the asynchronous setting. Consider equation (6.26) which defines the modification of coordinates $q$ in the asynchronous variable step size scheme. The scheme was designed as a sequence of drift and kick operators, the first performing a constant motion along a predefined time interval, the latter changing the momentum only. Obviously, the kick operator must not change the coordinates by definition. Therefore, $\partial s_i^j / \partial p_k = 0$.

  A $p$-dependent scaling function indeed leads to wrong results (as shown, for example, when applied to linear SDOF oscillators). To illustrate this, consider the existence of asynchronous potentials $V_i$, where one of them is always zero, i.e. $V_1(q) = 0$. Whenever $V_1$ is active one assumes that there will be no change to the phase space. Application of the arc-length parametrization in equation (6.34), however, will change the coordinates (but not the momentum) at all $t_1^j$. Therefore, $p$-dependent scaling functions may be applied to synchronous integrators, but not to AVIs.

- **Stability.** Explicit schemes are often applied with a time step close to the stability limit. In particular, since AVIs exhibit stability criteria which are difficult to derive, unconditional stability would be a desired side effect of the chosen time transformation.

- **Numerical efficiency** of the step size selection is of importance. Ideally, an explicit determination of the step size is preferred, as in equation (6.37). There is, however, no method known, which combines the logarithmic scaling with a time step length $s \propto 1/\nabla_q V(q)$. At least, any scaling function can be defined such that it is implicit in $p$ only. Then, the potential energy and its derivatives must be evaluated only once at the begin of each time step. This iterative nature is very different from standard implicit integrators, for example midpoint or Newmark, where the potential energy is evaluated in each iteration step.

The following step size functions seem suitable:

- simplified arclength with lower and upper bounds, equations (6.35)

$$s(q) = h_{\min} + \epsilon \left( \frac{2\left(H_0 - V(q)\right) + \nabla V(q)^T M^{-1} \nabla V(q)}{2H_0} + \left( \frac{\epsilon}{h_{\max} - h_{\min}} \right)^2 \right)^{-\frac{1}{2}}$$

(6.41)

## 6.7. Time step selection and solution

- reciprocal force with lower and upper bounds

$$s(q) = h_{\min} + \epsilon \left( \frac{\nabla V(q)^T M^{-1} \nabla V(q)}{H_0} + \left( \frac{\epsilon}{h_{\max} - h_{\min}} \right)^2 \right)^{-\frac{1}{2}} \quad (6.42)$$

- reciprocal potential energy with lower and upper bounds

$$s(q) = h_{\min} + \epsilon \left( \frac{V(q)}{H_0} + \left( \frac{\epsilon}{h_{\max} - h_{\min}} \right)^2 \right)^{-\frac{1}{2}} \quad (6.43)$$

- logarithmic scaling with lower and upper bounds, equation (6.37)

$$s(q) = h_{\min} + \epsilon \left( \frac{V(q)}{H_0} + \frac{\epsilon}{h_{\max} - h_{\min}} \right)^{-1} \quad (6.44)$$

The smaller the parameter $\epsilon$, the smaller is the time step in case of a large potential energy. $h_{\min}$ and $h_{\max}$ define lower and upper bounds to the time step. The time step bounds ensure that the denominator is nonzero. Notice, external forces and forces arising from constraint equations are included in $V(q)$ and $\nabla V_i$.

Equation (6.24) leads to the quadratic equation

$$0 = \frac{1}{2} p_k^T M^{-1} p_k \nabla s_i^j + p_k + p_k^0 \quad (6.45)$$

$$p_k^0 = -j_k + \left( w_k + V_i \left( q_k^+, \theta_k^+ \right) + \nabla_t V_i \left( q_k^+, \theta_k^+ \right) \right) \nabla s_i^j + s_i^j \nabla_q V_i \left( q_k^+, \theta_k^+ \right) \quad (6.46)$$

By introduction of the scalar variable $\beta = p_k^T M^{-1} p_k$ this equation can be solved efficiently. To do so, the variable $p_k$ is moved to the left hand side in equation (6.45). Then the $\| \circ \|_{M^{-1}}$-norm is taken and squared, i.e.

$$\beta = \left( \frac{1}{2} \beta \nabla s_i^j + p_k^0 \right)^T M^{-1} \left( \frac{1}{2} \beta \nabla s_i^j + p_k^0 \right)$$

$$= \frac{1}{4} \left( \nabla s_i^j \right)^T M^{-1} \nabla s_i^j \beta^2 + \left( \nabla s_i^j \right)^T M^{-1} p_k^0 \beta + \left( p_k^0 \right)^T M^{-1} p_k^0 \quad (6.47)$$

There generally exist two solutions. The solution which makes the absolute value of the energy error

$$|\mu_k| = \left| w_k + \frac{1}{2} \beta + V_i \left( q_k^+, \theta_k^+ \right) + \nabla_t V_i \left( q_k^+, \theta_k^+ \right) \right| \quad (6.48)$$

minimal is chosen if both are feasible. Negative $\beta$ are not allowed. In the implementation, the special case $\left( \nabla s_i^j \right)^T M^{-1} \nabla s_i^j = 0$ must be considered which makes equation (6.47) linear. If there exists no solution, then the nonsymplectic update (where one assumes $\nabla s_i^j = 0$) of the momentum is chosen.

When compared with the SEM Euler method, similar problems arise in the solution. In both cases, a scalar quadratic equation must be solved which is not possible under certain conditions. There are a few advantages of the regularization over the SEM approach, though:

1. If no solution is possible, there exists at least a reasonable guess for the time step length. The quadratic equation is used in SEM to obtain the time step length, but here it determines the perturbation to the Euler scheme which makes it symplectic in the presence of variable time steps.

2. Examples show that the number of nonsymplectic updates grows with the initial time step length. In SEM integration, the solvability depends on the phase space variables.

## 6.8 Examples

### 6.8.1 Synchronous integration of a single degree of freedom system with variable step sizes

**Time step regularization applied to a nonlinearly constrained oscillator**

This example applies the synchronous Euler method to a linear oscillator with Lagrange function

$$L(q,\dot{q}) = \frac{1}{2}\dot{q}^2 - \frac{1}{2}q^2 \tag{6.49}$$

with $q_0 = 0$ and $\dot{q}_0 = 1$. The critical time step is $h_{crit} = 2$. The system is subject to a unilateral constraint

$$g(q) = -q - 0.5 \leq 0 \tag{6.50}$$

which is associated with a quadratic penalty function

$$G(q) = \begin{cases} 0 & \text{if } q > -0.5 \\ \frac{1}{2}\rho g(q)^2 & \text{else} \end{cases} \tag{6.51}$$

being added to $L(q,\dot{q})$ and $\rho = 10$.

**Small time step and soft penalty**  Let test the accuracy given a small initial time step length $h_0 = 0.02$ and a simulation time $T = 100$. The adaptive methods are defined by $h_{max} = h_0$, $h_{min} = 10^{-5}h_{max}$, $\epsilon = 0.01$. Table 6.1 illustrates the maximum error in energy and the number of time steps being used. It also presents the actually used minimum and average time step lengths.

All variable time step methods improve the accuracy compared with constant time steps. Using the constant time step method with smaller step size ($h_0 = 0.01$, number of steps = 10000) still leads to a significantly larger error in energy of 0.0257. In this example, the inverse force norm (6.42) needs attention since it leads to a very small error while increasing the number of used time steps to a moderate size.

## 6.8. Examples

|  | constant | eq. (6.41) | eq. (6.42) | eq. (6.43) | eq. (6.44) |
|---|---|---|---|---|---|
| $\max_t |H(t) - H_0|/H_0$ | 0.0517 | 0.0070 | 0.0082 | 0.0147 | 0.0126 |
| number of steps | 5000 | 13062 | 10017 | 8031 | 9428 |
| $\min_t h(t)$ | 0.2 | 0.0033 | 0.0033 | 0.0089 | 0.0066 |
| mean $h(t)$ | 0.2 | 0.0089 | 0.0099 | 0.0124 | 0.0106 |

Table 6.1: Accuracy of variable synchronous time steps applied to a nonlinearly constrained oscillator $h_0 = 0.02$, $\rho = 10$

**Large time step and stiff penalty** Let us modify the example. A time step closer to the critical step size is used, $h_0 = 1$. The violation of the constraint can be reduced by increasing the penalty paremeter, $\rho = 100$. The constant-step size method turns out to be unstable. The error and efficiency results of the variable schemes are shown in table 6.2

In this case, the adaptation of the time step must ensure stability, not only accuracy. The only method which guarantees both is the simplified arc-length method. But this is done on cost of efficiency. Although its error is greater than in table 6.1, the number of time steps is increased. Interestingly, the maximum used time step length is not given by $h_{\max}$, but is much smaller. In practice, it may be difficult to tune the efficiency of the method. Sometimes one is looking for a stable method which requires a small number of time steps. What is common for all methods is that once the error is very large, the used time step is further decreased (one could, for example, use a time step $h_0 > h_{crit}$ and a larger initial velocity to verify this). This happens for the logarithmic scaling (6.44). Once numerical instability occurs, a further growth of the error is prevented by reducing the time step to very small values leading to an inefficient scheme. Still, the total simulation is very inaccurate due to the very large errors in the first time steps. A compromise of efficiency and accuracy is given by the scaling function (6.43).

|  | eq. (6.41) | eq. (6.42) | eq. (6.43) | eq. (6.44) |
|---|---|---|---|---|
| $\max_t |H(t) - H_0|/H_0$ | 0.1668 | 26.407 | 13.445 | 39.393 |
| number of steps | 13403 | 13055 | 5261 | 70463 |
| $\min_t h(t)$ | 0.0010 | 0.0002 | 0.0026 | 0.0003 |
| $\max_t h(t)$ | 0.0100 | 1 | 1 | 1 |
| mean $h(t)$ | 0.0074 | 0.0076 | 0.0190 | 0.0014 |

Table 6.2: Accuracy of variable synchronous time steps applied to a nonlinearly constrained oscillator $h_0 = 1$, $\rho = 100$

Using this example, the ability of variable time steps to improve accuracy, stability and efficiency when applied to highly nonlinear potential functions is shown. The compared step size functions exhibit a different behaviour with respect to these three properties.

**SEM Euler applied to a linear oscillator**

This example applies the synchronous SEM Euler method to a linear oscillator with Lagrange function

$$L(q, \dot{q}) = \frac{1}{2}\dot{q}^2 - \frac{1}{2}q^2 \tag{6.52}$$

with $q_0 = 0$ and $\dot{q}_0 = 1$. The critical time step is $h_{crit} = 2$, the initial time step is chosen to be $h_0 = 0.4$.

First, the standard SEM Euler method is tested, see equations (6.4) in section 6.2. Whenever the equations are not solvable or if an infeasible time step is obtained, then the initial time step is chosen. Surprisingly, this strategy is better than chosing, for example, some very small "minimal" time step length or $h_{crit}$ which may lead to an unstable scheme. The trajectory $q(t)$ and the computed time step sizes $h(t)$ are presented in figure 6.4.

The times where ill-conditioned equations appear are almost all times with $q(t) \to \max/\min$. They lead to an unstable time stepping scheme. The energy error passes $100\%$ at $t = 65$. Once, the time step selection computes a time step being larger than the critical time step (being projected to $h_{crit}$). At the end of the simulation the time step is almost constant because an unsolvable equation appears almost at each time step.

Let us improve the SEM Euler scheme. To this end, the multiplier $w_k$ is modified by a parameter $\epsilon_k$

$$w_k^{mod} = w_k + \epsilon_k \tag{6.53}$$

$\epsilon_k$ is chosen in such a way that the square root in equation (6.5) becomes zero. The modification is only applied if the square root would return a complex number.

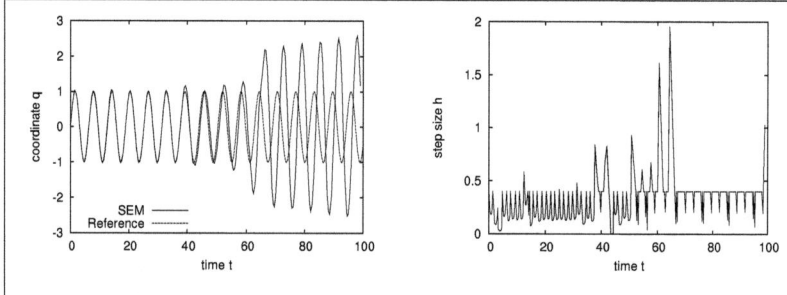

Figure 6.4: Synchronous SEM Euler applied to a linear oscillator. Left: $q(t)$, Right: $h(t)$.

## 6.8. Examples

The modification leads to an energy error in such time elements. It solves the issue of creating unsolvable equations by the time stepping scheme. It does not, however, solve the problem of negative time steps. The trajectory $q(t)$ and the computed time step sizes $h(t)$ are presented in figure 6.5. Obviously the time step selection and the error in $q$ and energy are greatly improved. Still, the energy error is monotonously too large at the end of the simulation. The pattern of time step selection also changes during the simulation.

This example illustrates effects of unsolvable equations in the SEM Euler method. Two strategies to deal with unsolvable equations are compared. Both exhibit long-term instabilities due to numerical energy growth. The modified SEM Euler improves situation considerably compared with the standard SEM scheme. Authors [158], however, report no problems regarding instabilities due to unsolvable equations. It may be assumed that the chance of the appearance of such critical points is reduced when applied to complex structures.

### 6.8.2 Asynchronous variable time steps applied to a linear oscillator

Let us now apply an asynchronous scheme to the example given in section 6.8.1. Two asynchronous potential energies are assumed and given by

$$V_1(q) = 0.4V(q), \quad V_2(q) = 0.6V(q) \tag{6.54}$$

An initial step size within the stable domain is chosen. The equivalent initial time steps for the individual potentials are obtained from an assumed initial synchronous time step as $h_{i,\max} = h_0 h_{i,crit}/h_{crit}$. For example, given an initial synchronous time step of $h_0 = 0.2$, one obtains equivalent step sizes for the two potentials $h_{0,1} = 0.316$ and $h_{0,2} = 0.258$. These numbers scale linearly for a different $h_0$. Furthermore, they define the maximum allowed time step $h_{\max}$ being used by the time step functions in equations (6.41), (6.42), (6.43). The lower bound for the time step size is defined by $h_{\min} = 10^{-5} h_{\max}$. The simulation time is $T = 100$.

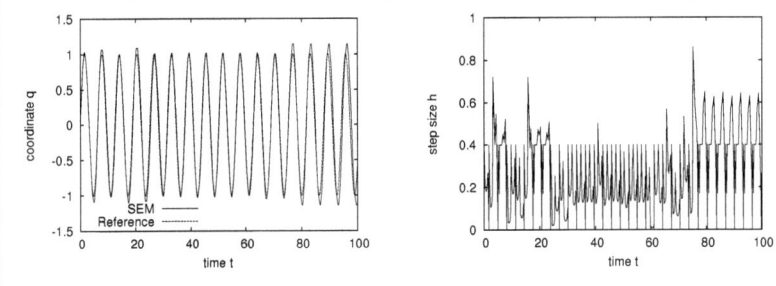

Figure 6.5: Synchronous modified SEM Euler applied to a linear oscillator. Left: $q(t)$, Right: $h(t)$.

An application of a scaling factor $\epsilon = 1$ in the step size functions in equations (6.41), (6.42), (6.43) leads to a scheme where the time step size is nearly constant. This is because the error in energy is quite small and the actual forces are relatively small compared with the total energy. In order to increase the influence of the state variables onto the time step selection, the parameter is changed to $\epsilon = 0.1$.

The different step size functions are implemented using a symplectic update, except the arc-length parameterization which is not applicable to the symplectic scheme. The number of kicks and the maximum error in $q(t)$ is given in table 6.3. Different starting step sizes are used, i.e. $h_0 = 0.2$, $h_0 = 0.4$, $h_0 = 1.2$. For larger time steps, one of the individual potentials uses a time step being unneggligably larger than the critical time step of the synchronous system and, therefore, the tested schemes become unstable.

The nonsymplectic arc-length method is listed to show that the effects of missing symplecticity may lead to large errors in asynchronous integration. In this case, it exhibits a very large error due to numerical damping and numerical energy growth.

There is no variable step size method available which outperforms the fixed-step size scheme. Almost all adaptive schemes lead to a significantly larger error for a similar number of time steps. Only the simplified arc-length method is comparable with constant step size. This happens because it renders nearly constant step sizes while all other schemes lead to huge differences in step size length. Furthermore, the simplified arc-length method is the only method which never reaches the upper limit of feasible time step sizes. This effect makes it very difficult to estimate the numerical efficiency prior the simulation. In fact, the measured number of time steps is nearly insensitive with respect to $h_0$. All the other methods start with $h_{\max}$ and reduce the time step at regular intervals. Therefore, once an error occured due to a large time step it cannot be reduced in the forthgoing simulation, but the efficiency will be reduced by the adaptation. Nevertheless, the adaptation itself seems to reduce accuracy since the fixed step size method starts with the same step sizes and leads to a smaller error. Furthermore, sometimes the quadratic equations are not solvable and require a few nonsymplectic steps as listed in table 6.3. Such incidences lead to an additional error.

The matter of accuracy can be approached by adjusting the parameter $\epsilon$ in the step size function. To illustrate this effect, the parameter is set to $\epsilon = 1$ for all time step functions. Repeating the tests improves accuracy/efficiency of the symplectic schemes compared with fixed step sizes. The variable step size schemes, however, still are less performant. One further observation is that the $q$-dependent arc length method returns step sizes which are equal for both potentials, i.e. leading to a synchronous scheme in this test case.

The main purpose of this example is to illustrate variable time steps applied to a standard case of structural dynamics: Many problems are governed by vibra-

## 6.8. Examples

|  | constant | eq. (6.41) | eq. (6.42) | eq. (6.43) | arc-length |
|---|---|---|---|---|---|
| $h_0 = 0.2$ | | | | | |
| $\max_t |H(t) - H_0|/H_0$ | 16.6% | 5.25% | 10.31% | 10.5% | 96.8% |
| max. error in $q(t)$ | 1.04% | 0.08% | 1.61% | 1.80% | 75% |
| number of steps | 2116 | 6010 | 3589 | 3577 | 919 |
| algorithmic errors | - | - | - | - | yes |
| $h_0 = 0.4$ | | | | | |
| $\max_t |H(t) - H_0|/H_0$ | 43.5% | 5.57% | 103.6% | 94.8% | 88.7% |
| max. error in $q(t)$ | 6.35% | 0.12% | 42.2% | 19.4% | 40% |
| number of steps | 1060 | 5722 | 3082 | 2125 | 646 |
| algorithmic errors | - | yes | - | - | yes |
| $h_0 = 1.2$ | | | | | |
| $\max_t |H(t) - H_0|/H_0$ | 526.4% | 5.91% | 344.7% | 344.1% | 513.1% |
| max. error in $q(t)$ | 90.8% | 0.10% | 110.7% | 110.6% | 82.6% |
| number of steps | 355 | 5632 | 3820 | 3898 | 592 |
| algorithmic errors | - | - | - | - | yes |

Table 6.3: Error and efficiency for different step size strategies in AVI applied to a linear oscillator. Top: $h_0 = 0.2$, center: $h_0 = 0.4$, bottom: $h_0 = 1.2$. Tested are fixed step size, $q$-dependent arc-length, inverse force, inverse potential energy and nonsymplectic arc-length. The table also shows for which settings algorithmic errors appeared which lead to nonsymplectic updates during the simulation.

tions and wave propagation without large nonlinearities. Although examples have shown that variable time steps may improve accuracy, stability and efficiency in case of very large nonlinearities, this is not true for linear oscillators. The example presented in this section, however, proves that asynchronous methods with symplectic variable time step selection may converge to the correct solution.

### 6.8.3 Limiting cases of variable time steps

**Symplectic versus nonsymplectic updates**

The example from section 6.8.1 is used and the arc length parametrization is applied as a time step function. The synchronous case is studied. The trajectory and the time step sizes are illustrated in figure 6.6. Two methods are tested: variable step sizes with and without symplectic update. The nonsymplectic version is simply the constant step size Euler where the step size $h_k$ is adapted at each time step. It can be seen that the nonsymplectic version continuously creates energy and, therefore, is unstable in long-term simulation. No unsolvable equations were generated by the symplectic scheme when applied to this example.

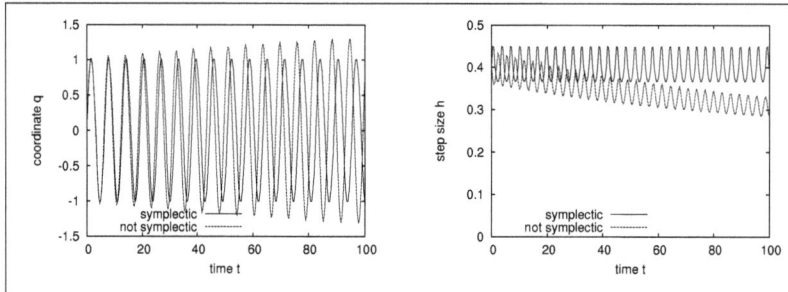

Figure 6.6: Synchronous arc length parameterization applied to a linear oscillator. Left: $q(t)$, Right: $h(t)$.

**Momentum dependent step size function in asynchronous integration**

Now apply an asynchronous scheme with arc length parametrization. Two potential energies are given by

$$V_1(q) = 0.4V(q), \quad V_2(q) = 0.6V(q) \tag{6.55}$$

Both potentials use the same initial time step $h_{1,2}^0 = 0.4$. The solution of the symplectic method is shown in figure 6.7. Clearly, the method does not converge. The reason is the dependency of the time step function on the momentum. The symplectic correction terms modify, therefore, the coordinates and the momentum. This happens at the end of the first period, first at $t = 3.8$ and then at $t = 5$ moving the trajectory to a wrong solution. Consider, for example, the case $V_1(q) = 0 \cdot V(q)$ and $V_2(q) = 1 \cdot V(q)$. One would assume, that $V_1$ does not contribute to the solution. The symplectic correction term of the arc length parameterization, however, will change $q_k$ even more than in this example.

Although the accuracy is bad due to asynchronicity, the nonsymplectic version of the same problem renders at least a nearly correct frequency, see figure 6.8.

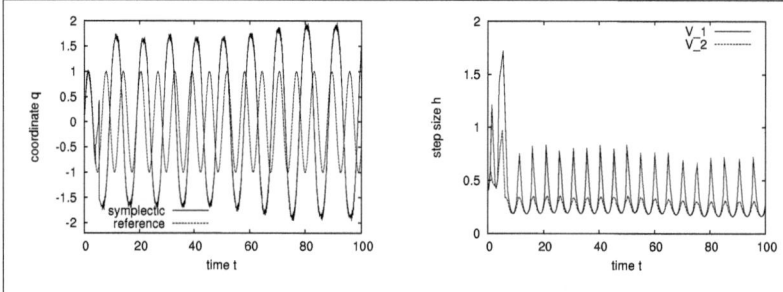

Figure 6.7: Asynchronous symplectic arc length parameterization applied to a linear oscillator. Left: $q(t)$, Right: $h(t)$.

## 6.8. Examples

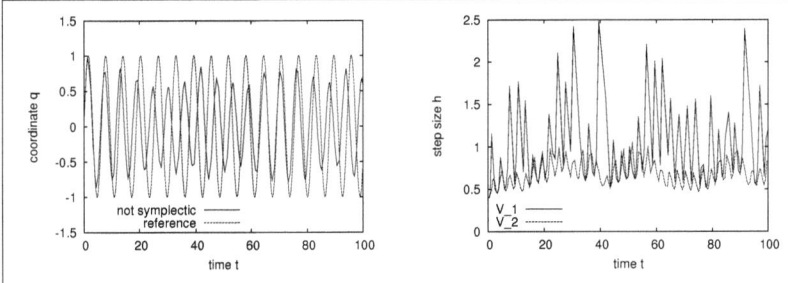

Figure 6.8: Asynchronous non-symplectic arc length parameterization applied to a linear oscillator. Left: $q(t)$, Right: $h(t)$.

**Stability for very large time steps**

Let us test the behaviour of different time step functions with respect to very large time steps. The nonsymplectic arc length parametrization is applied to the asynchronous setting with initial time steps $h_0 = 20$. The resulting trajectory is stable as shown in figure 6.9. The first time step leads to a very large error which is kept constant in the subsequent time steps. The computed time steps are within the stable interval $[0, 2]$.

The same setting is applied to the time step function in equation (6.42). The equivalent initial time steps for the individual potentials are obtained from an assumed initial synchronous time step as $h_{i,\max} = (h_0 \cdot h_{i,crit})/h_{crit}$ being $h_1^0 = 23.9$ and $h_2^0 = 36.5$. The resulting trajectory is stable as shown in figure 6.10. In opposite to the arc length method, the subsequent time steps are within the range of the initial step size. As a result, the error in the trajectory is much larger. The patterns of the time step size and of the displacement during the simulation are irregular. This is not what one would expect from a symplectic scheme. In fact, a non-symplectic update was chosen because equation (6.45) was often not solveable during the simulation for such large time steps.

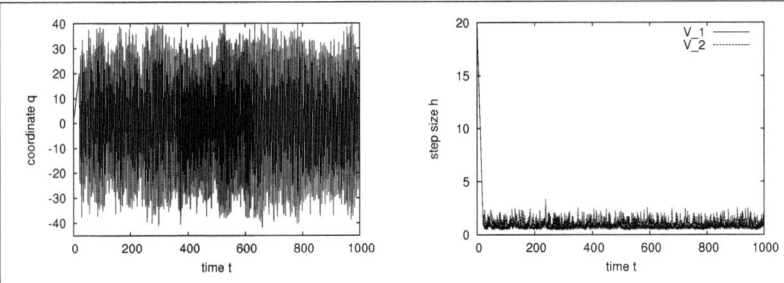

Figure 6.9: Stability of arc length parameterization applied to a linear oscillator. Left: $q(t)$, Right: $h(t)$.

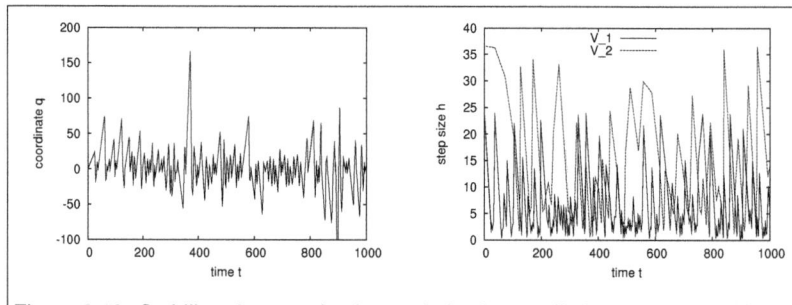

Figure 6.10: Stability of a symplectic regularization applied to a linear oscillator using step size function (6.42). Left: $q(t)$, Right: $h(t)$.

### 6.8.4 Variable step size integration of a cantilever beam

The example in section 5.6 is used to illustrate the performance of variable time steps in asynchronous integration of multiple degree of freedom systems. The mesh size parameter is chosen to be $n = 3$, see figure 5.10. Compared are: constant time step, SEM Euler (same procedure as in section 6.2) and one scaling function given by equation (6.43) (inverse energy).

Alternative time step functions turned out to be unsuitable: First, when trying equation (6.42) (inverse force), a test simulation with end time $T = 0.6 \times 10^{-3}$ was performed and the scaling parameter $\epsilon$ was adjusted such that approximately the same number of total time steps was required for AVIs with time step function (6.43). Extending the total simulation time to $T = 0.02$ leads to phase states with large strain energy and the generated time step sizes become extraordinary small. Furthermore, a symplectic mapping requires the assembly of tangential stiffness matrices at each time step which additionally decreases efficiency. Second, an adoption of the simplified arc length parametrization (6.41) leads to nearly constant time steps because the spatial weights of the individual potentials are small compared with the complete structure.

The total simulation time is chosen to be $T = 0.02$. The reference solution is given by the synchronous Euler method with the same time step length as the smallest one in the asynchronous simulation. When comparing the energy, it is evaluated as exact value $H(q(t), p(t))$ at 100 times within the simulation interval. This is in contrast to the energy condition in energy-preserving integration which is based on the discrete energy being evaluated at asynchronous times.

**Synchronous symplectic Euler**

The starting step size is chosen to be $h_0 = 0.5 h_{crit} = 4.94624 \cdot 10^{-7}$. The parameters of the scaling function are set to $h_{\max} = h_0$, $h_{\min} = 10^{-5} h_{\max}$ and $\epsilon = (h_{\max} - h_{\min})/5$. When using the SEM scheme, the initial time step is cho-

## 6.8. Examples

sen whenever an unsolvable equation is generated. The relative error in energy, required number of time steps and the used cpu time are presented in table 6.4. The variable step size methods increase both, cpu time and accuracy. The symplectic time step regularization exhibits a significantly smaller error in energy than the nonsymplectic method. The number of time steps is nearly equal, but the computational effort is increased in order to compute the symplectic correction terms. Even more interesting, the SEM scheme leads to an improved error while increasing the numerical cost insignificantly. This happens although the simulation is dominated by unsolvable equations which are neither symplectic, nor energy conserving.

A simulation with constant time steps and $h_0 = 1.3368216 \cdot 10^{-7}$ leads to $149609$ time steps and an energy error of $0.022\%$. These numbers are similar to those when using the variable step size strategy based on inverse energy. When using constant time steps with $h_0 = 3.4112 \cdot 10^{-7}$ one obtains $58631$ time steps and an energy error of $0.0648\%$. These numbers are are comparable with those of SEM.

Concluding, the tested variable step size methods converge to the correct solution, but do not improve the numerical efficiency (measured in accuracy related to numerical effort) when applied to free vibrations.

### Asynchronous variational integration

The choice of the weighting factors in the scaling function (6.43) is more complex than in the SDOF case. First, the $q$-dependent base term $V_i(q)$ is scaled by the factor $1/(w_i H_0)$ where $w_i$ is the relative spatial weight of the $i$-th potential $V_i(q)$ and $H_0$ the total energy. If the parameter $\epsilon$ is too large then the $q$-dependent base terms are too small compared with $\epsilon^2/(h_{\max} - h_{\min})^2$ leading to a nearly constant time step scheme. Therefore, the following parameter setting is used: $\epsilon = (h_{\max} - h_{\min})/10$. The magnitudes of the $q$-dependent base terms are, however, problem and state dependent.

Figure 6.11 presents the time series of total, kinetic and strain energy for different methods. The same figure shows the time series of the maximum displacement. Beside small deviations most tested methods describe the same dynamics. The variable time step method based on inverse energy turns out to be unsta-

|              | constant | eq. (6.43) (not sympl.) | eq. (6.43) (sympl.) | SEM      |
|--------------|----------|-------------------------|---------------------|----------|
| energy error | 0.12748% | 0.07297%                | 0.01936%            | 0.06665% |
| num. steps   | 40435    | 147680                  | 147474              | 57974    |
| cpu time     | 104.51s  | 369.02s                 | 554.8s              | 169.55s  |

Table 6.4: Synchronous variable step size integration of a cantilever beam. Energy error $\max_t |H(t) - H_0|/H_0$.

ble. The reference solution denotes the synchronous constant time step Euler method.

Figure 6.12 presents the total energy for the tested methods in more detail. Table 6.5 shows the maximum error in energy, the required number of kicks and the used cpu time. Obviously, the reference method exhibits the smallest energy error. The constant time step AVI follows in accuracy and is the fastest asynchronous method among the tested ones.

Interestingly, the modified SEM method exhibits a larger error in phase (see displacement plot) compared with the others. This is in agreement to the total error in energy being made in the interval $t \in [0, 0.009]$. Noteworthy, the modified SEM method was able to preserve the discrete energy exactly at verious times in the interval $t \in [0.01, 0.02]$ and exhibited a very small error in discrete energy during this period. Nevertheless, a small error in the end period may not reduce the error which was created during the beginning of the simulation (error is cumulative!).

The tested explicit time transformation turns out to be unstable. The inverse energy scaling function creates to continuous energy growth leading to a wrong solution in displacements. There is almost no difference between the symplectic and the non-symplectic versions of this time step regularization. Furthermore, changing the weights in the step size function (6.43) seems to change the rate in energy growth, but does not improve the stability qualitatively.

Concluding, variable step size methods based on time transformations are unstable in asynchronous integration of finite element structures while being stable and accurate in synchronous integration. Variable step sizes being based on SEM schemes are stable and accurate although symplecticity and energy conservation are enforced in a very small number of time steps during the simulation only. Due to the effort in implementation and computation, however, constant time steps may be preferred when applied to free vibrations.

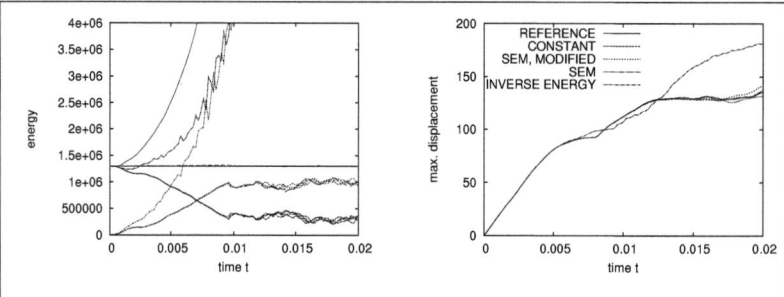

Figure 6.11: Energies (left) and max. displacement (right) for variable step size AVIs.

## 6.8. Examples

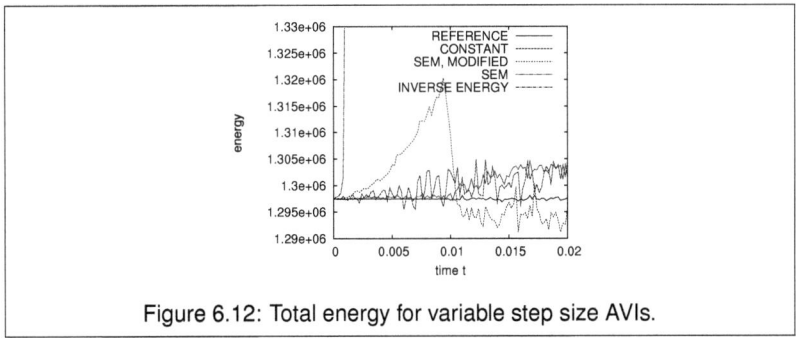

Figure 6.12: Total energy for variable step size AVIs.

|  | constant | eq. (6.43) | SEM | SEM (modified) |
|---|---|---|---|---|
| $\max_t |H(t) - H_0|/H_0$ | 0.5698% | 1484.89% | 0.6337% | 1.757% |
| number of kicks | 17519285 | 866602892 | 17782316 | 18385199 |
| cpu time | 36.59s | 3221.83s | 57.7s | 56.71s |

Table 6.5: Asynchronous variable step size integration of a cantilever beam

# Chapter 7

# Collision dynamics

## 7.1 Introduction

This chapter introduces contact algorithms to explicit asynchronous simulation of structural dynamics. When handling dynamic contact problems, there generally exist two approaches: penalty based and Lagrange multiplier methods. Penalty methods are simple to implement and penalty forces can be computed efficiently. But they are inaccurate allowing penetrations and may affect the critical time step. Lagrange multiplier methods often lead to iterative procedures. The possible large number of highly nonlinear constraints reduces the efficiency. Furthermore, redundant constraints may appear leading to singular systems of equations.

The application of an asynchronous collision integrator may eliminate some problems arising in Lagrange multiplier methods. Let the individual contact constraints be enforced at asynchronous times. If each spatial constraint is considered individually, the system of equation is simplified by two factors: (1) There is only a single constraint to be enforced at one time. (2) Furthermore, only a limited number of degrees of freedom is affected. The size of the equation system is, therefore, very small. By application of explicit collision integrators, see section 3.8, the equations are linear and the constraints can be enforced noniteratively. The operation only modifies the momentum and can, thus, be interpreted in terms of a KICK operator of an asynchronous variational integration algorithm, see section 6.4. Since each constraint is considered individually without affecting the critical time step, one may chose the time step size between two contact corrections according to local accuracy conditions, such as relative velocities and finite element sizes.

The subsequent paragraphs give an overview on the complete framework to asynchronous collisions including collision detection, spatial discretization, formulation of constraints and solution. Before start, section 7.2 recalls the mechanics of a continuuum subject to contact following standard text books [141, 273] with supporting derivations in appendix B. Section 7.3 introduces discrete distance fields. Distance fields are not entirely new to contact mechanics, but so far they were interpolated on Cartesian grids in conjunction with penalty methods. Instead of rectangular grids the finite elements themselves are used to provide an accurate interpolation. A plain interpolation is unstable when applied to structures with complex or thin-walled geometry. Two strategies for stabilization are proposed. It is shown how discrete distance can replace the impenetrability condition based on closest point projection. The resulting algorithm for approximating the closest point projection is simpler to implement, more robust and even more accurate when applied to discrete contact mechanics. The actual formulation and solution of the asynchronous collision response is derived in section 7.4. Some examples from structural dynamics verify convergence and performance of the presented method, section 7.5.

The spatial discretization of contact constraints is explained in section C of the appendix. Details on efficient collision detection schemes being suitable for distance fields and point-to-element integration are illustrated in appendix D.

## 7.2 Contact mechanics

### 7.2.1 Problem description

The displacement field $\mathbf{u}^{(i)}$ measures the distance from the reference to the deformed configuration.

$$\phi^{(i)}(\mathbf{X}, t) = \mathbf{x}^{(i)} = \mathbf{X}^{(i)} + \mathbf{u}^{(i)} \tag{7.1}$$

The deformation causes the bodies to contact and produce interactive forces $\mathbf{t}^{(i)}$ acting on portions of the subset $\partial_c \Omega^{(i)} \subset \partial \Omega^{(i)}$, see figure 7.1. Therefore, it is assumed

$$\partial \Omega^{(i)} := \partial_u \Omega^{(i)} \cup \partial_\sigma \Omega^{(i)} \cup \partial_C \Omega^{(i)} \tag{7.2}$$

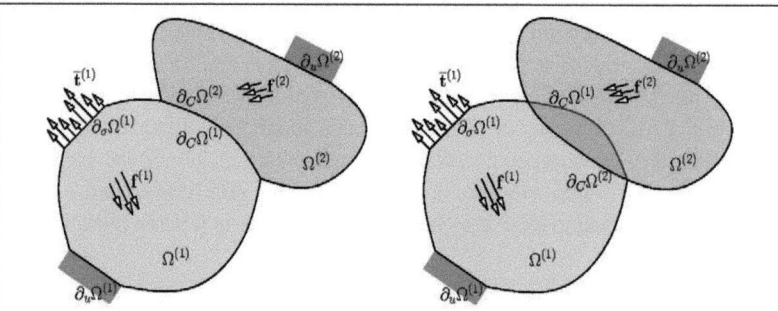

Figure 7.1: Two-body infinitesimal contact problem. A contact problem is presented showing two non-penetrating bodies in contact (left figure). During the simulation, any contact algorithm must handle trial steps where the bodies penetrate (right).

## 7.2. Contact mechanics

and
$$\partial_u\Omega^{(i)} \cap \partial_\sigma\Omega^{(i)} = \partial_C\Omega^{(i)} \cap \partial_\sigma\Omega^{(i)} = \partial_u\Omega^{(i)} \cap \partial_C\Omega^{(i)} = \emptyset \qquad (7.3)$$

where $\partial_u\Omega^{(i)}$ and $\partial_\sigma\Omega^{(i)}$ are the parts of the surface where displacements and tractions are prescribed:

$$\phi^{(i)}|_{\partial_u\Omega^{(i)}} = \bar{\phi}^{(i)}, \quad \sigma^{(i)}|_{\partial_u\Omega^{(i)}}\mathbf{n}^{(i)} = \bar{\mathbf{t}}^{(i)} \qquad (7.4)$$

and $\mathbf{n}^{(i)}$ is the normal on the boundary. The set of trial solutions $\mathcal{S}_t^{(i)}$ at time $t$ is defined through the $C^0$ continuous Sobolev space $H^1(\Omega^{(i)})$

$$\mathcal{S}_t^{(i)} = \{\phi^{(i)}(\cdot,t) : \overline{\Omega}^{(i)} \to \mathbb{R}^{n_d} |\, \phi^{(i)} \in H^1(\Omega^{(i)}),\, \phi^{(i)}|_{\partial_u\Omega^{(i)}} = \bar{\phi}^{(i)}\} \qquad (7.5)$$

Correspondingly, the space of kinematically possible variations (virtual displacements) is defined

$$\mathcal{V}_t^{(i)} = \{\delta\phi^{(i)} : \overline{\Omega}^{(i)} \to \mathbb{R}^{n_d} |\, \delta\phi^{(i)} \in H^1(\Omega^{(i)}),\, \delta\phi^{(i)}|_{\partial_u\Omega^{(i)}} = 0\} \qquad (7.6)$$

For each body $(i)$ the balance of virtual work is given through

$$0 = G^{(i)}(\phi^{(i)}, \delta\phi^{(i)}) = G_{dyn}^{(i)}(\phi^{(i)}, \delta\phi^{(i)}) - \int_{\partial_C\Omega^{(i)}} \mathbf{t}^{(i)} \cdot \delta\phi^{(i)} \mathrm{d}\Gamma \qquad (7.7)$$

wherein $G_{dyn}(\phi, \delta\phi)$ denotes the virtual work of the problem without contact portions, i.e. from internal and (prescribed) external forces balancing the virtual work of the contact tractions $\mathbf{t}^{(i)}$. Introducing the notation $\phi$ for a collection of the motions $\phi^{(i)}$, the contact virtual work writes

$$G_c(\phi, \delta\phi) := -\int_{\partial_C\Omega^{(1)}} \mathbf{t}^{(1)} \cdot \delta\phi^{(1)} \mathrm{d}\Gamma - \int_{\partial_C\Omega^{(2)}} \mathbf{t}^{(2)} \cdot \delta\phi^{(2)} \mathrm{d}\Gamma \qquad (7.8)$$

and adding the variational quantities of both bodies yields the global variational problem: Find $\phi \in \mathcal{S}$, such that

$$G_{dyn}(\phi, \delta\phi) + G_C(\phi, \delta\phi) = 0 \quad \forall \delta\phi \in \mathcal{V},\, \phi \in \mathcal{S} \qquad (7.9)$$

where the displacements and tractions are subject to certain contact constraints. Most importantly, the Signorini impenetrability condition [239] must be satisfied. The Signorini condition is enforced by imposing normal forces on the contacting boundaries. Other constraints evolve from friction laws.

### 7.2.2 Kinematics

Assume the existence of a contact frame $C$ being defined as a surface which serves for parametrizing the contacting domain. It may be located intermediate the two contact boundaries, may be identical to one of the contact boundaries are is an independent surface.

All quantities refering to the contact frame are denoted by $(\hat{\cdot})$. Points in $C$ are labeled by the position vector $\hat{\mathbf{x}} \in C$. Particles of both interacting bodies which are coupled through the contact frame are labeled by $\hat{x}^{(i)}$. The underlying parameterization of the contact frame is given by the set $\mathcal{A} \in \mathbb{R}^{n_d-1}$ with points denoted $\hat{\xi} \in \mathcal{A}$ and the invertible map $\hat{\Psi}_t : \mathcal{A} \to \mathbb{R}^{n_d-1}$, i.e.

$$C_t = \hat{\Psi}_t(\mathcal{A}) \tag{7.10}$$

Therefore, $C$ is a $(n_d - 1)$-dimensional manifold in $\mathbb{R}^{n_d}$. Material points in the reference and in the current configuration are denoted $\hat{\mathbf{X}} = \hat{\Psi}_0(\hat{\xi})$ and

$$\hat{\mathbf{x}} = \hat{\Psi}_t(\hat{\xi}) = \hat{\phi}\left(\hat{\Psi}_0(\hat{\xi})\right) \tag{7.11}$$

A basis for $C$ is given by its directional derivatives of the parametrization $\hat{\Psi}$. Assuming sufficient smoothness of the contact frame it is differentiable and the basis vectors for the reference configuration (7.12) are

$$\hat{\mathbf{T}}_\alpha(\hat{\xi}) := \hat{\Psi}_{0,\alpha}(\hat{\xi}) := \frac{\partial \hat{\Psi}_0(\hat{\xi})}{\partial \hat{\xi}_\alpha} \tag{7.12}$$

and for the spatial configuration

$$\hat{\tau}_\alpha(\hat{\xi}) := \hat{\Psi}_{t,\alpha}(\hat{\xi}) = \hat{\mathbf{F}}_t \cdot \hat{\Psi}_{0,\alpha}(\hat{\xi}) = \hat{\mathbf{F}}_t \cdot \hat{\mathbf{T}}_\alpha(\hat{\xi}) =: \hat{\phi}_{,\alpha}(\hat{\mathbf{x}}) \tag{7.13}$$

defining the tangent plane to $C$ at $\hat{\xi}$, whereby the tensor $\hat{\mathbf{F}}_t$ denotes the deformation gradient. The normal vector (if $n_d = 3$) is

$$\hat{\mathbf{n}} = \frac{\hat{\tau}_1 \times \hat{\tau}_2}{\|\hat{\tau}_1 \times \hat{\tau}_2\|} \tag{7.14}$$

Since these basis vectors are generally non-orthogonal a dual basis is introduced such that

$$\hat{\tau}^\beta \cdot \hat{\tau}_\alpha = \delta^\beta_\alpha \tag{7.15}$$

where $\delta^\beta_\alpha$ is the Kronecker delta. To perform transformations between the bases, a metric tensor $\hat{\mathbf{m}}$ is used having the components

$$\hat{m}_{\alpha\beta} = \hat{\tau}_\alpha \cdot \hat{\tau}_\beta \tag{7.16}$$

Analogously, a metric tensor $\hat{\mathbf{M}}$ may be defined for the reference configuration. For the contact surfaces $C^{(i)}$ identical basis vectors and metrics may be found as well, see figure 7.2. The transformations are given by

$$\hat{\tau}^\alpha = \hat{m}^{\alpha\beta}\hat{\tau}_\beta, \quad \hat{\tau}_\alpha = \hat{m}_{\alpha\beta}\hat{\tau}^\beta \tag{7.17}$$

with the contravariant metric $\hat{m}^{\alpha\beta} = (\hat{m}_{\alpha\beta})^{-1}$.

Concluding, the contact constraints are defined on the surface $C$. It becomes necessary to define additional mappings to describe the interaction within the contact frame and within the global frame

$$\hat{\phi}^{(i)} : C \to \mathbb{R}^n \tag{7.18}$$
$$\hat{\mathbf{t}}^{(i)} : C \to \mathbb{R}^n \tag{7.19}$$

## 7.2. Contact mechanics

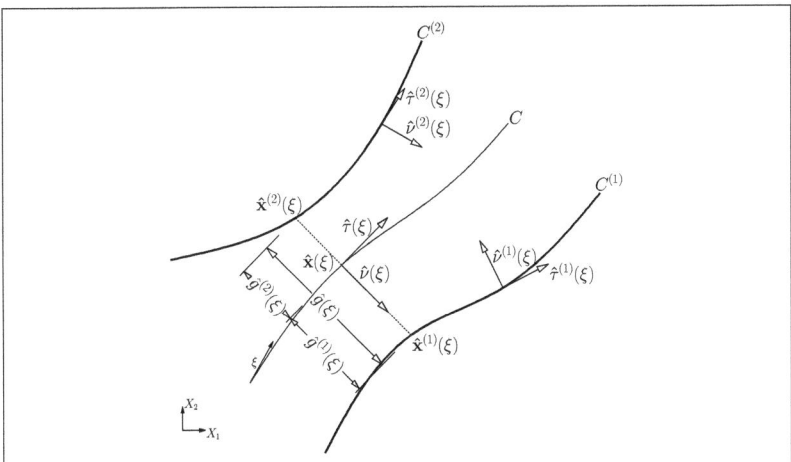

Figure 7.2: Parameterization within the contact frame. The non-penetrating case is illustrated.

which represent mappings of the displacements $\phi^{(i)}$ and tractions $\mathbf{t}^{(i)}$ applied to the bodies $\Omega^{(i)}$ onto the contact surfaces $C^{(i)}$. The contact constraints are evaluated in the contact frame and, thus, the contact virtual work can be replaced

$$G_c(\phi, \delta\phi) \leftarrow \hat{G}_c(\hat{\phi}, \hat{\delta\phi}) \tag{7.20}$$

It is assumed that there exist one-to-one mappings of points on the contact surfaces to points on the contact frame

$$\bar{\mathbf{x}}^{(i)} : \hat{\mathbf{x}}^{(i)} \rightarrow \hat{\xi}^{(i)} \tag{7.21}$$
$$\bar{\mathbf{X}}^{(i)} : \hat{\mathbf{X}}^{(i)} \rightarrow \hat{\xi}^{(i)} \tag{7.22}$$

such that the position of a point $\hat{\mathbf{x}}^{(i)}$ on the contact surface $C^{(i)}$ can be obtained from its coupled point on the contact frame, i.e. $\hat{\mathbf{x}}^{(i)} = \bar{\mathbf{x}}^{(i)}(\hat{\xi}^{(i)})$.

### 7.2.3 Gap function

The gap function $\hat{g}$ defines the signed distance between any point $\hat{\mathbf{x}}^{(1)}$ to the contact boundary $C^{(2)}$ of the target body $\Omega_t^{(2)}$ in the current configuration. It defines three possible states:

$$\begin{array}{lll} \hat{g} < 0 & \hat{\mathbf{x}}^{(1)} & \text{is outside of } \overline{\Omega}_t^{(2)} \\ \hat{g} = 0 & \hat{\mathbf{x}}^{(1)} & \text{is on the boundary of } \overline{\Omega}_t^{(2)}, \mathbf{x} \in C^{(2)} \\ \hat{g} > 0 & \hat{\mathbf{x}}^{(1)} & \text{is inside of } \overline{\Omega}_t^{(2)} \text{ (infeasible)} \end{array} \tag{7.23}$$

The gap function is defined by the closest point projection between points on the contact surface $C^{(1)}$ of the contactor body $\Omega_t^{(1)}$

$$\hat{g}(\hat{\xi}) = \left(\hat{\mathbf{x}}^{(2)}(\hat{\xi}) - \hat{\mathbf{x}}^{(1)}(\hat{\xi})\right) \hat{\nu}(\hat{\xi}) \tag{7.24}$$

where the vector $\hat{\nu}$ is the normal to the tangential plane at the projection point $\hat{x}^{(2)}$ and pointing outwards of the boundary.

The variation of the gap function is given by

$$\delta \hat{g} = \hat{\nu} \cdot \left( \delta \hat{\phi}^{(2)} - \delta \hat{\phi}^{(1)} \right) \tag{7.25}$$

while the variation of the tangential glide path is

$$\delta \hat{\xi}^\beta = \left( \delta \hat{\phi}^{(2)} - \delta \hat{\phi}^{(1)} \right) \cdot \hat{\tau}^\beta \tag{7.26}$$

see appendix B for more details.

### 7.2.4 Contact integral

The contact integral is given through equations (7.8) and (7.20)

$$\hat{G}_c(\hat{\phi}, \delta \hat{\phi}) = - \int_C \left( \hat{t}^{(1)} \cdot \delta \hat{\phi}^{(1)} + \hat{t}^{(2)} \cdot \delta \hat{\phi}^{(2)} \right) d\Gamma \tag{7.27}$$

Balance of linear momentum across the contact frame requires $\hat{t}^{(1)} d\Gamma = -\hat{t}^{(2)} d\Gamma$ for all $\hat{x} \in C$. Hence, (7.27) writes

$$\hat{G}_c(\hat{\phi}, \delta \hat{\phi}) = - \int_C \hat{t}^{(2)} \cdot \left( \delta \hat{\phi}^{(2)} - \delta \hat{\phi}^{(1)} \right) d\Gamma \tag{7.28}$$

The traction $\hat{t}^{(2)}$ can be decomposed into normal and tangential components

$$\hat{t}^{(2)} = \hat{t}_N \hat{\nu} + \hat{t}_\alpha \hat{\tau}_\alpha \tag{7.29}$$

The variations $\delta \hat{\phi}^{(i)}$ are expressed in terms of the the gap function (7.25) and the glide path (7.26). Using these notations, equation (7.28) can be expressed through

$$\hat{G}_c(\hat{\phi}, \delta \hat{\phi}) = - \int_C \left( \hat{t}_N \delta \hat{g} + \hat{t}_\alpha \delta \hat{\xi}^\alpha \right) d\Gamma \tag{7.30}$$

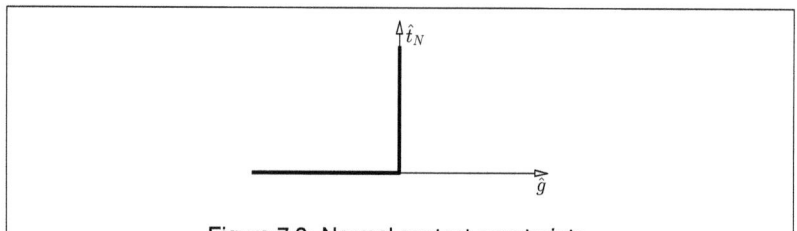

Figure 7.3: Normal contact constraints.

## 7.2. Contact mechanics

### 7.2.5 Constraints

**Normal contact conditions**

Impenetrability of the two bodies is satisfied if $\hat{g}(\hat{\xi}) \leq 0$. The gap function and the corresponding contact pressure $\hat{t}_N$ (Lagrange multiplier) are related through the Karush-Kuhn-Tucker conditions of the associated constrained variational problem

$$\hat{t}_N(\hat{\xi}) \geq 0 \tag{7.31}$$
$$\hat{g}(\hat{\xi}) \leq 0 \tag{7.32}$$
$$\hat{t}_N(\hat{\xi})\hat{g}(\hat{\xi}) = 0 \tag{7.33}$$
$$\hat{t}_N(\hat{\xi})\dot{\hat{g}}(\hat{\xi}) = 0 \tag{7.34}$$

which are known as Signorini conditions. Equations (7.31) to (7.33) state that the pressure must be nonnegative and the gap function nonpositive. The complementary condition (7.33) forces the pressure to be nonzero only if the bodies are in contact, i.e. the inequality constraint is active ($\hat{g}(\hat{x}) = 0$), see figure 7.3.

Equation (7.34) is a persistency condition specifying that when the contact pressure $\hat{t}_N$ is non-zero, the rate of separation between the contact surfaces must be zero at $\hat{\xi}$.

**Frictional contact conditions**

The frictional response is characterized by the relative velocity between points on the contact surfaces. The relative velocity is decomposed using the dual basis

$$\dot{\hat{\phi}}^{(1)} - \dot{\hat{\phi}}^{(2)} = \hat{\mathbf{v}}_N + \hat{\mathbf{v}}_T$$
$$= \hat{v}_N \hat{\nu} + \hat{v}_{T\alpha}^b \hat{\tau}^\alpha \tag{7.35}$$

The contact traction can be resolved into normal and tangential components as well

$$\hat{\mathbf{t}} := \hat{\mathbf{t}}^{(2)}(\hat{\mathbf{x}}) = -\hat{\mathbf{t}}^{(1)}(\hat{\mathbf{x}})$$
$$= \hat{\mathbf{t}}_N + \hat{\mathbf{t}}_T$$
$$= \hat{t}_N \hat{\nu}^{(2)} + \hat{t}_{T\alpha}^b \hat{\tau}^\alpha \tag{7.36}$$

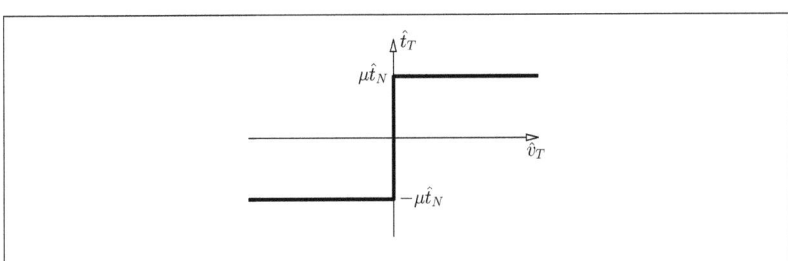

Figure 7.4: Tangential contact constraints for Coulomb friction.

For convenience, the Coulomb model of friction is used and incorporated into the Karush-Kuhn-Tucker conditions

$$\Phi := \left\|\hat{\mathbf{t}}_T\right\| - \mu \hat{t}_N \leq 0 \tag{7.37}$$

$$\hat{\mathbf{v}}_T - \zeta \frac{\hat{\mathbf{t}}_T}{\left\|\hat{\mathbf{t}}_T\right\|} = 0 \tag{7.38}$$

$$\zeta \geq 0 \tag{7.39}$$

$$\Phi\zeta = 0 \tag{7.40}$$

Equation (7.37) restricts the magnitude of the tangential traction to the product of the friction coefficient $\mu$ and the contact pressure. Equations (7.38) and (7.39) force the relative velocity (slip) to be collinear with the tangential traction, and (7.40) permits slip only when $\left\|\hat{\mathbf{t}}_T^b\right\| = \mu \hat{t}_N$, see figure 7.4.

## 7.3 Distance field

### 7.3.1 Level sets

The level set method is primarily used to implicitly describe propagating interfaces which are usually discretized on Cartesian meshes, see [209, 234]. The level set equation describes the evolution of a scalar field $d(\mathbf{x}, t)$ through

$$\dot{d}(\mathbf{x}, t) + v(\mathbf{x}, t) \left\|\nabla_{\mathbf{x}} d\right\| = 0, \quad \forall (\mathbf{x}, t) \in \Omega \times \mathcal{T} \tag{7.41}$$

where $v(\mathbf{x}, t)$ is the advection velocity. This partial differential equation is used to describe the motion of an interface by associating its geometry with the zero-iso contour of $d$. Typically, $d$ is initialized as the signed distance to the interface, satisfying $\left\|\nabla_{\mathbf{x}} d\right\| = 1 \Rightarrow \nabla_{\mathbf{x}} \left\|\nabla_{\mathbf{x}} d\right\| = 0$. The advection velocity $v$ is often replaced by a corresponding extensional velocity $v_e$ such that $\left\|\nabla_{\mathbf{x}} d\right\| = 1$ is maintained. The levelset equation (7.41) of a distance field can then be recast as

$$\dot{d}(\mathbf{x}, t) + v_e(\mathbf{x}, t) = 0 \tag{7.42}$$

Consider a moving interface $\mathcal{F}$ which divides a domain $\Omega$ into two disjoint subsets $\Omega^+$ and $\Omega^-$ to either side of $\mathcal{F}$. $\mathcal{F}$ is parametrized at the zero level set [209]

$$\mathcal{F}(t) = \{\mathbf{x} : d(\mathbf{x}, t) = 0\} \tag{7.43}$$

where the velocity field is $v(\mathbf{x}, t) = \mathbf{v}(\mathbf{x}, t) \nabla_{\mathbf{x}} \cdot d(\mathbf{x}, t)$ on $\mathcal{F}$. If the function $d$ is a signed distance function [196] it becomes

$$d(\mathbf{x}, 0) = \begin{cases} \min_{\mathbf{y} \in \mathcal{F}} \left\|\mathbf{x} - \mathbf{y}\right\| & \forall \mathbf{x} \in \Omega^+ \\ -\min_{\mathbf{y} \in \mathcal{F}} \left\|\mathbf{x} - \mathbf{y}\right\| & \forall \mathbf{x} \in \Omega^- \end{cases} \tag{7.44}$$

## 7.3. Distance field

with a velocity

$$v(\mathbf{x}, t) = v_e \, \forall \mathbf{x} \in \mathcal{F} \qquad \nabla_\mathbf{x} d(\mathbf{x}, t) \cdot \nabla_\mathbf{x} v(\mathbf{x}, t) = 0 \, \forall \mathbf{x} \in \Omega \qquad (7.45)$$

which preserves the signed distance function, i.e. the length of the distance gradient is constant since $\frac{\partial}{\partial t} \|\nabla_\mathbf{x} d\|^2 = 0$.

The level set $d(\mathbf{x}, t)$ is usually discretized and interpolated using a Cartesian grid, see figure 7.5. It implicitly defines the position of the surface $\mathcal{F}$. The dynamics of the levelset field is generally specified by Eulerian coordinates, i.e. the grid coordinates are constant in time. The motion of the surface $\mathcal{F}$ in time can be described using time stepping methods, i.e. the position of $\mathcal{F}$ can be determined at times $t > 0$.

In the context of collision detection [59, 96] the surface $\mathcal{F}$ becomes the contact boundary $\partial_C \Omega$. Instead of propagating the dynamics of the boundary in time, the distance field is usually recomputed for regions of interest. Eulerian distance fields provide the following properties when applied to contact:

+ Eulerian distance fields provide a trivial strategy for inside-outside tests and can be used directly in collision tests.

+ The rectangular grid can be easily coupled with hierarchical collision detection procedures, for example octrees, see section D.1.

+ By interpolation on the grid, one obtains an approximation of the gap function $\hat{g}$ and the normal vector $\hat{\nu}$.

- High sampling rates are required to represent objects with fine detail. Accuracy is limited due to the nature of the Cartesian grid.

- In many applications, only some parts of given geometries require a fine resolution. In turn, distance fields may generate a lot of data slowing down data processing.

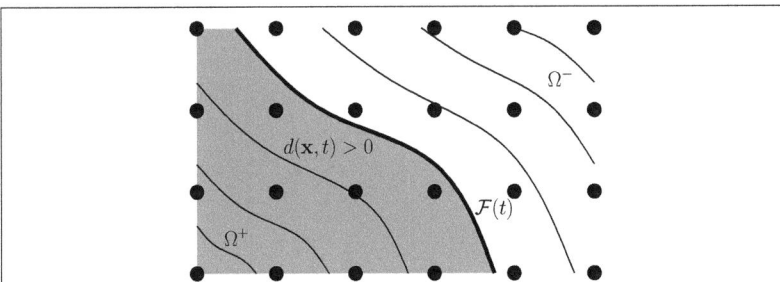

Figure 7.5: Level set on a grid. Implicit representation of the surface $\mathcal{F}$ by the level set interpolated on a rectangular grid.

## 7.3.2 Discrete distances

Let the distance field be described in terms of Lagrangian coordinates, i.e. the distance $d$ is expressed with respect to the reference configuration

$$d = d(\mathbf{X}) \tag{7.46}$$

with material coordinate $\mathbf{X}$. It describes the signed distance between $\mathbf{X}$ and the boundary $\partial_C \Omega$ of the considered body.

The distance field is now discretized with respect to the finite element nodes

$$d^h(\xi) = \sum_A N_A(\xi) d_A \tag{7.47}$$

with local material coordinate $\xi$, node index $A$ and finite element shape function $N_A$. The discrete values $d_A$ are stored at the nodes. The discrete field has the following properties:

- It is not defined outside of the finite element mesh, i.e.

$$\begin{aligned} d^h(\mathbf{X}) &= 0 \quad \text{if } \mathbf{X} \in \Omega \\ d^h(\mathbf{X}) &> 0 \quad \text{if } \mathbf{X} \in \partial_C \Omega \end{aligned} \tag{7.48}$$

- $d^h$ is $C^0$-continuous.

- The condition $\left\| \nabla_\mathbf{X} d^h(\mathbf{X}) \right\| = 1$ is generally not satisfied.

- The normal vector $\nu$ being perpendicular to the tangential plane on the boundary at $\mathbf{X} \in \partial_C \Omega$ becomes

$$\nu(\mathbf{X}) = -\frac{\nabla_\mathbf{X} d^h(\mathbf{X})}{\left\| \nabla_\mathbf{X} d^h(\mathbf{X}) \right\|}, \ \forall \{\mathbf{X}, d^h(\mathbf{X}) = 0\} \tag{7.49}$$

- The interpolation returns reasonable values only in the case where the nodal values differ in at least one node of each element, i.e. $d_i \neq d_j$ (if $i \neq j$). This case must be intercepted by the mesh generator. Particularly, at least one node must not be part of the surface what may be circumvented by subdividing elements. If all nodes are associated with zero distances, the discrete field returns zero distances in the finite elements interior which is wrong, see section 7.3.4 for more details.

## 7.3.3 Computing discrete distances

The computation of a distance field may be a very expensive application. To be accurate, for each discrete point of the mesh the distance to all parts of the related surface must be computed and the smallest is chosen, yielding a complexity of $\mathcal{O}(n \cdot m)$ (number of nodes $n$, number of faces $m$). The complexity of this approach

## 7.3. Distance field

can be reduced by first identifying the closest discrete points on the related surface for each mesh point and then computing the distance to the adjacent faces of these closest points only. When interpreting distance fields as level sets, they may be efficiently computed using level set methods like Fast Marching [235].

The Fast Marching Method was originally developed for Cartesian grids, see figure 7.6. The idea is that distances of points lying on (or near) the reference surface are already known. Considering a point with an unknown value, its distance may be approximated if the distances of neighbouring points are known. The level set value of an arbitrary point is, therefore, not the exact distance to the surface, but a good approximation while only the information of a few neighboring points is used, yielding an algorithm of complexity $O(n \cdot \log(n))$ [235]. In [59] the distance field of a finite element structure was computed by creating a Cartesian grid occupying the same space as the finite element mesh. The actual distance values are then obtained by interpolating the values of the underlying Cartesian grid. The reason for this approach is, that although stable modifications of the Fast Marching Method for acute triangular meshes [7] exist, there is no stable version for arbitrary tetrahedral meshes in three dimensions. The nodal distance to be determined in each step depends on the discrete distances of the current marching front. As a result, an error is induced in each step which grows during the propagation.

Instead of propagating the distance through the body, one could remember the surface patch which is closest to a considered mesh point, obtaining the Closest Feature Fast Marching (CFFM) method. The approach was presented in [186] and first applied to arbitrary tetrahedral meshes in [184]. A closest feature denotes either the point, the edge or the surface patch of the elements defining the surface which is closest to the considered point. The algorithm requires a division of all spatial points into 3 mutually exclusive sets: ALIVE, ACTIVE and FAR. Additionally, each point stores two properties, the distance to the surface as well as a reference to the closest feature. At initialization, all points are marked FAR, except those being on the body's surface and their neighbours. The points on the surface are marked ALIVE, their distance is set to zero and their closest feature becomes the set of surface patches they belong to. The adjacent points

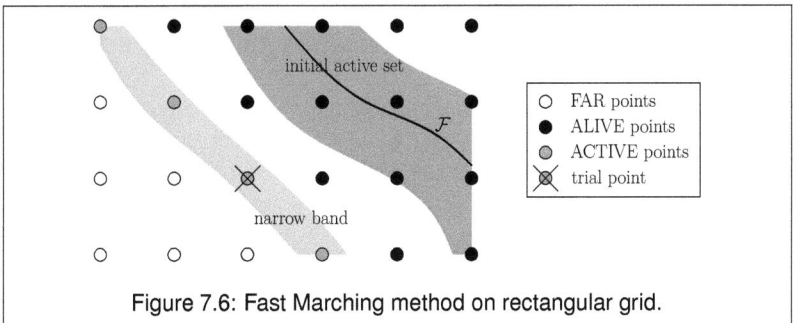

Figure 7.6: Fast Marching method on rectangular grid.

of the initial ALIVE points are set being ACTIVE; they form the so called narrow band. The narrow band is then propagated through the body, starting from its boundary and moving towards the interior skeleton until no FAR points exist. The procedure applied to finite continuum elements is summarized in algorithm 7.7. The resulting method is of complexity $\mathcal{O}(n \cdot \log(n))$. It is more accurate than the original Fast Marching Method because it computes the accurate distances to the boundary instead of propagating approximate values. Errors may appear if a node stores the wrong closest feature it is referring to. Still, the nodal distance may be a good approximation of the exact value. Furthermore, the error can be corrected at deeper nodes during the propagation.

In this work, the distance field is precomputed in the reference configuration [96]. Not all finite element nodes are associated with a discrete distance. Starting from the bounding surface, the two top layers of finite elements are identified, see figure 7.8. Only the nodes of these elements are considered during the distance field computation. This saves time and memory. Furthermore it reduces the effort in collision detection. Since trial steps of a time stepping scheme should not exhibit deep penetrations, this assumption is feasible for many cases.

If the complete boundary of a body is used as reference surface $\mathbf{F}$, then the distance gradient is not perpendicular to the boundary when measured at points close to corners and edges. Depending on the interpolation functions, some kind of smoothing appears. There are applications, however, where smoothing is not desired. This can be approached by careful definition of the reference surface $\mathbf{F}$. In [94] a strategy for a safe partial distance field update is proposed. Therein, the boundary of one body (contactor) is intersected with the volume of the target. In the next step, all intersecting target surface patches are identified. These surface patches are used to define the initial set in CFFM. The discrete distances of all nodes in the overlapping domain are recomputed at each deformed configuration. When evaluating the distance on the intersected contactor surface, it is measured with respect to the intersected target surface patches and the accuracy of the distance gradient can be improved, see figure 7.9. Other approaches exist to partial distance field updating which are more efficient, but less robust [59, 184].

### 7.3.4 Stable interpolation and assumed distance gradients

When using standard finite element shape functions, the representation of distances in the interior of elements may be erroneous. This happens whenever points of maximum distance are located in the considered element's interior. The maximum in the interior is eliminated if low order finite element shape functions are used for interpolation. Such situations always appear for elements which are located at a body's skeleton, see figure 7.10 (let the body's skeleton be the set of those points where the closest point projection returns multiple projection points with identical minimal distance). But these elements are often not crucial since

## 7.3. Distance field

> Given is a set of faces and a set of finite elements.
> **for all** finite element nodes $A$ being part of the given elements **do**
>     $\text{state}_A = \text{FAR}$
> **end for**
> Create a graph which associates each finite element node to neighbouring nodes whereby the adjacency is defined through the elements.
> Create an empty set of active nodes (the narrow band) being sorted by their distance.
> **for all** nodes $A$ which are part of the surface **do**
>     $d_A := 0$, $\text{state}_A := \text{ACTIVE}$
>     Determine the associated feature. Take all surface patches this node is part of, triangulate them and collect the triangle patches to a feature.
>     Add node $A$ to the narrow band.
> **end for**
> Define the function $\text{proj}(A, f)$ which returns the (nonnegative) smallest distance between node $A$ and the triangles of feature $f$.
> **while** narrow band is not empty **do**
>     Find the active node $A$ with smallest distance and remove it from the narrow band.
>     $\text{state}_A = \text{ALIVE}$
>     **for all** neighbours $B$ of node $A$ **do**
>         **if** $\text{state}_B == \text{ACTIVE}$ **then**
>             Compute $g = \text{proj}(B, \text{feature}_A)$
>             **if** $g < d_B$ **then**
>                 $d_B := g$, $\text{feature}_B := \text{feature}_A$
>                 Resort the set of active nodes.
>             **end if**
>         **else if** $\text{state}_B == \text{FAR}$ **then**
>             $d_B := \text{proj}(B, \text{feature}_A)$, $\text{feature}_B := \text{feature}_A$, $\text{state}_B := \text{ACTIVE}$
>             Insert node $B$ into the set of active nodes.
>         **end if**
>     **end for**
> **end while**
> If there are any nodes $A$ left with $\text{state}_A == \text{FAR}$ then $A$ belongs to another body than the given surface patches.
>
> Figure 7.7: Closest Feature Front Marching

one is generally not interested in measuring very deep distances in collision detection. They are important, however, if the skeleton is close to the boundary, for example if more than one element face is part of the boundary or if individual nodes are on the boundary, but not the element faces they are part of. Some critical situations are illustrated in figure 7.11 for two and in figure 7.12 for three

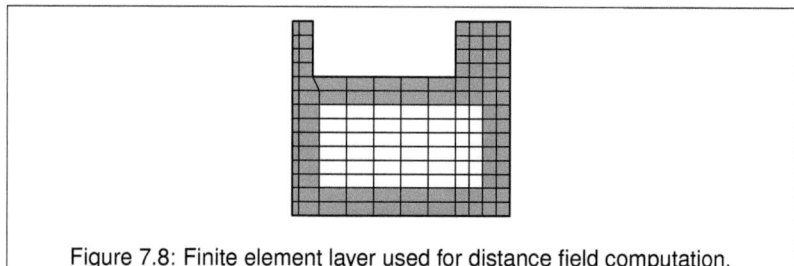

Figure 7.8: Finite element layer used for distance field computation.

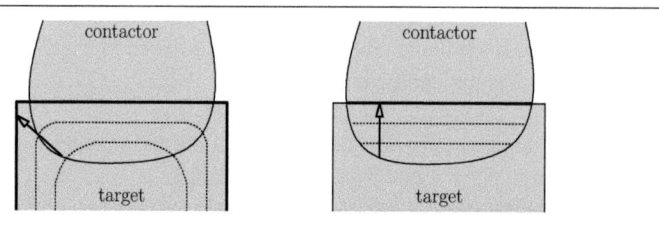

Figure 7.9: Safe partial distance field update. Left: Distance with respect to the target's boundary. Right: Distance refers to actual intersection of the target's boundary with the contactor. Dotted lines: contours. Arrows: approximated projection vectors to the boundary.

dimensions. Obviously, an enrichment of the interpolation may stabilize the distance approximation. The function space must contain at least edge, face and bubble shape functions. For hexahedral elements, a quadratic tensor-product interpolation with 27 nodes seems sufficient. For tetrahedra one requires at least 15 support points, i.e. a 10-noded quadratic tetrahedron enriched by 1 bubble and 4 area functions with supports in the element center and on the element faces.

An enrichment of the shape function space may stabilize the interpolation of the distance $d^h$, but the interpolation of the distance gradient $\nabla_{\mathbf{x}} d^h$ may still be insufficient. Consider, for example, a 2nd order quadrilateral element. All element faces are on the boundary. Only the bubble shape function contributes nonzero terms to the distance field. The shape function gradient is zero at the interior support point (which is acceptable) and at the finite element nodes. The latter may lead to a wrong collision response. The matter of nearly zero gradients may be approached by normalizing the gradient to unit length. This strategy decreases accuracy due to round-off errors and possible ill-conditioning of the normalization. Another strategy would be to enrich the elements by piecewise linear polynomials. This is equivalent to subdividing finite elements into linear tetrahedra. The distance gradient will be nonzero everywhere, but discontinuous, see figure 7.13.

A stabilization of the gradient field can be realized through assumed gradients, see figure 7.13. The presented strategy extends ideas from [196]. Given

## 7.3. Distance field

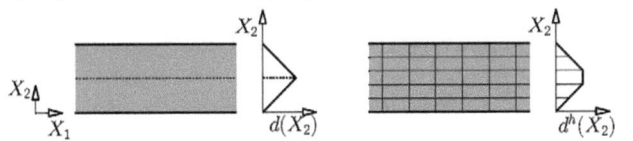

Figure 7.10: Representation of skeletons. Left: The dotted line denotes the true skeleton. Right: The exact position of the skeleton can not be measured by the (linear) interpolation.

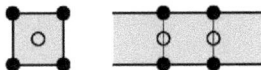

Figure 7.11: Unstable discrete distance field in 2D, exemplified for a single 1st order quadrilateral element. Left: If all finite element faces are on the boundary, an additional support point in the element center stabilizes the interpolation. Right: If two opposite element faces are on the boundary, additional support points on the interior element edges stabilize the interpolation.

a distance field based on finite element shape functions, the distance gradient $\mathbf{G} = \nabla_\mathbf{X} d$ is interpolated independently by

$$\mathbf{G}^h(\mathbf{X}) = \sum_A N_A(\mathbf{X})\mathbf{G}_A, \quad \mathbf{G}_A \in \mathbb{R}^3 \qquad (7.50)$$

For finite element nodes which are not on the boundary, the discrete gradient $\mathbf{G}_A$ is determined from a least square problem which minimizes the error of the equivalence condition $\mathbf{G}^h(\mathbf{X}) \approx \nabla_\mathbf{X} d^h$:

$$\min_{\mathbf{G}_A} \int_\Omega \left\| \sum_A N_A(\mathbf{X})\mathbf{G}_A - \nabla_\mathbf{X} d^h \right\|^2 dV \qquad (7.51)$$

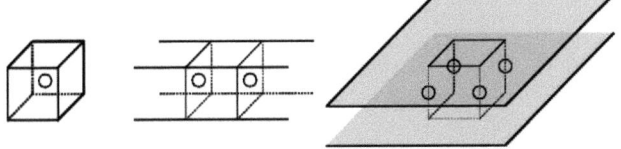

Figure 7.12: Unstable discrete distance field in 3D, exemplified for a single 1st order hexahedral element. Left: If all finite element faces are on the boundary, an additional support point in the element center stabilizes the interpolation. Center: If only two opposite element faces are not on the boundary, additional support points on the remaining element faces stabilize the interpolation. Right: If two opposite element faces are on the boundary, additional support points on the interior element edges stabilize the interpolation.

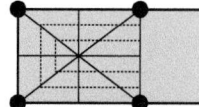 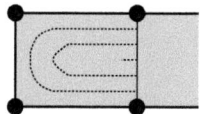

Figure 7.13: Stable distance gradient for a quadrilateral element. Left: Piecewise linear interpolation/subdivision into simplices. Dotted lines: contours of distance interpolation. Right: Assumed gradient interpolation. Dotted lines: contours of a distance function being associated with assumed gradients.

Deriving by $\mathbf{G}_A$ and assuming the mass lumping condition

$$\int_\Omega N_A(\mathbf{X})N_B(\mathbf{X})dV = \delta_A^B \int_\Omega N_B(\mathbf{X})dV \tag{7.52}$$

with Kronecker delta $\delta_A^B$ leads to

$$\mathbf{G}_A = \sum_B \frac{\int_\Omega N_A(\mathbf{X})\nabla_\mathbf{X} N_B(\mathbf{X})d_B dV}{\int_\Omega N_A(\mathbf{X})dV} \tag{7.53}$$

For finite element nodes which are part of the boundary, this strategy may lead to erroneous values because the distance field $d^h$ itself contains insufficient data. The discrete gradients $\mathbf{G}_A$ of boundary nodes are, therefore, the average of the surface normals of all surface patches $F$ being adjacent to the considered node

$$\mathbf{G}_A = \frac{\sum_{F\in A}-\nu_F W_F}{\sum_{F\in A} W_F}, \quad A \notin \partial_C\Omega \tag{7.54}$$

Therein, $\nu_F$ denotes the normal vector of surface $F$ at node $A$, which can be easily obtained from triangulation, and $W_F$ represents a weighting factor, for example the area of the surface patch $F$. A similar procedure using average nodal normal vectors on surface nodes was used by [218, 264]. The special treatment of boundary nodes is not required if enriched shape functions are used.

The distance gradient in the deformed configuration $\gamma(\mathbf{X})$ can be obtained from

$$\gamma(\mathbf{X})_\alpha = \frac{\partial}{\partial \mathbf{x}_\alpha}d(\mathbf{X}) = \frac{\partial \mathbf{X}_\beta}{\partial \mathbf{x}_\alpha}\frac{\partial}{\partial \mathbf{X}_\beta}d(\mathbf{X})$$
$$\gamma(\mathbf{X}) = \mathbf{F}^{-T}(\mathbf{X})\mathbf{G}(\mathbf{X}) \tag{7.55}$$

One can, therefore, store the nodal gradients in the reference configuration and map them into the deformed space during the simulation.

### 7.3.5 Computing the closest point projection

The distance field can be used to efficiently compute the closest point projection of a point $\mathbf{x}$ to the boundary. Let this projection be $\mathbf{y} = \mathbf{x}+\mathbf{p}$, where $\mathbf{p}$ denotes the projection/distance vector from $\mathbf{x}$. The boundary is defined through $d(\mathbf{x}) = 0$. The

## 7.3. Distance field

closest point projection defines the projection vector **p** to be the shortest vector to the surface, i.e.

$$\min_{\mathbf{p}} \|\mathbf{p}\|, \text{ subject to } d(\mathbf{x} + \mathbf{p}) = 0 \tag{7.56}$$

which is transformed into an equivalent problem using the Lagrangian

$$L(\mathbf{p}, \lambda) = \|\mathbf{p}\|^2 + \lambda \cdot d(\mathbf{x} + \mathbf{p}) \to \min_{\mathbf{p}} \max_{\lambda} \tag{7.57}$$

The objective function is quadratic, but the distance may be highly nonlinear, the surface may be nonconvex and nonsmooth.

An approximate solution is obtained by replacing the distance function by a first order Taylor expansion

$$d(\mathbf{x} + \mathbf{p}) = d(\mathbf{x}) + \nabla_{\mathbf{x}} d(\mathbf{x})^T \mathbf{p} + \mathcal{O}(\mathbf{p}^2) \tag{7.58}$$

Substituting $\gamma(\mathbf{x}) = \nabla_{\mathbf{x}} d(\mathbf{x})$, one obtains

$$\mathbf{p} = -\frac{\gamma(\mathbf{x})}{\|\gamma(\mathbf{x})\|^2} d(\mathbf{x}) \tag{7.59}$$

The expansion $d(\mathbf{x} + \mathbf{p})$ approximates the tangential plane in $\mathbf{x} + \mathbf{p}$. This plane is perpendicular to **p**, since the scalar product of **p** and any vector to a point $\hat{\mathbf{y}}$ on the approximated surface $0 = d(\mathbf{x}) + \nabla_{\mathbf{x}} d(\mathbf{x}) \cdot (\hat{\mathbf{y}} - \mathbf{x})$ is zero.

The advantages of using the distance field for closest point projection compared with exact projections are

1. Once the material coordinate **X** is found, the distance function $d(\mathbf{X})$ can be easily evaluated by interpolation. The evaluation time is independent of the complexity of the body's shape.

2. For small penetrations, the approximation is usually sufficiently accurate.

3. Directions are always feasible, independent from nonsmooth geometries, nonconvex surface features, neighbourhood of corners and wedges, see figure 7.14.

4. The discrete distance field is $C^0$-continuous. When used in conjunction with assumed gradients, it is $C^1$-continuous.

### 7.3.6 Replacing the gap function

The structure of the projection vector given through the distance field in equation (7.59) is similar to equation (7.24), i.e.

$$\frac{d(\mathbf{x})}{\|\gamma(\mathbf{x})\|} \frac{(-\gamma(\mathbf{x}))}{\|\gamma(\mathbf{x})\|} = \mathbf{p} \quad \longleftrightarrow \quad \hat{g}(\hat{\xi}) \cdot \hat{\nu}(\hat{\xi}) = \hat{\mathbf{x}}^{(2)}(\hat{\xi}) - \hat{\mathbf{x}}^{(1)}(\hat{\xi}) \tag{7.60}$$

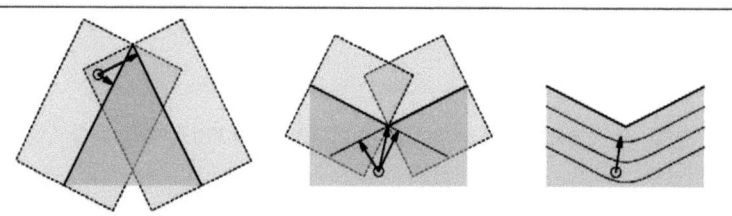

Figure 7.14: Distance field and closest point projection. The highlighted rectangles around the surface patches are bounding volumes (halos) in which the search takes place. Left: Closest point projection at sharp corners. The point lies outside, but the right face returns a positive distance. Center: Closest point projection at nonconvex corners. The point is outside of the halos and may not be used in detection. Multiple projections appear. Special considerations of edges and corners are required. Right: Distance field simply evaluates a distance and a smoothed gradient which automatically includes handling of sharp or nonconvex geometries.

The definitions of the gap function and the surface normal can, therefore, be replaced by the distance function and its gradient. Let the gap function be the sum of the distances of a point $\hat{x}$ on the contact frame to the boundaries of the two bodies, see figure 7.2

$$\hat{g}(\hat{\xi}) = \sum_{i=1,2} \frac{d^{(i)}(\hat{\mathbf{X}}(\hat{\xi}))}{\left\|\gamma^{(i)}(\hat{\mathbf{X}}(\hat{\xi}))\right\|} \qquad (7.61)$$

Variation leads to

$$\delta\hat{g} = \sum_{i=1,2} \left( \frac{\gamma^{(i)}(\hat{\mathbf{X}}(\hat{\xi})) \cdot \left(\delta\hat{\phi}^{(i)} + \hat{\tau}_\beta^{(i)} \delta\xi^{(i)\beta}\right)}{\left\|\gamma^{(i)}(\hat{\mathbf{X}}(\hat{\xi}))\right\|} - \frac{d^{(i)}(\hat{\mathbf{X}}(\hat{\xi}))\delta\left\|\gamma^{(i)}(\hat{\mathbf{X}}(\hat{\xi}))\right\|}{\left\|\gamma^{(i)}(\hat{\mathbf{X}}(\hat{\xi}))\right\|^2} \right) \qquad (7.62)$$

Due to perpendicularity of the basis it is $\gamma^{(i)} \cdot \hat{\tau}_\beta^{(i)} = 0$. The last term is assumed being zero since $\left\|\gamma^{(i)}(\hat{\mathbf{X}}(\hat{\xi}))\right\| = 1$ is satisfied by the continuous distance function. In feasible contact, $\hat{g} = 0$, both tangential planes are identical and one has

$$\hat{\nu}(\hat{\xi}) = -\frac{\gamma^{(1)}(\hat{\mathbf{X}}(\hat{\xi}))}{\left\|\gamma^{(1)}(\hat{\mathbf{X}}(\hat{\xi}))\right\|} = \frac{\gamma^{(2)}(\hat{\mathbf{X}}(\hat{\xi}))}{\left\|\gamma^{(2)}(\hat{\mathbf{X}}(\hat{\xi}))\right\|} \qquad (7.63)$$

Then equation (7.62) leads to

$$\delta\hat{g} = \hat{\nu} \cdot \left(\delta\hat{\phi}^{(2)} - \delta\hat{\phi}^{(1)}\right) \qquad (7.64)$$

which is identical to the previous result (7.25).

The tangential velocity (7.35) is obtained from the decomposition

$$\hat{\mathbf{v}}_T = (\mathbf{I} - \hat{\nu} \otimes \hat{\nu})\,\hat{\mathbf{v}} \qquad (7.65)$$

## 7.3. Distance field

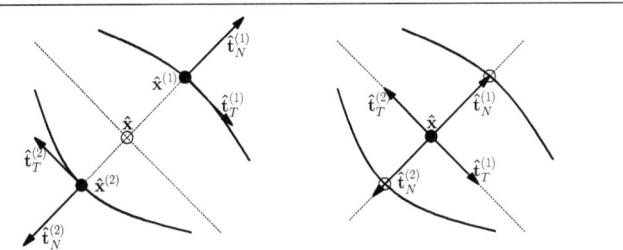

Figure 7.15: Momentum preservation. Left: tractions applied to projection points $\hat{x}^{(i)}$ in case of $\hat{g} > 0$. Right: tractions applied to material points $\hat{X}^{(i)}$ located at the same spatial coordinate $\hat{x}$ on contact frame.

Hence, no assumptions on the smoothness of the contact surface are required for the computation of the tangential basis, see section 7.2.2.

When applied to contact problems, distance fields have some advantageous properties compared with closest point projection:

- **Exact momentum preservation** In frictionless contact, both formulations preserve the balance of momentum: The normal tractions are applied to spatial points being located on the intersections of the two contact surfaces and a line along the normal vector $\hat{\nu}$ and the normal tractions $\hat{t}_N$. This is different for the friction forces in the standard approach: If the normal contact condition $\hat{g} = 0$ is not exactly enforced by the contact algorithm, the angular momentum is generally not preserved, see figure 7.15 on the left. When using distance fields, the contact tractions are applied to material points with the same deformed coordinate. Therefore, balance of linear and angular momentum is satisfied even if $\hat{g} \neq 0$, see figure 7.15 on the right.

- **Numerical efficiency** The computation of the closest point projection involves the solution of the optimization problem

$$\|\mathbf{y}(\xi) - \hat{\mathbf{x}}\| \to \min, \quad \mathbf{y}(\xi) = \sum_A N_A^F(\xi)\mathbf{x}_A \qquad (7.66)$$

which finds the point $\mathbf{y}$ on some finite element face which is closest to a given point $\hat{\mathbf{x}}$ on the contact frame. $\xi$ are two-dimensional coordinates defining the parameterization on the finite element face. The face is interpolated using the shape functions $N_A^F$. The optimality conditions lead to

$$0 = \sum_{A,B} N_A^F(\xi)\nabla_\xi N_B^F(\xi)\mathbf{x}_A \cdot \mathbf{x}_B - \sum_A \nabla_\xi N_A^F(\xi)\mathbf{x}_A \cdot \hat{\mathbf{x}} \qquad (7.67)$$

which is solved iteratively.

The application of the distance field involves the identification of the local finite element coordinate $\xi \in \mathbb{R}^3$ for a given spatial coordinate $\hat{\mathbf{x}}$. In case of isoparameric continuum elements, this leads to

$$0 = \mathbf{y}(\xi) - \hat{\mathbf{x}}, \quad \mathbf{y}(\xi) = \sum_A N_A(\xi)\mathbf{x}_A \qquad (7.68)$$

with finite element shape function $N_A$. When comparing equations (7.67) and (7.68), the latter is generally faster to solve by iterative methods. Consider, for example, distorted first order hexahedral elements or distorted second order elements. Then the degree of nonlinearity is higher in (7.67). Hence, one usually requires more iterations in order to obtain similar accuracy.

Furthermore, in one usually must check more faces being close to $\hat{x}$ than elements being intersected by $\hat{x}$. The complexity of projection finding is increased if the contacting boundaries are replaced by smooth surface representations and if nonconvex or nonsmooth boundaries are present.

## 7.4 Asynchronous collisions

### 7.4.1 Asynchronous collision detection

The focus of asynchronous collision detection lies on updating the spatial data structures of global contact search with minimal effort. In synchronous schemes, the spatial hierarchy is updated for all nodes and elements at one time. In the asynchronous context, only a few nodes are affected by a single KICK event or DRIFT phase. In AVIs it is more crucial than in standard methods that the update of the data structures with respect to the affected nodes is a numerically cheap operation.

The priority queue is an ordered set of KICK operators. Some of these KICK operators are "contact elements" each representing the collision of a single contactor node. Other KICK operators represent the response due to the restoring forces of individual spatial integration points. The latter are associated with a set of affected finite elements. Spatial integration points of the strain energy may influence the degrees of freedom of a single finite element (if they are located in the element's interior) or of multiple finite elements (if they are located on the element's boundary, for example in the nodes). Every finite element knows if it is subject to collision detection. In this case, it is associated with an AABB in the spatial data structure. Whenever a strain energy based KICK appears, it applies a DRIFT to all nodes which are influenced by the KICK. It further checks if any associated finite element is subject to collision detection. In this case, all associated AABBs are updated with respect to the current coordinates. The coordinates of a single AABB are not necessarily synchronously. Assuming that the time step of strain energy KICKs is generally very small, the spatial data structures can be considered being sufficiently accurate with respect to the target elements.

Whenever a collision KICK is taken from the priority queue, it first drifts the contactor node to the current time. The position code of the node is updated in

## 7.4. Asynchronous collisions

the spatial hierarchy. Then one can perform a global collision detection for the associated node.

After finding the collision pair candidates in the global search phase, all involved nodes are drifted to the contactor node time. This will improve the accuracy of the local collision detection. Furthermore, it allows the accurate computation of the deformation gradient of the considered target element. This is important since the prediction of an accurate normal vector is crucial. In addition, the conservation of linear and angular momentum is only guaranteed, if the collision response is applied to the contactor point and the target element at the same time and spatial coordinate.

Subsequently, the local contact search is applied, i.e. the considered contactor node is accurately intersected with the target element and the local finite element coordinate $\xi^{(2)}$ within the target element is obtained.

## 7.4.2 Normal contact

Given the local finite element coordinate $\xi^{(2)}$ of contactor node $A$ in the target element one can assemble the discrete gap gradient from the variation

$$\delta \hat{g}_A = \frac{\gamma_A}{\|\gamma_A\|} \cdot \left( \delta \mathbf{u}_A - \sum_B N_B^{(2)}(\xi^{(2)}) \delta \mathbf{u}_B \right) \qquad (7.69)$$

which can be obtained from equations (7.64) and (C.9). Therein, $\gamma_A$ denotes the distance gradient at the contactor node $A$, $N_B^{(2)}$ the finite element shape function of the target element, $\mathbf{u}_A$ the displacements of node $A$.

The computation of the collision response follows the procedure presented in section 3.8.5, page 66. The gap rate is obtained

$$\dot{\hat{g}}_A = (\nabla \hat{g}_A)^T \mathbf{v}^- \qquad (7.70)$$

with discrete gap gradient $\nabla \hat{g}_A$ and global velocity vector $\mathbf{v}$ which is related to the discrete momentum $\mathbf{j}$, equation (3.2), through $\mathbf{v} = \mathbf{M}^{-1} \mathbf{j}$. The constraint is considered active if

$$\dot{\hat{g}}_A > 0 \qquad (7.71)$$

else the contact pair will be skipped.

Direct application of equation (3.151) corresponds to the modification of velocities

$$\mathbf{v}^+ = \mathbf{v}^- - 2\mathbf{M}^{-1} \left( \nabla \hat{g}_A \left[ (\nabla \hat{g}_A)^T \mathbf{M}^{-1} \nabla \hat{g}_A \right]^{-1} \dot{\hat{g}}_A \right) \qquad (7.72)$$

Assuming a diagonal mass matrix and nodal masses $m_A$, this update can be computed efficiently. The gap gradient is sparse and, hence, only a few components of $\mathbf{v}$ are affected by the update. Notice, the asynchronous nature of the collision response leads to a sequential collision response involving simple equations with scalar Lagrange multipliers (3.148). This is contrary to synchronous algorithms where equation (3.151) requires the solution of a system of linear equations due to coupling terms among individual contact constraints.

Application of a coefficient of restitution $\kappa$ [36], $0 \leq \kappa \leq 1$, leads to

$$\mathbf{v}^+ = \mathbf{v}^- - \frac{(2-\kappa) \dot{\hat{g}}_A}{(\nabla \hat{g}_A)^T \mathbf{M}^{-1} \nabla \hat{g}_A} \mathbf{M}^{-1} \nabla \hat{g}_A \qquad (7.73)$$

## 7.4.3 Normal contact with nodal restraints

If nodal restraints are defined, see section 5.4, the computation of the contact response is not so easy since not only $\dot{\hat{g}}_A \leq 0$ must be enforced. The contact projection must preserve the restraint conditions $\mathbf{G}^T \mathbf{u} = 0$, equation 5.20. An additional projection as in equation (5.22) is not possible since it does not obey the energy preservation properties of the contact response.

## 7.4. Asynchronous collisions

The system of equations (3.148) is, therefore, extended by the restraints. Nevertheless, the problem can still be solved efficiently since at most three restraints can be defined per node and each subset of restraints influences only the displacements which belong to the same node. Furthermore, the equality constraints can be handled in the same manner as inequality constraints if the hidden restraint $\mathbf{G}^T \mathbf{v}^- = 0$ is satisfied prior the collision.

The Lagrange multiplier (3.148) becomes a vector consisting of the restraint multipliers $\lambda$ and the contact multiplier $\mu$

$$\begin{pmatrix} \lambda \\ \mu \end{pmatrix} = \left[ \begin{pmatrix} \mathbf{G} & \nabla \hat{g}_A \end{pmatrix}^T \mathbf{M}^{-1} \begin{pmatrix} \mathbf{G} & \nabla \hat{g}_A \end{pmatrix} \right]^{-1} \begin{pmatrix} \mathbf{G}^T \mathbf{v}^- \\ \dot{\hat{g}}_A \end{pmatrix} = \mathbf{S} \begin{pmatrix} \mathbf{G}^T \mathbf{v}^- \\ \dot{\hat{g}}_A \end{pmatrix} \quad (7.74)$$

with

$$\mathbf{S} = \begin{pmatrix} \mathbf{A} & \mathbf{B} \\ \mathbf{C} & \mathbf{D} \end{pmatrix}^{-1}$$

$$\mathbf{A} = \mathbf{G}^T \mathbf{M}^{-1} \mathbf{G}$$
$$\mathbf{B} = \mathbf{G}^T \mathbf{M}^{-1} \nabla \hat{g}_A$$
$$\mathbf{C} = \mathbf{B}^T$$
$$D = (\nabla \hat{g}_A)^T \mathbf{M}^{-1} \nabla \hat{g}_A \quad (7.75)$$

The matrix S always exists: The restraints are linearly independent by definition. Only the contact gradient may be linearly dependent on the restraints. This case is very unlikely. It can be checked easily and may only appear in erroneous generated meshes.

The update of the velocities is then computed from

$$\mathbf{v}^+ = \mathbf{v}^- + \Delta \mathbf{v}_N$$

$$\Delta \mathbf{v}_N = -(2-\kappa) \mathbf{M}^{-1} \begin{pmatrix} \mathbf{G} & \nabla \hat{g}_A \end{pmatrix} \begin{pmatrix} \lambda \\ \mu \end{pmatrix} \quad (7.76)$$

In order to solve the system of equations one has to find a simple expression for

$$\mathbf{S} = \begin{pmatrix} \mathbf{S}_{11} & \mathbf{S}_{12} \\ \mathbf{S}_{21} & S_{22} \end{pmatrix} \quad (7.77)$$

S is given by

$$\mathbf{S}_{11} = \mathbf{A}^{-1} + (\mathbf{A}^{-1} \mathbf{B}) S_{22} \mathbf{B}^T \mathbf{A}^{-1}$$
$$\mathbf{S}_{12} = -\mathbf{A}^{-1} \mathbf{B} S_{22}$$
$$\mathbf{S}_{21} = \mathbf{S}_{12}^T$$
$$S_{22} = \frac{1}{D - \mathbf{B}^T \mathbf{A}^{-1} \mathbf{B}} \quad (7.78)$$

Using these definitions and assuming that the restraint rates are kept zero during the simulation, the Lagrange multipliers simplify to

$$\lambda = S_{12} \mathring{g}_A$$
$$\mu = S_{22} \mathring{g}_A \quad (7.79)$$

When implementing the computation of the multipliers, one has to provide fast mappings between node indices, indices of global degrees of freedom and the local DOF indices at nodes. The mass is diagonal and constant for all DOFs of the same node. The blocks of matrix $\mathbf{A}^{-1}$ (without mass, i.e. $(\mathbf{G}^T\mathbf{G})^{-1}$) and the blocks of matrix $\mathbf{G}$ are stored at each node. During each collision response, one first computes the vector $\mathbf{B}$ and the scalar $D$. Then, the vector

$$\mathbf{E} = \mathbf{A}^{-1}\mathbf{B} \quad (7.80)$$

is computed and temporarily stored. The scalar $S_{22}$ is computed by

$$S_{22} = \frac{1}{D - \mathbf{B}^T\mathbf{E}} \quad (7.81)$$

For the multipliers one obtains

$$\lambda = -\mathbf{E} S_{22} \mathring{g}_A$$
$$\mu = S_{22} \mathring{g}_A \quad (7.82)$$

The product (7.80) can be obtained efficiently by using the block structure of matrix $\mathbf{A}$

$$\mathbf{A} = \begin{pmatrix} \mathbf{A}_1/m_1 & 0 & 0 & \cdots & 0 \\ 0 & \mathbf{A}_2/m_2 & 0 & \cdots & 0 \\ 0 & 0 & \mathbf{A}_3/m_3 & & \vdots \\ \vdots & \vdots & & \ddots & 0 \\ 0 & 0 & \cdots & 0 & \mathbf{A}_n/m_n \end{pmatrix} \quad (7.83)$$

$$\mathbf{A}^{-1} = \begin{pmatrix} m_1\mathbf{A}_1^{-1} & 0 & 0 & \cdots & 0 \\ 0 & m_2\mathbf{A}_2^{-1} & 0 & \cdots & 0 \\ 0 & 0 & m_3\mathbf{A}_3^{-1} & & \vdots \\ \vdots & \vdots & & \ddots & 0 \\ 0 & 0 & \cdots & 0 & m_n\mathbf{A}_n^{-1} \end{pmatrix} \quad (7.84)$$

wherein $\mathbf{A}_i = \mathbf{G}_i^T \mathbf{G}_i$ are precomputed constant symmetric, positive definite matrices of maximum dimension $3 \times 3$ per node $i$ and $m_i$ denotes the respective nodal mass. For the computation of $\mathbf{E}$ one does not need all nodes, since the sparse vector $\mathbf{B}$ only contains elements which belong to the nodes which are affected by the considered collision. For the construction of $\mathbf{B}$ and $\mathbf{E}$ it is helpful to determine an ordered set of the involved nodes. Then one iterates through the set of involved nodes and adds three components (three DOFs per node) to $\mathbf{B}$ and $\mathbf{E}$. In fact, it is not necessary to generate $\mathbf{B}$ at all. The contributions of each node can be added to $\mathbf{E}$ and $S_{22}^{-1}$ during the loop through the involved nodes set.

## 7.4.4 Coulomb friction

The collision response requires the computation of the normal velocity increment $\Delta v_N$, equation (7.76). The tangential velocity increment is then computed from the decomposition (7.65). Therefore, another set of constraints must be satisfied serving as a predictor step, i.e.

$$0 = \dot{\bar{g}}_A = \mathbf{v}_A - \sum_B N_B^{(2)}(\xi^{(2)})\mathbf{v}_B \qquad (7.85)$$

These are three additional constraints which enforce a zero relative velocity of both contacting material points, one constraint along the direction of each Cartesian axis. The gradient matrix $\nabla \bar{g}_A$ is assembled from the variation

$$\delta \bar{g}_{A\alpha} = \delta u_{A\alpha} - \sum_B N_B^{(2)}(\xi^{(2)})\delta u_{B\alpha}, \quad \alpha = 1\ldots 3 \qquad (7.86)$$

The computation of the corresponding velocity increment $\Delta \bar{\mathbf{v}}$ is similar to the normal response $\Delta \mathbf{v}_N$, but this time there are three instead of a single contact constraint. In the presence of nodal restraints, the solution involves the following steps:

$$\begin{aligned}
\Delta \bar{\mathbf{v}} &= -(2-\kappa)\mathbf{M}^{-1}\begin{pmatrix} \mathbf{G} & \nabla \bar{g}_A \end{pmatrix}\begin{pmatrix} \bar{\lambda} \\ \bar{\mu} \end{pmatrix} \\
\dot{\bar{g}}_A &= (\nabla \bar{g}_A)^T \mathbf{v}^- \\
\bar{\mathbf{B}} &= \mathbf{G}^T \mathbf{M}^{-1} \nabla \bar{g}_A \\
\bar{\mathbf{D}} &= (\nabla \bar{g}_A)^T \mathbf{M}^{-1} \nabla \bar{g}_A \\
\bar{\mathbf{E}} &= \mathbf{A}^{-1} \bar{\mathbf{B}} \\
\bar{\mathbf{S}}_{22} &= \left[\bar{\mathbf{D}} - \bar{\mathbf{B}}^T \bar{\mathbf{E}}\right]^{-1} \\
\bar{\lambda} &= -\bar{\mathbf{E}}\bar{\mathbf{S}}_{22}\dot{\bar{g}}_A \\
\bar{\mu} &= \bar{\mathbf{S}}_{22}\dot{\bar{g}}_A
\end{aligned} \qquad (7.87)$$

A predictor of the tangential velocity increment can be obtained by

$$\Delta \mathbf{v}_T^{pred} = \Delta \bar{\mathbf{v}} - \Delta \mathbf{v}_N \qquad (7.88)$$

The friction law, equations (7.37)-(7.40), is applied to the tangential velocity by checking the Coulomb yield surface for each predicted nodal tangential traction. The nodal contact tractions are the momentum changes

$$\hat{\mathbf{t}}_{T,A}^{pred} = m_A \Delta \mathbf{v}_{T,A}^{pred}/\Delta t, \quad \hat{\mathbf{t}}_{N,A} = m_A \Delta \mathbf{v}_{N,A}/\Delta t \qquad (7.89)$$

with node index $A$ and nodal mass $m_A$. Obviously, the nodal mass and the time step length can be eliminated from equation (7.37) and the velocity increments can be used directly. The tangential velocity change for node $A$ is then

$$\mathbf{v}_{T,A} = \begin{cases} \mathbf{v}_{T,A}^{pred} & \left\|\mathbf{v}_{T,A}^{pred}\right\| \leq \mu \left\|\mathbf{v}_{N,A}\right\| \\ \left(\mu \left\|\mathbf{v}_{N,A}\right\| / \left\|\mathbf{v}_{T,A}^{pred}\right\|\right) \mathbf{v}_{T,A}^{pred} & \text{else} \end{cases} \qquad (7.90)$$

The post collision velocity is obtained from

$$\mathbf{v}^+ = \mathbf{v}^- + \Delta\mathbf{v}_N + \Delta\mathbf{v}_T \tag{7.91}$$

### 7.4.5 Time step selection

**Sequential synchronous collisions**

The asynchronous collision reponse can be used to improve explicit synchronous contact algorithms. One reason for the inefficiency of those algorithms is the factorization of the matrix $\left[(\nabla\hat{g})^T\mathbf{M}^{-1}\nabla\hat{g}\right]$ in equation (3.151) when multiple contacts are active. Although the matrix itself is very sparse, its inverse may be dense. Furthermore, its dimension is the number of nodes being in contact. The computation of the projection is, therefore, computationally expensive. The contact gradients $\nabla\hat{g}$ change with time and, therefore, the factorization must be repeated at each time step. Furthermore, the matrix may be singular if spatial discretizations equivalent to the two-pass node-to-segment integration are used.

The sequential procedure, on the other hand, may be less accurate in regions with complex geometries. Using a smaller time step may, however, improve the accuracy. A sequential response is equivalent to the collision procedure presented in [58] without computing the actual collision times $\alpha_c$, section 3.8.5.

A synchronous sequential procedure may be more efficient than a completely asynchronous collision response. This is because the data structures used in global collision detection must be updated only at synchronous times. The number of these times is much smaller than in the asynchronous setting, but then the complete structure is affected instead of a small spatial region.

**Asynchronous adaptive time step selection**

Asynchronous integration targets at problems with spatially varying mesh densities. Then there may exist regions on the contact boundaries with very fine meshing and other domains with very coarse meshing. In such cases the time step of a synchronous collision response is tied to the smallest mesh size on the boundary. A time step too large may invoke interpenetrations which are not detected. Situation can be improved by assigning smaller time steps to surface patches of smaller size. Furthermore, the time step sizes can be arbitrarily chosen. There exists no critical time step to the collision response while the only restriction is given by the accuracy of the contact detection.

Define a representative quantity for the "length" $l_A$ of a finite element node $A$, e.g.

$$l_A = \left(\frac{m_A}{\rho_A}\right)^{\frac{1}{3}} \tag{7.92}$$

## 7.5. Examples

with nodal mass $m_A$ and nodal mass density $\rho_A$. Then every single collision response may be associated to an individual kick event $i$ with kick time $t_i^j$ which is kept in the priority queue of the asynchronous integrator. After each response, the next kick time is computed

$$t_i^{j+1} = t_i^j + \Delta t_i^j \tag{7.93}$$

and the kick is reinserted into the priority queue. The time step is

$$\Delta t_i^j = \min\left(\Delta t_i^{\max}, \frac{\alpha_C l_i}{\|\mathbf{v}_i^+\|}\right) \tag{7.94}$$

such that the time step length depends on the size of the node and the current absolute nodal velocity.

Ideally, a quantity for the relative velocity to close contact candidates should be chosen, but this is not available in most cases. Using a two-pass strategy, however, a conservative estimation can be obtained: If two nodes with distance $\Delta x$ approach each other with velocities $v_A$ and $v_B$, then the collision time will be at least $\Delta t_C \geq 0.5\Delta x / \max(v_A, v_B)$. It is sufficient, when the faster node handles the collision detection. In a two-pass strategy, where both nodes are associated to collision detection times, it is irrelevant which node has the faster velocity.

When choosing the constant $\alpha_C$ one must ensure that the next collision detection must take place, before the considered node penetrates the target too deep (or before the target penetrates the contactor too deep when taking the characteristic length of the contactor side).

## 7.5 Examples

### 7.5.1 Two elastic bars

This example is excerpted from [36, 72]. A longitudinal impact of two elastic bars is considered, see figure 7.16. The geometry of each bar is given by $L = 10$, $H = B = 1$. The material is linear elastic with Young's modulus $E = 1$, Poisson's ratio $\nu = 0$, mass density $\rho = 1$. The impact is full elastic. Before the collision, the velocities are $v^{(1)} = -v^{(2)} = 0.1$. The bar tips should remain in contact for $t < 20$.

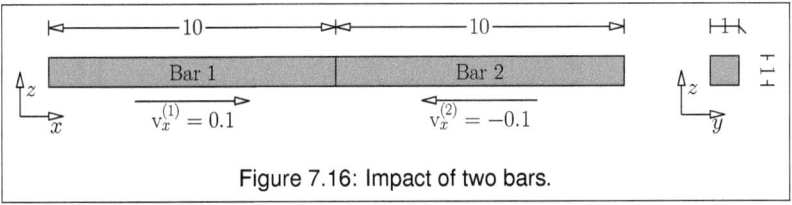

Figure 7.16: Impact of two bars.

The asynchronous integrator is used with a time step ratio $\beta = 0.5$ related to the critical time step. The step size parameter for the collision kicks is given by $\alpha_C = 0.75$ with maximum collision step time being the average critical time step. As a reference solution serves Velocity Verlet with midpoint collision, see section 3.9.8 page 88. The time series history variables are determined for both at certain points at time. The total simulation time is $T = 50$; the number of save intervals is 5000. When measuring the required cpu time, the evaluation of history variables is turned off, because it may affect the performance.

For both integrators the same contact detection algorithms are applied. For Velocity Verlet the sequential response, see section 7.4.5 page 204, is used for a fair comparison. A general contact methodology is applied, that means all bounding faces (and the element layer underneath) are subject the collision detection and distance field computation. Although the normal vectors at the corners are not accurate due to averaging, the direction of the response is computed nearly accurately: The motion of the two bars is parallel to the longitudinal axis.

The mesh of the first bar consists of $5n \times n \times n$ 8-noded brick elements C3D_8N_27C, see section 4.8.5. The mesh size parameter $n$ controls the number of elements per side. The element sizes along the $y$ and $z$ direction are uniform. Along the $x$ axis the position of nodes is chosen to be

$$x_i = \frac{L}{n}\frac{i}{5n} + \left(L - \frac{L}{n}\right)\left(\frac{i}{5n}\right)^2, \quad 0 \leq i \leq n \tag{7.95}$$

That means, the smallest elements are located at the contact interface. Along the $x$ axis, the element size grows linearly. In order to enforce a nonconforming mapping at the contact interface, the right bar is discretized by $5(n+1) \times (n+1) \times (n+1)$ elements with equivalent node positions, see figure 7.17.

The tip displacements for mesh size parameter $n = 4$ are illustrated in figure 7.18. The used time steps are $h = 0.0174$ for Verlet and for AVI $h_{min} = 0.0129$, $h_{max} = 0.0485$, $h_{average} = 0.0264$. The tip velocities are presented in figure 7.19. In order to get single quantities for the tip nodes, the values of all nodes located at the tip are averaged. The displacements are in good agreement. There exist spurious oscillations in the velocities during the persistent impact phase. Such were also reported by others applying explicit collisions [36]. They result from the explicit representation of the discrete potential action $V_d$. The oscillations can be reduced by decreasing the time step length and by using another finite element formulation. For example, taking standard isoparametric elements C3D_8N instead of C3D_8N_27C nearly doubles the magnitude of the oscillations. The spurious oscillations are greater if the collisions are applied asynchronously.

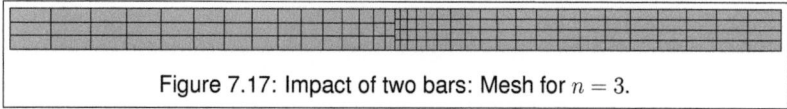

Figure 7.17: Impact of two bars: Mesh for $n = 3$.

## 7.5. Examples

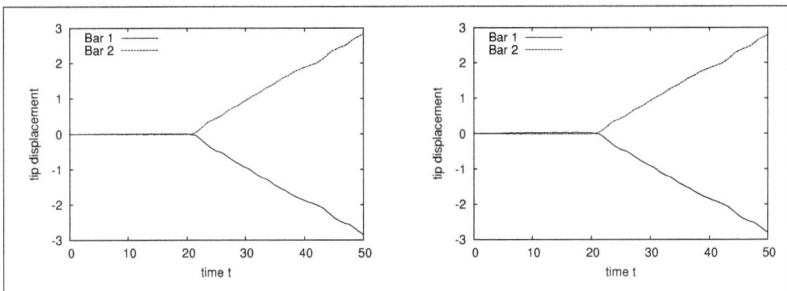

Figure 7.18: Impact of two bars: Tip displacements over time. Left: Verlet. Right: AVI. $n = 4$

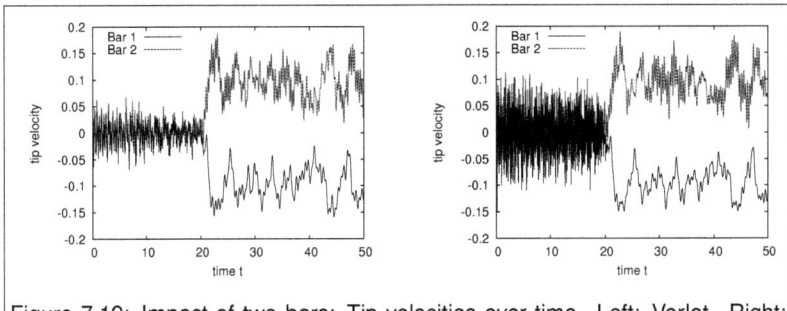

Figure 7.19: Impact of two bars: Tip velocities over time. Left: Verlet. Right: AVI. $n = 4$

The energy balance is presented in figure 7.20. Both algorithms nearly preserve the total energy. There is a very small energy drift in the asynchronous integrator, however, which seems to be subject to the asynchronicity of the strain energy evaluations as noted in section 5.6.

The cpu times are presented in figure 7.21. It illustrates the total cpu time and the time which was spent by the contact algorithm. The latter is measured

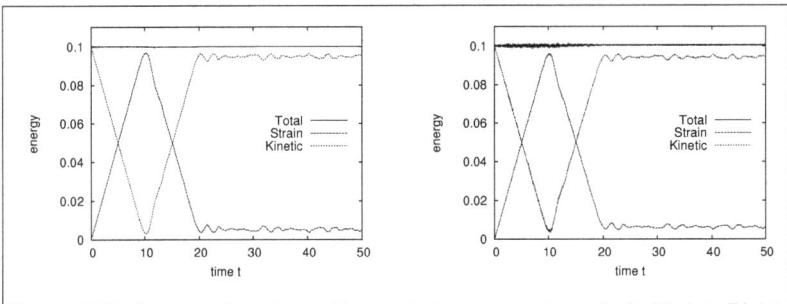

Figure 7.20: Impact of two bars: Energy balance over time. Left: Verlet. Right: AVI. $n = 4$

as the difference of the total cpu times of two simulations, one with and one without contact. Obviously, the numerical cost grows significantly slower in the asynchronous case.

### 7.5.2 Elastic block sliding on rigid obstacle

This example serves to test Coulomb friction and nodal restraints. A block with dimensions $1 \times 1 \times 1$ is sliding on a rigid plane subject to gravitation. The block is discretized by $3 \times 3 \times 3$ C3D_8N_27C elements. The obstacle is discretized by $5 \times 5 \times 1$ elements and has the dimension $5 \times 2 \times 0.2$. All nodes of the basement are fixed through nodal restraint conditions. The material is linear elastic with Young's modulus $E = 100$, Poisson's ratio $\nu = 0$, mass density $\rho = 1$. The body is subject to a constant vertical body force $F = 1$. The initial horizontal velocity is $v_x = 1$. The Coulomb parameter of friction is $\mu = 0.5$. The total simulation time is $T = 3$. The time step ratio related to the critical time step is $\beta = 0.5$. The step size parameter for the collision kicks is given by $\alpha_C = 0.5$ with maximum collision step time being the average time step

Figure 7.22 illustrates the geometry at the beginning and at the end of the simulation. Figure 7.23 presents the horizontal displacements and velocities at the block's bottom. A single value of the displacements is obtained by averaging the nodal values at the bottom surface. The results are in good agreement with the analytical solution of a rigid block: The displacements describe a parabola with end displacement $u_x = 1$ while the velocity decreases linearly until time $t = 2$. Figure 7.24 presents the energy balance. The energy dissipated by the friction grows until almost no energy is left in the system. The total energy is nearly preserved by the algorithm.

The material parameters are changed to $E = 10$ and $\nu = 0.2$. Due to the reduced stiffness, the block starts to roll on the interface. Some configurations are presented in figure 7.25. The energy balance is shown in figure 7.26.

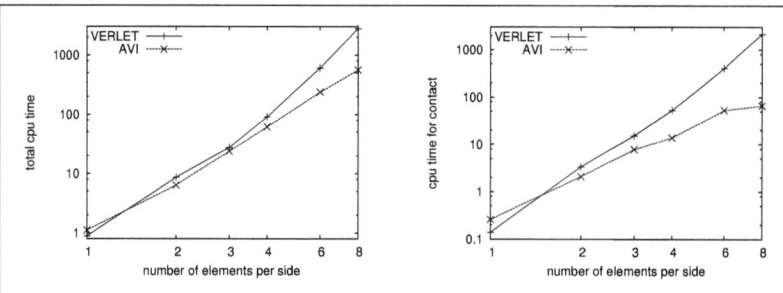

Figure 7.21: Impact of two bars: CPU time for different mesh sizes. Left: total cpu time. Right: cpu time due to contact algorithm.

## 7.5. Examples

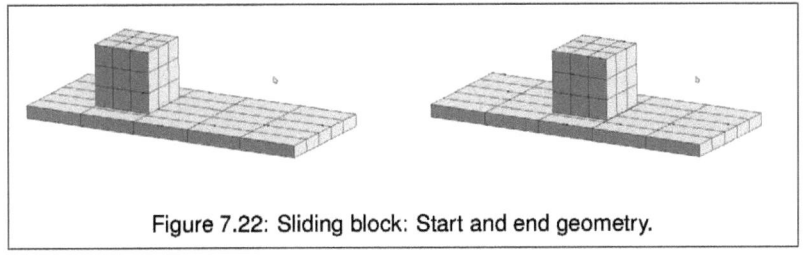

Figure 7.22: Sliding block: Start and end geometry.

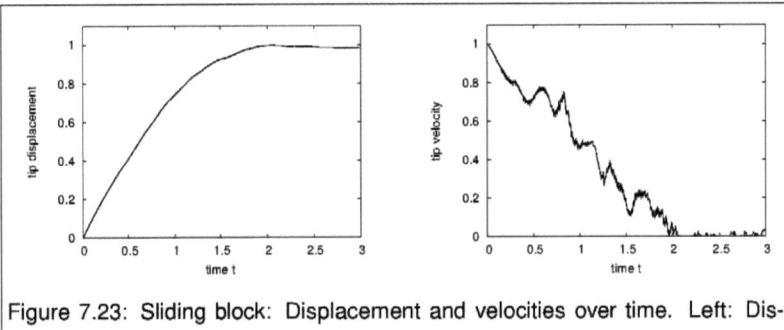

Figure 7.23: Sliding block: Displacement and velocities over time. Left: Displacement. Right: Velocity.

### 7.5.3 Block assembly

This example illustrates the asynchronous collision procedure applied to a rather complex problem. An assembly of 18 cubes is hit by another moving cube. The geometry of the initial frame is presented in figure 7.27. All cubes are of dimension $1 \times 1 \times 1$. The hitting cube is rotated around its center by the axis $(1/\sqrt{2}, -1/\sqrt{(2)}, 0)$ and the angle $\pi/4$. Its center is defined by $(-0.5, -0.5, 2)$ whereby the origin $(0, 0, 0)$ is defined in one of the bottom corners of the block assembly. Every cube is discretized by either first order tetrahedra or first order hexahedra on a regular grid, each cube with different element sizes. C3D_4N_1I and C3D_8N_27C elements are used. The material is linear elastic with Young's

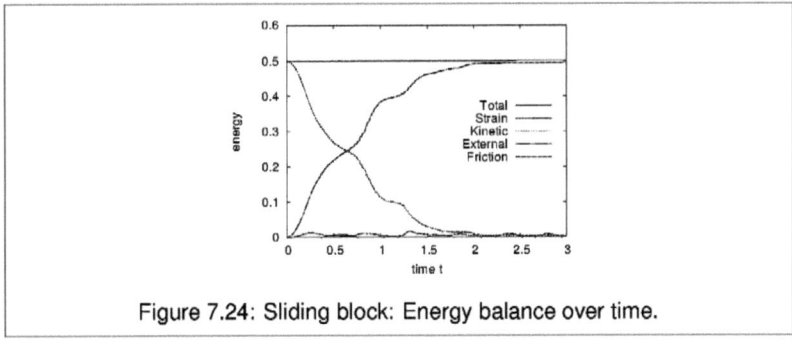

Figure 7.24: Sliding block: Energy balance over time.

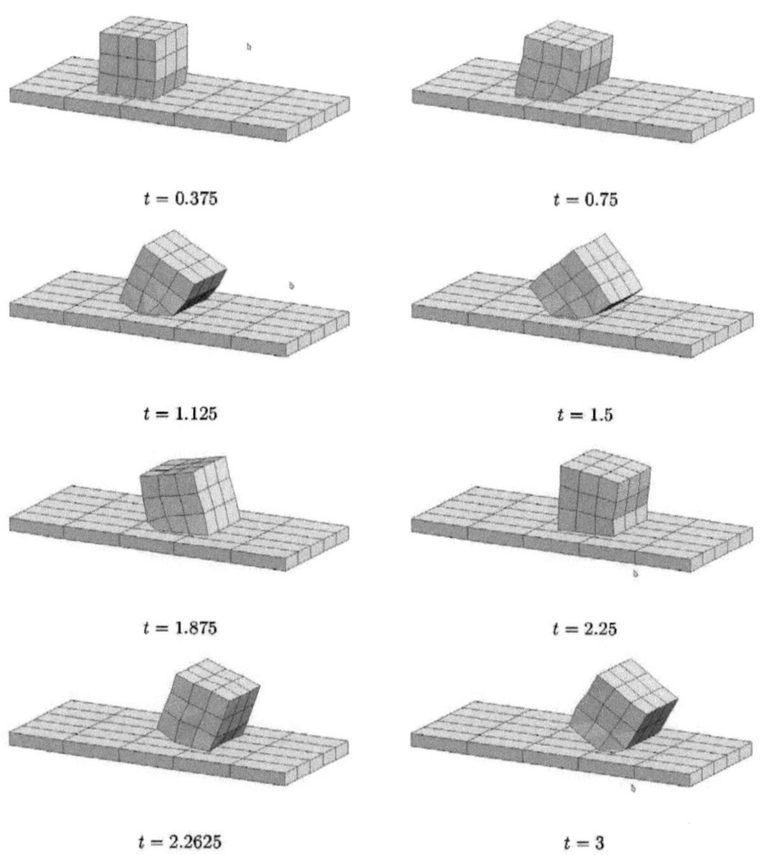

# Chapter 7. Collision dynamics

| | |
|---|---|
| $t = 0.375$ | $t = 0.75$ |
| $t = 1.125$ | $t = 1.5$ |
| $t = 1.875$ | $t = 2.25$ |
| $t = 2.2625$ | $t = 3$ |

Figure 7.25: Soft sliding block. Geometry at different times.

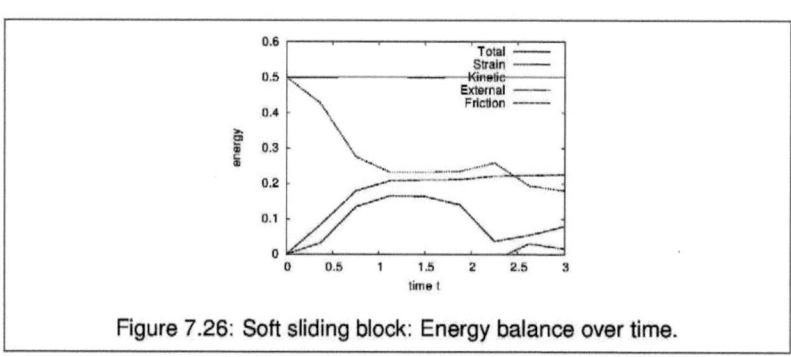

Figure 7.26: Soft sliding block: Energy balance over time.

## 7.5. Examples

modulus $E = 1000$, Poisson's ratio $\nu = 0.3$ and mass density $\rho = 1$. The initial velocity of the hitting cube is 2 in $x$ and $y$ direction.

The total simulation time is $T = 4$. The time step ratio related to the critical time step is $\beta = 0.5$. The step size parameter for the collision kicks is given by $\alpha_C = 0.25$ with maximum collision step time being twice the average time step. The minimal critical time step in the system was identified as $h_{crit}^{min} = 0.476 \times 10^{-3}$, the maximum critical time step as $h_{crit}^{max} = 2.469 \times 10^{-3}$ and the average being $h_{crit}^{average} = 0.957 \times 10^{-3}$. Figure 7.27 illustrates the energy over time. The total energy error does not exceed 1%. Figure 7.28 presents the geometry at various times during the simulation.

When analyzing the geometries at discrete times, small interpenetrations can be observed. These have three reasons:

- The **spatial density** of the integration points in the contact search is too small. During the first impact between the hitting cube and the block assembly, one cube is strongly deformed at its corner, the other in the center of its face. Some interpenetration can not be detected, because subsequent collisions between contactor edges and target elements are not found by the contact search. To improve the accuracy, one needs to add more integration points to the contactor surface or apply additional search strategies for edges and faces, see for example [36].

- **Cumulative effects**. The algorithm tries to prevent collisions by changing the velocity. The momentum change is applied at times, where already a collision was detected. Even if a correct velocity change was computed, subsequent events may increase the interpenetration until the next collision reponse takes place. A predictor-corrector algorithm could improve this, but is not effcient in explicit analysis, in particular in asynchronous simulation. In [36] the velocity change was combined with a nonsymplectic coordinate change which tries to eliminate existing penetrations. Such a strategy is, however, more complex in asynchonous integration.

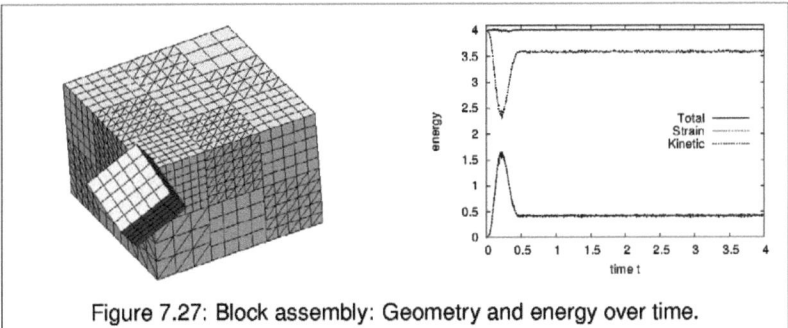

Figure 7.27: Block assembly: Geometry and energy over time.

- **Too large collision time steps.** One reason for the superiority of asynchronous collisions with respect to cpu time is that less contact detections take place. In this example, 24884 spatial integration points for the strain energy, and 3752 spatial integration points for the contact integral are used. During the simulation, $238,944,839$ strain energy kicks and $7,837,942$ collision kicks were performed. This is equivalent to average time steps of $0.416 \times 10^{-3}$ for the strain energy and $1.915 \times 10^{-3}$ for the collision detection.

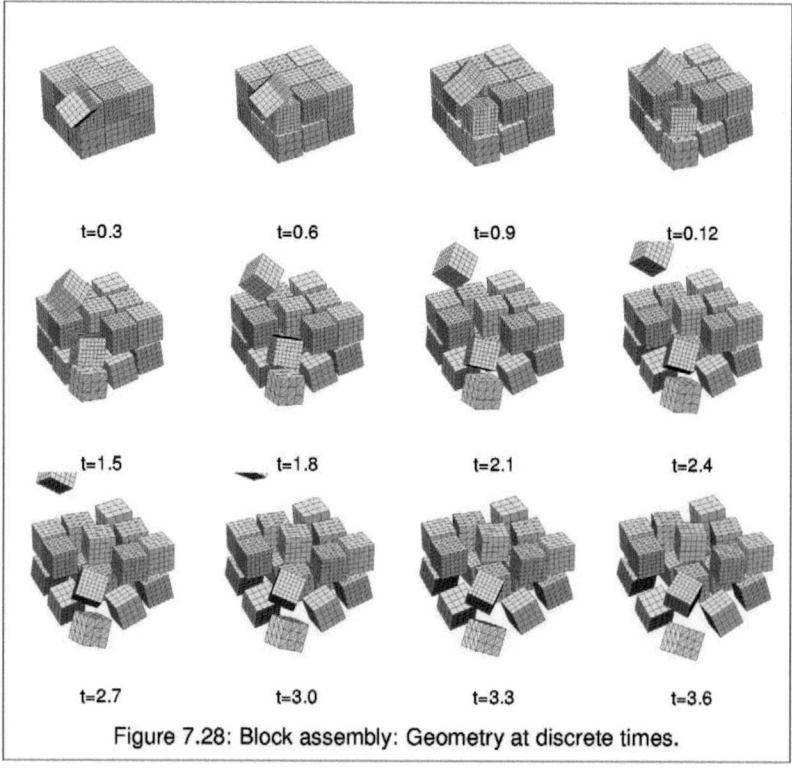

Figure 7.28: Block assembly: Geometry at discrete times.

# Chapter 8

# Summary

## 8.1 Longer time steps in explicit dynamics

This thesis considered variational integrators. Variational integrators can be derived by a strict procedure and automatically inherit some properties of continuous conservative systems, i.e. symplecticity and preservation of momentum, which make them favorable to other time stepping schemes.

Aside standard integrators such as symplectic Euler, Verlet, Newmark and Midpoint, the variational principle allows the derivation of schemes which make use of special properties of the considered mechanical system. For example, if the response of a system is dominated by linear forces, one can apply various types of multiscale integrators. The thesis studied some variational integrators which make use of a dominant linear response, such as r-RESPA, mollified impulse, exponential integration and implicit-explicit integration, see section 3.7. These schemes are non-iterative and increase the accuracy (and, therefore, the stability) regarding the linear parts of the response. Using model problems, various methods were tested on stability regarding an increasing time step size, see sections 3.9.1 to 3.9.3.

Processed and composition methods increase the order of a numerical integrator and, thus, usually the critical time step as well. If one is interested in numerical efficiency, however, they may not be the first choice. Composition methods increase the number of force evaluations (at least by the factor three), but do not extend the critical time step by the same factor. Rowlands's method requires the computation of the tangential stiffness matrix at every discrete point in time. In structural dynamics, this assembly may impair the numerical efficiency even more than the additional matrix-vector products.

Multiple time stepping was not used in its optimal field of application in this research. It has been presented to show the difficulties finding regions without resonances. In elasticity, multiple time stepping is generally not used to superimpose a linear system and a nonlinear perturbation as done in this article. It has been successfully applied to improve numerical efficiency when different time steps are used for different spatial regions.

Exponential integrators target at improving the accuracy when the linear forces are dominant. They require the computation of a matrix exponential. This operation limits their application to problems of small to moderate size. Otherwise, approximations to the matrix exponential must be used. A certain approximation is given by an implicit treatment of the linear forces. Although it is less accurate in treating the linear forces, it was shown to be more robust with respect to perturbations.

By application of the mollified impulse method the stability of IMEX integrators can be improved. This is achieved through replacing the original system by a model which filters the nonlinear forces. The improvement of stability happens on cost of accuracy. The 'nonlinear forces' are defined as a perturbation to an assumed dominant linear part. As long as the perturbations are small, the integrator is sufficiently accurate and stable. With growing perturbation the filtered model becomes less accurate and less stable.

For small and moderate nonlinearities the solution of the modified system was shown to be similar to the reference solution of the original system. There was even no stability limit to be observed. For larger nonlinearities the modified system represented a very inaccurate approximation to the original one. Then the critical time step was still significantly larger compared with Verlet's method. The magnitude of nonlinearity depends on the initial conditions which, therefore, have a huge impact on stability. This is a very interesting result since stability of explicit integrators is usually tied to the largest natural frequency regardless of the magnitude of excitation.

The mollified impulse method filters, however, parts of the nonlinear forces even in case of very small time steps. As a result, the accuracy may be unacceptably bad even if a time step size being smaller than the critical time step of Verlet is used. The mollified IMEX method should, therefore, be chosen carefully. It targets at approximate solutions using very large time steps without iteration. The accuracy of the approximation is reasonable for problems with dominant linear response only. Since it requires the inversion of a linear combination of the mass and the stiffness matrix, its numerical efficiency may be reduced for problems with many degrees of freedom. Possible candidate problems are those from model order reduction which usually satisfy both constraints, i.e. a small number of unknowns and dominant linear part. More studies are required to analyse the accuracy and stability for various applications.

## 8.2 Time discretization of nonlinear constraints

The treatment of linear and nonlinear holonomic constraints in conjunction with collisions must be included in the considerations on an overall concept of the formulation of a mechanical integrator. Lagrange multiplier methods are considered as solution methods which can be easily transformed into penalty and Augmented Lagrange regularizations. The thesis presented the well known methods SHAKE for holonomic constraints in a Lagrangian framework, RATTLE for holonomic constraints in a Hamiltonian framework and DCR for explicit treatment of unilateral constraints in a Hamiltonian framework.

Examples discuss the treatment of nonlinear holonomic constraints by alternative methods. Based on the discretization of SHAKE, a constraint enforcement using the implicit midpoint method is studied in section 3.9.5. It is identified being unstable. An illustrative explanation for the instabilities was found, i.e. enforcing the constraints at the time step's midpoint can generally not avoid oscillations at the time step end points leading to an hourglass-like interpolation of the discplacements. The implicit midpoint method is related to the trapezoidal Newmark. By transforming the midpoint scheme into Newmark's equations the instabilities can be identified as spurious oscillations in the velocities in Newmark's scheme. Such oscillations were reported by other authors [48, 125].

Stable time discretization schemes of nonlinear constraints in a Hamiltonian framework are usually based on RATTLE [3, 180]. RATTLE leads, however, to unsymmetric systems of equations. Given a too large time step, the missing symmetry may lead to convergence problems in iterative solvers, see sections 3.9.6 and 3.9.7. It also requires the solution of two sets of constraints: the displacement based constraints and the constraint rates, i.e. hidden velocity constraints. Furthermore, symmetric systems of equations are generally faster to solve. This is why the direct application of Newmark and Midpoint is tempting. A simple strategy is proposed which tries to reduce the oscillations in the midpoint scheme and which may be applied to Newmark as well. The constraint rates are enforced using an additional projection which is essentially identical with DCR. The resulting scheme is symplectic, momentum preserving and symmetric.

After the successful application of DCR to stabilize the implicit midpoint scheme, the procedure is further simplified by application of DCR as a standalone procedure to the treatment of constraints. The system of equations is symmetric and explicit, the method energy preserving and symplectic. The method should be used, however, with care. It does not enforce the feasibility of the constraints, but only the constraint rates. Only as long as the time step is reasonable small, the violation of the constraints remains acceptable. Furthermore, DCR is usually applied in conjunction with explicit time discretizations of the potential energy. In combination with an implicit treatment of potential forces one may need to apply the Chawla algorithm [142] instead of DCR.

DCR was applied to a simple contact problem in section 3.9.8. The violation of the impenetrability constraints was further reduced by assigning the projection to the time step's midpoint in conjunction with Velocity Verlet instead of a projection between two time steps. The same example is used to study the stability of DCR compared with the penalty method. The resulting algorithm is the basis of asynchronous collision response introduced in section 7.4.

## 8.3 Continuous assumed gradient method

In recent years various assumed gradient methods were developed improving the accuracy of standard finite elements. A special class among these methods is based on interpolations of assumed deformation gradients which are continuous along finite element interfaces. Various schemes were developed in this research starting from nodal integration. Nodal integration was interpreted by a continuous strain field with finite element nodes as support points of the interpolation. Then the instabilities were analyzed and some penalty based approaches were presented which stabilize the solution. Stable interpolation schemes were developed and their behavior of these schemes with respect to accuracy, convergence and efficiency was compared with other assumed gradient methods and standard finite elements in static and dynamic problems.

Assumed gradient fields have shown the ability to improve the accuracy of isoparametric finite elements without adding any additional degrees of freedom. It is possible to design efficient assumed gradient interpolation schemes which are stable without the need of a stabilizing penalty energy function. Basically all schemes that are able to capture all local nonzero energy deformation modes are possible. These schemes always are of higher polynomial order than the underlying finite element shape functions. The interpolation is not necessarily of tensor-product structure. Partial higher order polynomials may be sufficient as being used in the bubble-supported and edge-supported elements.

In this research, the following new numerical schemes were provided for three-dimensional finite elements: (1) Support of the assumed gradient field with points in the nodes and in the element interior, (2) support with points on the edges only and (3) interpolation with higher order complete-tensorial support. All of them were shown to improve accuracy compared with NICE/NS-FEM and FS-FEM.

The advance in accuracy mainly results from the assumed continuity of strains along element interfaces and, therefore, the assumed smoothness of the finite elements. The major accuracy improvements are observed in the interior while the representation of strains at a body's boundary is not much improved. The overall increased accuracy is the reason why the new method behaves better in nearly incompressible problems. It does not, however, eliminate volumetric locking. Additional strategies to avoid incompressible locking are recommended.

## 8.3. Continuous assumed gradient method

When extending the comparison to problems of dynamic analysis, questions which usually arise are: (1) What are the limitations of the schemes compared with others, in particular stability? (2) How large is the benefit in accuracy? (3) Does the additional numerical effort (when compared with standard elements) pay out the improved accuracy? (4) How are the new schemes applicable to complex problems involving boundary conditions, domain decomposition, contact/impact or arbitrarily complex meshes with different element types?

All of the CAG elements were shown to improve accuracy in structural dynamics - except nodal integration (without stabilization) which does poorly converge in modal analysis (with applications in system identification, model order reduction, etc.) and exhibits a bad accuracy in examples of time integration. When applied to contact problems, the amount of spurious oscillations when using explicit collision integrators is reduced compared with isoparametric FEM.

The stable CAG methods require additional numerical effort when compared with standard elements. This is because they use a slightly greater number of integration points and lead to a greater number of nonzero elements in the stiffness matrix. The actual difference in number of integration points related to standard FEM is dependent on the mesh topology. The more averagings can take place at element interfaces, the less the number of integration points and the greater is the accuracy. The beam example is a structure which is unfavorable for CAG methods. Bold structures (such as blocks) lead to less integration points.

Nevertheless, surprisingly well results were found regarding efficiency and accuracy. Although the tetrahedron with nodal-interior support C3D_4N_1I contains almost twice as many integration points, the computational time is reduced in explicit codes when compared with standard FEM. At the same time the accuracy is improved. This is because the maximum frequency is reduced and a larger time step can be used in time integration. Previously developed assumed gradient methods, i.e. FS-FEM, did not lead to such significant improvements. Situation is different for hexahedral elements: All stable CAG methods require more integration points and computation time. The maximum frequency does not differ much compared with standard FEM and does not balance the numerical effort of a single restoring force evaluation. The hexahedral CAG element which improves accuracy at most is the C3D_8N_27C element with complete higher order support.

An important observation is made by comparison of the displacement solution of the C3D_8N_27C and C3D_4N_1I elements with their quadratic standard FEM counterparts. Using the same number of nodes, both solutions are equivalent. One could replace the quadratic elements by first-order CAG elements without losing accuracy. On the other hand one has more flexibility in mesh generation. Boundary conditions arising from contact/impact and mortar methods are easier to implement.

A problem when using both element types in the same mesh is meeting the regularity conditions. The advantages of continuous strains can only be used if an averaging can actually take place at finite element interfaces. While the C3D_4N_1I element has support points in the nodes (and interior), the C3D_8N_27C has support in the nodes, in the interior, and on faces and edges. The latter two can not be averaged when connected to C3D_4N_1I elements. Therefore, accuracy is reduced unless transition elements are developed.

The presented schemes are similar to those of SFEM/SPIM/NICE, but there exist a few advantages:

- The presented approach separates the deformation field from the numerical integration scheme. This is one of the reasons why nodal integration is not well understood.

- Only by assuming a continuous strain field, it can be intuitively explained why certain interpolation schemes exhibit spurious modes.

- Most SPIM/SFEM methods are developed in two dimensions. At the moment, three-dimensional implementations are limited to linear tetrahedrons and only a few smoothing cell schemes. One reason could be that the creation of smoothing cells is more complex in three dimensions than in two. In the presented approach, no smoothing cells are required.

- The illustrated method intuitively extends to arbitrary finite element types by using the same methodology as for 1st order elements.

- Heterogeneous meshes with different finite element types can be used. Regularity conditions were explained. It was discussed which effects appear if regularity conditions are not satisfied. The FS-FEM was shown not to be an ideal candidate to meet the regularity conditions in heterogenious meshes.

**Benchmark of higher order elements** The modal analysis of the 2nd order tetrahedron leads to surprisingly good results for the nodally integrated element with bubble stabilization. It would be interesting to test the performance of 2nd order elements and various assumed gradient schemes in detail.

**Incompressible locking** The method should be accompanied with strategies to avoid volumetric locking. A possible candidate was implemented in FS-FEM [203] where the volumetric strain was nodally interpolated. F-bar methods being based on nodally supported Jacobians [46] also naturally fit into the presented framework. It would be interesting to test if a nodally supported volumetric strain field exhibits the same instabilities as the plain assumed gradient field which leads to the spurious modes in nodal integration.

## 8.4. Asynchronous variational integration

**Stability** Although an illustrative explanation for the appearance of spurious modes was given in this article, a theoretical justification is not yet available. The only available test of a formulation on temporal stability is modal analysis. Examples have shown, however, that one must check various types of geometry and one must test even very high eigenmodes. For example, the 1st order nodally integrated hexahedron with one additional support point in its interior appears to be stable in many examples and even in modal analysis. But when applied to the compressible block example, then instabilities appeared.

**Integration weights and support point locations** It would be interesting if it is possible to derive optimal integration weights. The same holds for the position of the support points of the gradient field. One constraint is stability which must not be destroyed when 'tuning' the weights of those integration points which are responsible for the stabilization. Such a strategy could be seen in the $\alpha$-FEM/PIM approaches which superpose an instable and a stable strain field and try to find optimal weight factors for both. The weights are chosen to obtain an overall error in energy of zero. In the opinion of the author, this strategy may destabilize the solution.

**Stress continuity** The presented approach relies on the idea of improving the quality of kinematic interpolation by assuming continuity of the strains (or the deformation gradient) at finite element interfaces. This assumption holds in homogenious media. An even more natural choice would be to assume continuity of stresses.

## 8.4 Asynchronous variational integration

Asynchronous variational integrators were presented. The treatment is initially independent from the physical problem, i.e. it may be applicable to other fields such as molecular dynamics. The subsequent space-time discretization allows the fully asynchronous case with individual time step sizes per integration point, the original AVIs with individual time step sizes per finite element, r-RESPA with integer ratios between the different time step sizes and synchronous time stepping (symplectic Euler or Velocity Verlet).

The asynchronous procedure was extended to treat linear nodal restraint conditions at runtime. This is achieved through filtering the momentum change of a KICK event (due to the restoring force) such that the constraints are satisfied at all times. The projection is a specialisation of RATTLE to AVIs yielding an extraordinary simple scheme. Although an elimination of restrained degrees of freedom prior the simulation would be possible as alternative, the projection based on internal Lagrange multipliers is better suited when combined with collisions and variable time steps.

Of particular interest was the estimation of the critical time steps. When applied to CAG elements, it is not easy to determine the critical time step. In the original AVI method this is done using the CFL condition which uses elemental stiffness and mass matrices. In the CAG method, no equivalent to the elemental stiffness matrix exists due to the continuity of the strain interpolation. Two approaches to the CFL condition were proposed which seem suitable for CAG elements. Examples illustrated that one of them renders the time stepping scheme sufficiently stable and efficient. Problems regarding resonances as observed in r-RESPA or in AVI applied to molecular dynamics did not appear in the numerical tests.

Examples compare the performance of AVI and synchronous Velocity Verlet. In case of irregularly dense meshes there exist great potential of AVI to save computing time. A worst case scenario illustrated the numerical overhead of AVI related to standard Verlet. If all elements are of identical size then no benefit can be found because the individual time step sizes do not vary within the model. Hence, the computing times were much longer.

## 8.5 Variable time steps

In many dynamic explicit codes, the time step may be adopted if it endangers the stability or the accuracy of the solution, for example if a finite element collapses or becomes too stiff to ensure stability. But these corrections may happen only a few times among a total number of a million time steps. When adopting the time step size at every time step, one has to preserve the multisymplectic form. Some correction terms must be added to the time stepping scheme in order to stay symplectic. Depending on the choice of the time step function, this may lead to an iterative or even implicit procedure. Emphasis of this research lies on time stepping schemes leading to explicit, non-iterative equations.

The flow of fixed-step size AVI was interpreted by a sequence of KICK and DRIFT elements, i.e. by events which modify the momentum only and time elements which perform a motion with constant velocity. The time step function must be selected in such a way that the symplectic correction terms do only lead to a velocity update and not to a change of displacements. Hence, the time step function must depend on the displacements only and not on the momentum. In order to allow individual step sizes for each integration point when starting from a synchronous initial condition, every integration point obtained its individual time counter which serves as additional degree of freedom. The symplectic correction leads to a system of equations which is quadratic in the momentum and linear in all other variables. A procedure, which transforms such equations to a scalar

## 8.6. Asynchronous collisions

quadratic form in synchronous variable step size integration, was adopted to the asynchronous case. Various known time step functions were presented.

SDOF examples showed the potential of variable time steps when applied to highly nonlinear problems. They further illustrated accuracy and convergence of individual step size strategies. The application to linear problems, however, did not improve accuracy or efficiency when compared with constant time steps. On the contrary, most symplectic step size methods required the assembly of the tangential Hessian (stiffness matrix in elasticity) rendering the schemes inefficient. Furthermore, the tuning of the parameters of step size functions in order to obtain an efficient scheme turns to be complex and problem dependent. The stability of explicit integrators may be improved, but once the critical time step is passed the integration error becomes inadmissible.

The examples clearly illustrate the need of symplectic updates in favour of adhoc adaptation of the step sizes. Furthermore, the numerical trajectories converged to the correct solution in the presented examples. The long-term instability of SEM is not inherited by variable step size methods based on time transformations.

The problems of SEM integration were illustrated in a small SDOF example. Therein, the appearance of unsolvable equations may render the subsequent time integration inaccurate and even unstable. The problems are less obvious in MDOF systems of elasticity. Therein, a quite large energy error appeared at the beginning of the simulation while the energy is kept nearly constant in subsequent time steps. Nevertheless, the large energy error at the beginning of the simulation is unaccepable.

MDOF examples from elasticity also compare time step functions with and without symplectic correction in the synchronous context and in AVI. The asynchronous schemes turn out to be unstable. Even when applied to synchronous time steps, there is no benefit in accuracy or numerical efficiency when simulating a linear oscillator. Only in the presence of extremely nonlinear effects, variable time steps may stabilize and improve the solution.

Since there are no benefits of using variable time steps in linear elasticity, one should try to improve the treatment of strong nonlinearities in a different manner. For example, instead of using penalty contact forces a discontinuous momentum change may lead to similar accuracy without the need of a smaller time step during the impact.

## 8.6 Asynchronous collisions

One component of the presented contact algorithm are discrete distance fields. An overview on the development of this concept was presented. The existing approaches are based on supplementary rectangular grids on which the distance field is interpolated and on penalty contact forces applied to first order tetrahedral finite elements. The approach using distance fields was extended to arbitrary finite element types and a Lagrange multiplier formulation. No supplementary grid is required. Instead, the distances are interpolated on the same finite element mesh. This formulation requires a stabilization for some geometries. An approach based on assumed distance gradients was proposed to achieve this. Furthermore, assumed gradients lead to some kind of 'edge smoothing' near corners and allow the treatment of general contact problems. No user input is required where the contact pairs are defined before the simulation.

The discrete distance field approximates the closest point projection. Hence, it provides a robust and unique quantity being used as impenetrability constraint. It does not assume any smoothness requirements on the boundary as required by closest point projection. The collision detection is simple to implement and robust. Its accuracy only depends on the density of the integration points in the contact integral. Furthermore, the contact tractions always balance linear and angular momentum. This is not satisfied for procedures based on closest point projection in the presence of friction.

The distance field was used in conjunction with a spatial node-to-element integration scheme in asynchronous time stepping. The idea is that in synchronous contact detection, the smallest surface patch decides on the time step size between two collision detections. In the presence of small and large finite elements in the same mesh, the frequency at which each element is tested on collision can be adopted to the element size. The approach assumes that the collision response does not affect the critical time step which is true for DCR. Furthermore, DCR only changes the momentum and, thus, can be interpreted as a KICK event in the asynchronous procedure. An additional procedure which directly eliminates spurious interpenetrations seems to be very difficult in AVI.

DCR was applied to AVI termed Asynchronous Collision Response. The treatment of collisions includes inelastic impacts, Coulomb friction and the presence of nodal restraints. All of them can be solved by very small systems of equations since only one constraint is treated at one time. Hence, the algebraic solution is very fast compared with synchronous contact where the coupling of degrees of freedom leads to large matrices to be factorized. The asynchronous treatment further eliminates the appearance of overconstrained configurations where the two-pass node-to-element integration generates two equivalent constraints. The asynchronous collision reponse provides three interpretations in the implementation: (1) An adaptation of individual time steps for each boundary node with respect to accuracy (velocity and element sizes). (2) a fixed step size strategy where the individual step sizes are adjusted to the critical time step of adjacent finite elements. (3) synchronous collisions where each contact constraint is processed sequentially.

## 8.6. Asynchronous collisions

Examples verified the accuracy and efficiency of the asynchronous collision response method and the robustness of the discrete distance field. Normal contact and frictional contact were tested. Energy and momentum preservation was excellent. Furthermore, the potential to save computing time turned out to be even greater than in asynchronous integration of the strain energy. This was because the asynchronous procedure allows the contact detection being less frequent than the evaluation of the strain energy.

# Appendix A

# Verification of CAG elements

The meshes of the following examples are generated using a uniform cartesian grid of hexahedron elements which are either scaled or transformed into polar coordinates, depending on the example geometry. Each mesh is described by the number of hexahedral elements per edge, where 'edge' denotes a boundary edge along one of the coordinate directions. For reproduction purposes, the meshes with tetrahedral elements are based on hexahedral background meshes. Figure A.1 illustrates how simplicial meshes are created: Any given brick element can be subdivided into at least five simplices. By chosing 6 simplices a regular mesh can be created.

The nodally integrated C3D_4N_NI is equivalent to the formulations of Bonet [21], NICE [133] and NS-FEM [169]. C3D_8N_NI is equivalent to NS-FEM and NICE in the locations of the support points (and similar in coefficients). The face based tetrahedron C3D_4N_1F is equivalent to the three-dimensional ES/FS-FEM [203]. It is implemented for comparison. An implementation of hexahedra with facial support is not tested due to its limited usability. Further results refer to the F-Bar method based on nodal patches, F-Bar-SN-Q and F-Bar-SN-T [46]. The 10-noded tetrahedron elements are implemented in order to test if the presented strategy stabilizes nodal integration of higher order elements as well. They are used in modal analysis, but not in the benchmark examples.

All elements were implemented straight-forward without any strategies to avoid volumetric locking. When comparing the FS-FEM, then the implementation without separation of volumetric and deviatoric strains is considered in the examples [203].

## A.1 Patch test

The patch test verifies the ability to represent constant strain fields. To do so, a linear deformation field is applied to a structure with irregular element geometries. In figure A.2 a structure consisting of $3 \times 3 \times 3$ 8-noded hexahedron elements is

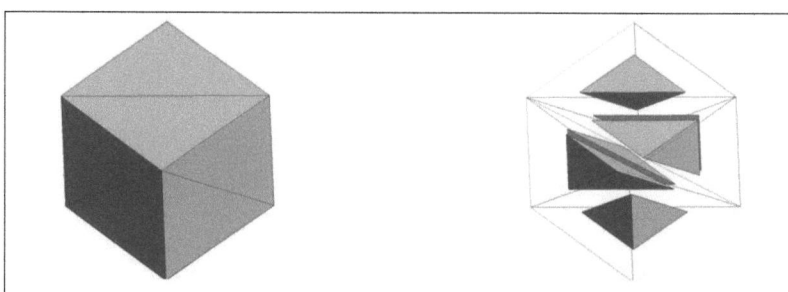

Figure A.1: Mesh generation: Creating 6 simplex elements by subdividing a given hexahedron. Left: surface and boundary edges. Right: subdivided simplex elements, being shrinked around their centers by factor 0.5.

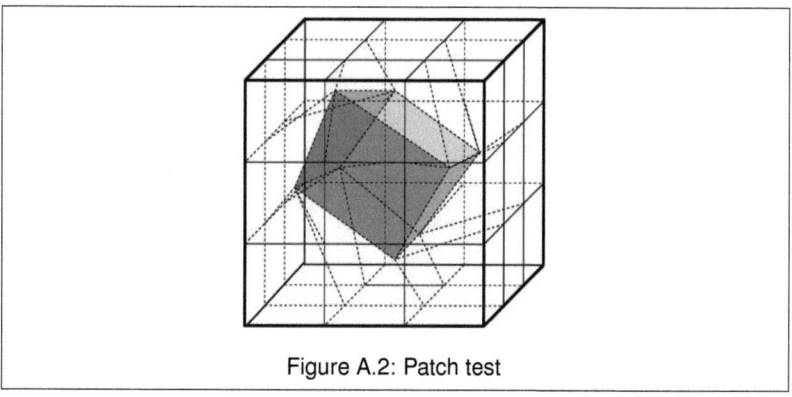

Figure A.2: Patch test

shown, where the nodal coordinates of the center element are randomly chosen. Application of a displacement field given through the nodal displacements

$$u_{A\alpha} = u_{0\alpha} + a_\alpha X_{A\alpha} \tag{A.1}$$

and measurement of the deformation gradients at all integration points proves the patch test. The test may be expanded to other element types by inserting additional nodes (for example for C3D_27N elements) or subdividing the hexahedra into tetrahedral elements.

## A.2 Cantilever beam

This example illustrates the performance of assumed gradient methods when applied to bending in linear elasticity. A cantilever beam as shown in figure A.3 with dimensions $L = 10m$ and $b = h = 1m$ is subject to a vertical parabolic area load on its right edge. The load is of total $P = -100N$. The material is defined by

## A.2. Cantilever beam

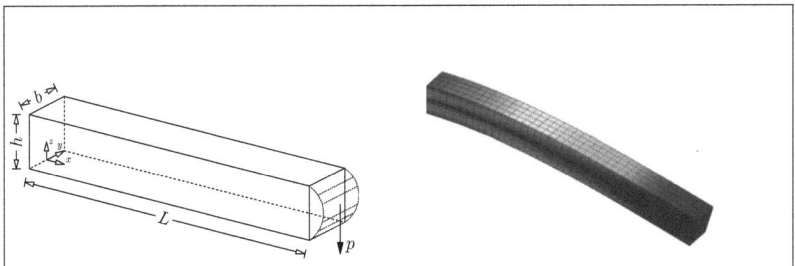

Figure A.3: A cantilever beam. Left: Geometry and loading. Right: von Mises stress field using C3D_8N_27C elements, displacement is scaled by 150.

$E = 30 \times 10^6 kPa$ and $\nu = 0$. From plane stress theory for thin beams one can approximate the reference solutions [257]

$$u_z = \frac{P}{6EI}\left(3\nu z^2(L-x) + (4+5\nu)\frac{h^2 x}{4} + (3L-x)x^2\right) \tag{A.2}$$

$$\sigma_{xz} = \frac{P}{2I}\left(\frac{h^2}{4} - z^2\right) \tag{A.3}$$

where $I$ is the moment of inertia $I = h^3/12$, $u_z$ the vertical displacement of the center axis and $\sigma_{xz}$ the shear stress.

The vertical deflection along the neutral line ($y = z = 0$) and the shear stress along the midplane ($x = L/2$, $y = 0$) are plotted together with the reference solutions in figures A.4 to A.7, individually for 1st order hexahedron and tetrahedron elements. The maximum deflection and shear stress is illustrated using different mesh refinements, i.e. $10e \times e \times e$ where $e$ denotes the number of elements per edge. The shear and deflection distributions are illustrated for $e = 8$.

The shear stress distribution is unsymmetric for the simplex elements due to unsymmetric meshing. When comparing the edge and face supported elements and standard FEM elements, the stress distribution is artificially worsened in the example. The plotted values are nodally averaged values of an assumed interpolation of the stress inside the elements. In particular, the face supported simplex element exhibits nodally averaged stresses being even worse than those of the standard simplex element, although the displacement distribution is reasonable good.

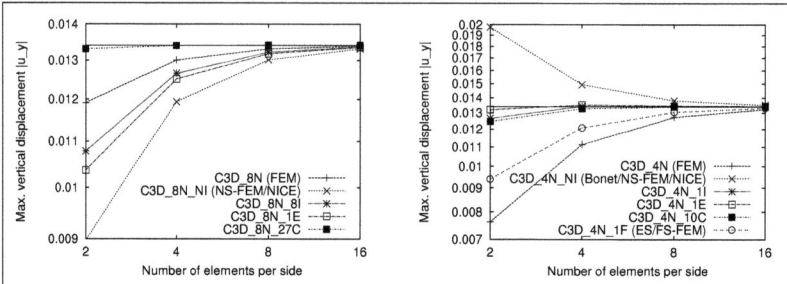

Figure A.4: Maximal displacement of a cantilever beam for different meshes. Left: brick elements. Right: simplex elements.

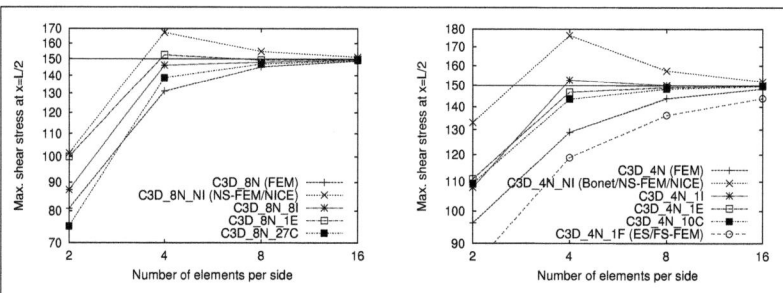

Figure A.5: Shear stress at $x = L/2$, $y = z = 0$ of a cantilever beam for different meshes. Left: brick elements. Right: simplex elements.

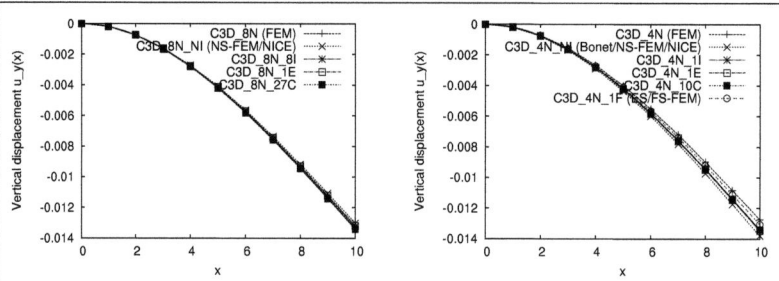

Figure A.6: Vertical displacement distribution of a cantilever beam at $y = z = 0$ for $e = 8$. Left: brick elements. Right: simplex elements.

## A.2. Cantilever beam

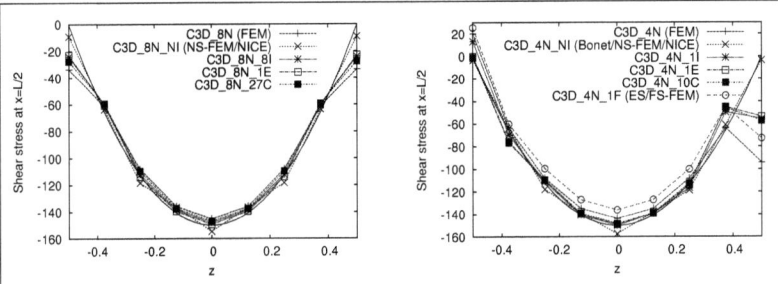

Figure A.7: Nodal shear stresses of a cantilever beam at $x = L/2$, $y = 0$ for $e = 8$. Left: brick elements. Right: simplex elements.

## A.3 Lamé problem

The Lamé problem of linear elasticity is given by a hollow sphere with inner radius $a$ and outer radius $b$ subject to internal pressure $p$, as shown in figure A.8. For this example, the analytical solution is available in polar coordinates (see [257]), i.e.

$$u_r = \frac{Pa^3 r}{E(b^3 - a^3)}\left((1-2\nu) + (1+\nu)\frac{b^3}{2r^3}\right) \tag{A.4}$$

$$\sigma_r = \frac{Pa^3(b^3 - r^3)}{r^3(a^3 - b^3)} \tag{A.5}$$

$$\sigma_\theta = \frac{Pa^3(b^3 + 2r^3)}{2r^3(b^3 - a^3)} \tag{A.6}$$

where $r$ is the radial distance from the centroid of the sphere to the point of interest.

Due to the symmetry, only one eighth is modelled. Symmetry conditions are imposed on the three planes as $u_x = 0$ for all nodes with $x = 0$, $u_y = 0$ if $y = 0$ and $u_z = 0$ if $z = 0$. The model parameters are given by the material $E = 1$, $\nu = 0.25$, geometry $a = 1$, $b = 4$ and the loading $p = 1$. The mesh generator creates equidistant nodes along the radius $r$, the angle $\theta_z$ and the angle $\theta_r$, see figure A.8 (right). The computed nodal displacements and stress along the $y$-axis are plotted in figures A.9 to A.14 together with the reference solution. The stresses and radial displacement at the inner boundary were computed for different mesh refinements, where $1.25 e \times e \times e$ elements were used along each polar coordinate direction. The distribution of displacement and stress within the sphere is given for a mesh parameter $e = 8$.

The solutions of 1st order tetrahedra converge to values which are slightly different from the reference solution. This effect is due to unsymmetric meshing. It is clearly seen that the stresses and, in particular, the displacements oscillate in the interior domain for the instable nodally integrated elements, see figures

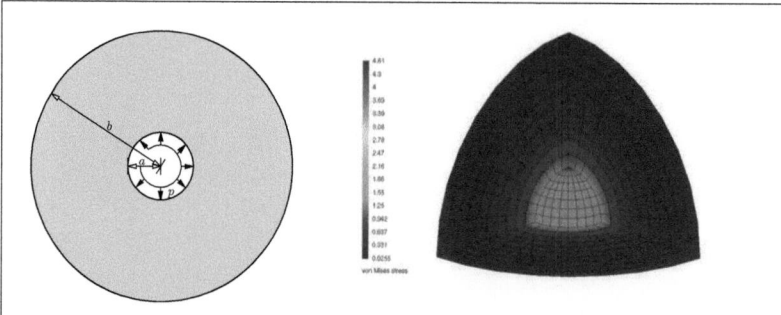

Figure A.8: Lamé problem. Left: Geometry and loading. Right: von Mises stress field using C3D_8N elements ($e = 8$), undeformed.

A.14 and A.12. The boundary stresses are badly approximated by 1st order elements - the accuracy of inner stresses is better. The boundary stresses are badly approximated by 1st order elements - the accuracy of inner stresses is better. Therefore, the accuracy of the displacements given in figure A.9 may be different from the accuracy in stress given in A.10 and A.11. These two figures illustrate the different behaviour of the individual methods to represent boundary stresses.

## A.4 Cook's tapered panel

This example is a commonly used problem, see [26,53], to test combined bending and shear for compressible and nearly incompressible materials. The geometry is presented in figure A.15. A panel is clamped on its left side and subjected to a uniform vertical shear load $F = 100 N/mm$ on its right side. The quantity of interest is the vertical displacement $u_z$ of the top right corner $A$. Its convergence with respect to the number of elements per edge can be used to measure the performance of the used element types. If tetrahedral elements are used, then the number of elements per edge is the number of corresponding hexahedral elements, each subdivided into six tetrahedrons. The constitutive relation is a geometrically nonlinear linear-elastic material with elastic modules $E = 240.565 N/mm^2$. The compressible case is defined by Poisson's ratio $\nu = 0$, the nearly incompressible case by $\nu = 0.4999$. The example is originally a two dimensional plane strain problem. Since the assumed gradient field has been implemented in three dimensions, an equivalent model is defined by constraining the motion along the $y$-axis being perpendicular to the plane. Along direction $y$ only one element is used.

In figure A.16 the displacement $u_z$ is presented for various 1st order hexahedron elements, in figure A.17 for various linear tetrahedrons. All assumed gradient elements show an improved accuracy compared with standard FEM. Notice the remarkably well behaving unstable nodally integrated elements C3D_8N_NI and C3D_4N_NI and the partially stabilized nodally integrated hexahedron C3D_8N_1I

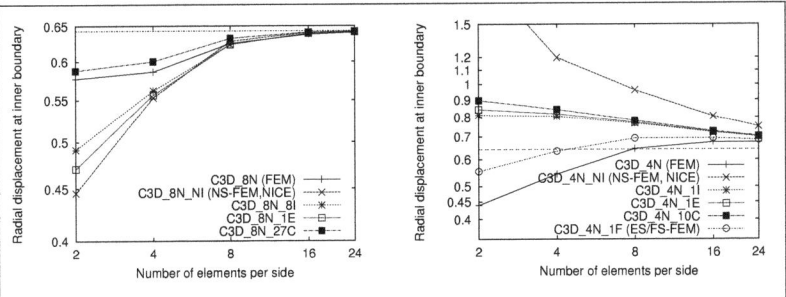

Figure A.9: Inner radial displacement for the Lamé problem for different meshes. Left: brick elements. Right: simplex elements.

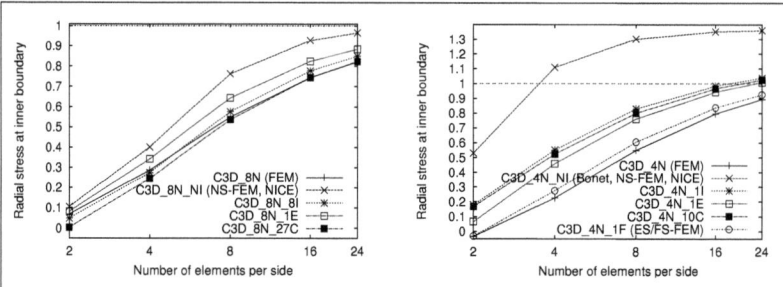

Figure A.10: Inner radial stress for the Lamé problem for different meshes. Left: brick elements. Right: simplex elements.

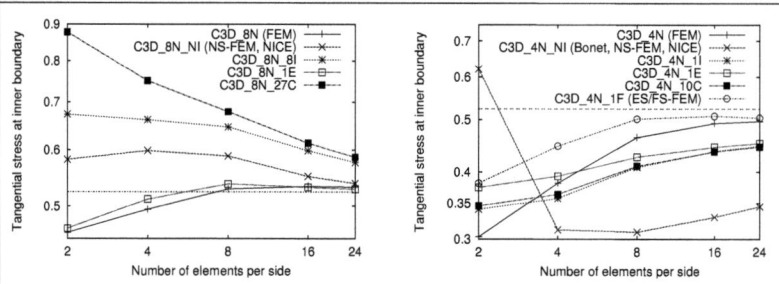

Figure A.11: Inner tangential stress displacement for the Lamé problem for different meshes. Left: brick elements. Right: simplex elements.

with one integration point in the interior. As being noted by several authors [168, 169, 280], the tetrahedron C3D_4N_NI exhibits an overly-soft behavior, i.e. the displacement converges from too large values. Volumetric locking is nearly eliminated in the incompressible case by the nodally integrated elements. The stable assumed gradient hexahedral elements (particularly with edge-based and higher order assumed gradients) reduce the incompressible locking, but do not eliminate it. The stable assumed gradient tetrahedra exhibit strong volumetric

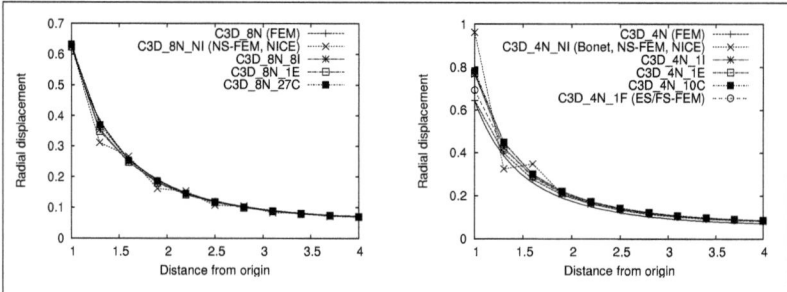

Figure A.12: Radial displacement distribution (nodal values) for the Lamé problem at $e = 8$. Left: brick elements. Right: simplex elements.

## A.4. Cook's tapered panel

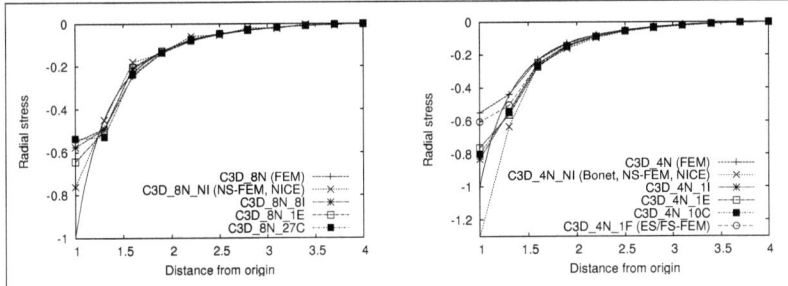

Figure A.13: Radial stress distribution (nodal values) for the Lamé problem at $e = 8$. Left: brick elements. Right: simplex elements.

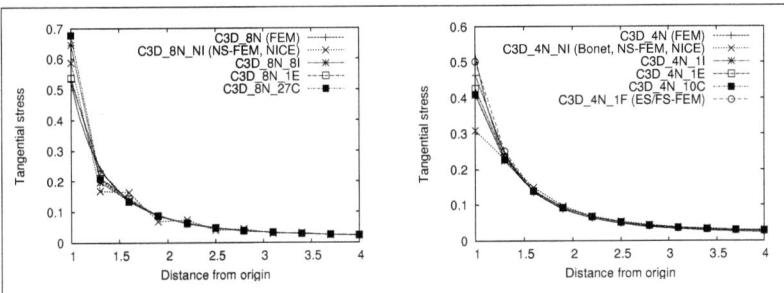

Figure A.14: Tangential stress distribution (nodal values) for the Lamé problem at $e = 8$. Left: brick elements. Right: simplex elements.

locking. The C3D_4N_NI element leads to worse accuracy when compared with the (identical) NICE-T4 element. This is due to different meshes being used in this study and in the article [133] where the values are excerpted from.

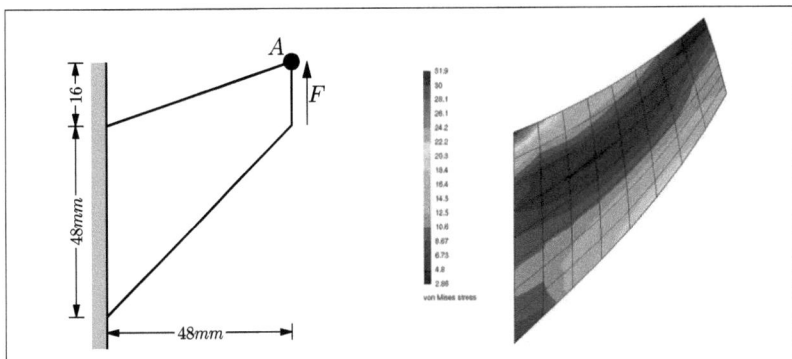

Figure A.15: Cook's panel. Left: Geometry, loading and boundary conditions. Right: $C_0$-continuous von Mises stress field using C3D_8N_27C elements for $\nu = 0$.

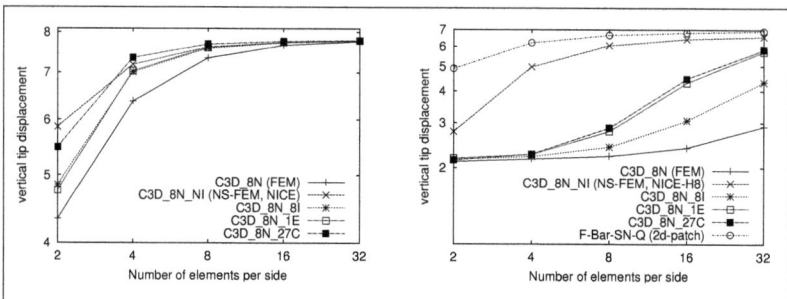

Figure A.16: Cook's panel: vertical tip displacement $u_z$ for 1st order hexahedron elements. Left: $\nu = 0$. Right: $\nu = 0.4999$.

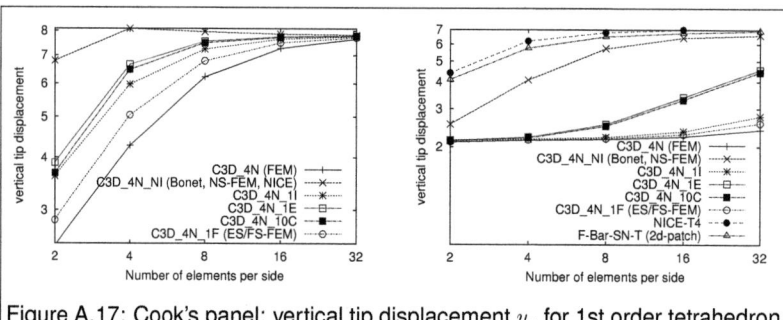

Figure A.17: Cook's panel: vertical tip displacement $u_z$ for 1st order tetrahedron elements. Left: $\nu = 0$. Right: $\nu = 0.4999$.

## A.5 Compressed block

The next example is often used to demonstrate the performance of assumed gradient fields and enhanced elements, see [26, 53, 223] in case of near incompressibility and large strains. The structure is a three-dimensional cuboid block. Due to symmetry along the $x - z$ and $y - z$ planes only one quarter is modeled. The

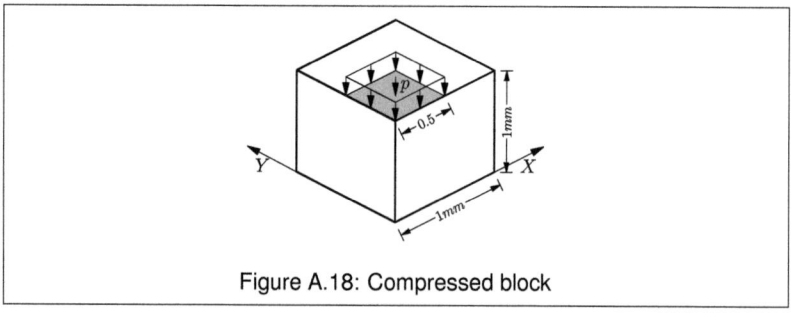

Figure A.18: Compressed block

## A.6. Performance of distorted elements

geometry and loading conditions are given in figure A.18. The material model is a nearly-compressible Neo-Hookean with strain energy density function

$$U^d(J, \mathbf{C}) = \frac{1}{2}\mu \left(\text{tr}(\mathbf{C}) - 3\right) - \mu \ln J + \frac{1}{2}\lambda \left(\ln J\right)^2 \tag{A.7}$$

with $J = \det(\mathbf{F})$ and $\mathbf{C} = \mathbf{F}^T\mathbf{F}$. The parameters are given by $\lambda = 400889.806 N/mm^2$ and $\mu = 80.194 N/mm^2$. Different loading levels $p = \alpha p_0$ may be considered with reference load $p_0 = 4N/mm^2$. Herein, the load factor $\alpha = 80$ was used. The surface load $p$ is considered as "dead load", i.e. it is assumed to act as a constant vertical load specified in the reference configuration. The Dirichlet boundary conditions are $u_z = 0$ for all nodes on the bottom surface and $u_x = u_y = 0$ for all nodes on the top surface (if these conditions would be applied to all nodes on the bottom surface, some iterative solution methods may fail). Due to symmetry, the model is subject to the constraints $u_x = 0$ for all nodes on the $y - z$ plane and $u_y = 0$ for all nodes on the $x - z$ plane. The vertical compression of the upper center point will be studied with respect to the load factors and the number of elements per edge.

The results for the pressure ratio $p/p_0 = 80$ are given in figure A.19. For the selected mesh sizes up to 16x16x16 no convergence was obtained for all tested elements, although the assumed gradient considerably improved accuracy when compared with standard finite elements C3D_8N and C3D_4N. The unstable nodal integrated elements diverged during the Newton iteration. Furthermore, the nodal integrated hexahedron with one interior point, C3D_8N_1I, turned out to be unstable when being applied to this example. Here, spurious deformation modes were activated, see figure A.20. Generally the edge-based and the higher order assumed gradient fields perform better than the nodal-based field with bubble stabilization.

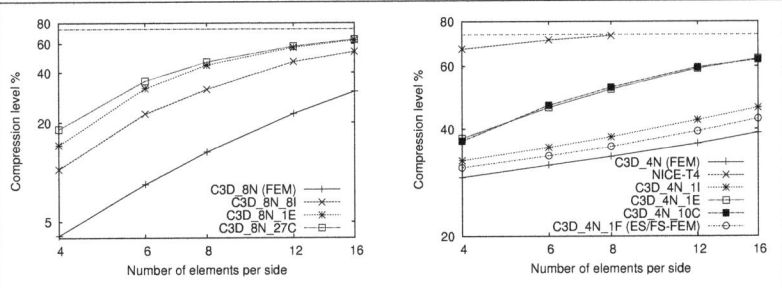

Figure A.19: Compression level in % versus number of elements per side for the three-dimensional block for load level $p/p_0 = 80$. Left: 1st order hexahedron elements. Right: 1st order tetrahedron elements.

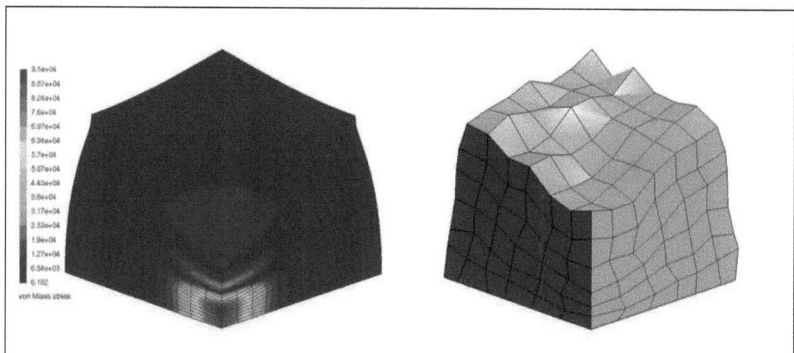

Figure A.20: Block under compression, deformed configuration $p/p_0 = 80$. Left: $C_0$-continuous von Mises stress field using C3D_8N_27C elements. Right: Deformed configuration due to instabilities when using C3D_8N_1I elements

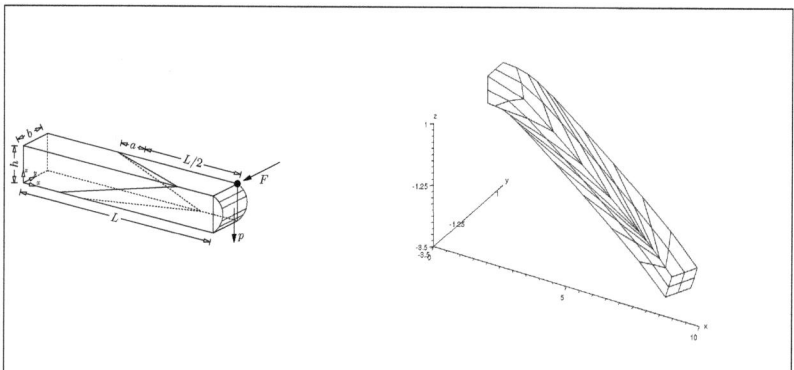

Figure A.21: Beam with distorted mesh. Left: Geometry and loading. Right: Displacement field using C3D_8N_27C elements.

## A.6 Performance of distorted elements

The purpose of this example is to test the sensitivity of assumed gradient elements with respect to mesh distortion. A similar test was proposed in [222]. The geometry and loading of a cantilever beam is presented in figure A.21. The beam is subject to a parabolic vertical area load on its free edge. A horizontal point load is acting on one of the vertices on its free end. The material is linear elastic. Geometrical nonlinearities are considered. The model parameters are $L = 10$, $h = b = 1$, $E = 3 \times 10^7$, $\nu = 0$, $P = 100 \times 10^3$, $F = 100 \times 10^3$. The maximum deflection is compared for different element types and for different values of the mesh distortion parameter $a$. The percentages denote the deviation of the actual value with respect to the undistorted mesh ($a = 0$). For a $10 \times 2 \times 2$ mesh the results are listed in table A.1. For a $20 \times 4 \times 4$ mesh see table A.2.

## A.7 Convergence of natural frequencies

This example is used to test the convergence and accuracy of dynamic eigenvalues. The geometry is presented in figure A.22. A beam with square cross section is clamped on its left side. The quantity of interest are the 1st, 2nd and the 200th eigenvalues. The constitutive law is a linear elastic material with elastic modulus $E = 30000 N/mm^2$, Poisson ratio $\nu = 0$ and mass density $2400 kg/m^3$. The convergence is measured with respect to the number of elements per edge, whereby the number of elements along the longitudinal axis are 10 times as much as along the other directions. The reference values have been computed using a 27-noded hexahedron element with $100 \times 10 \times 10$ elements. Due to symmetry the 1st two bending modes must be equal and are averaged in case of the tetrahedral meshes in order to obtain a unique representative value.

In figure A.23 the first natural frequency for various families of finite element types and $\nu = 0$ are shown. The stable assumed gradient schemes exhibit a significantly improved accuracy for 1st order tetrahedra. When considering 1st order hexahedra, only C3D_8N_27C improves the accuracy of the first eigenvalues, while the other schemes behave worse. Then all assumed gradient schemes perform better than standard FEM. Furthermore, higher eigenvalues are well approximated by all stable assumed gradient schemes, see figure A.24 for the 200th eigenvalue. Notice, the instable nodal integrated elements converge to wrong solutions.

Table A.1: Maximum displacement and deviation for mesh distortion parameters $a$ and a mesh with $10 \times 2 \times 2$ elements per edge

| $a$ | C3D_8N | C3D_8N_NI | C3D_8N_8I | C3D_8N_1E | C3D_8N_27C | |
|---|---|---|---|---|---|---|
| 0 | 4.68976 | 4.72447 | 4.79279 | 4.77723 | 5.07468 | |
| 3 | 48% | 29% | 49% | 55% | 38% | |
| 4.9 | 68% | 59% | 77% | 81% | 61% | |
| $a$ | C3D_4N | C3D_4N_NI | C3D_4N_1I | C3D_4N_1E | C3D_4N_10C | C3D_4N_1F |
| 0 | 3.26678 | 5.35719 | 4.3898 | 4.96579 | 4.90264 | 3.94155 |
| 3 | 69% | 1% | 46% | 44% | 48% | 63% |
| 4.9 | 90% | 23% | 88% | 88% | 89% | 89% |

# Appendix A. Verification of CAG elements

Table A.2: Maximum displacement and deviation for mesh distortion parameters $a$ and a mesh with $20 \times 4 \times 4$ elements per edge

| $a$ | C3D_8N | C3D_8N_NI | C3D_8N_8I | C3D_8N_1E | C3D_8N_27C | |
|---|---|---|---|---|---|---|
| 0 | 5.08042 | 5.09454 | 5.11962 | 5.11819 | 5.19704 | |
| 3 | 23% | 8% | 16% | 16% | 10% | |
| 4.9 | 53% | 29% | 53% | 57% | 45% | |
| $a$ | C3D_4N | C3D_4N_NI | C3D_4N_1I | C3D_4N_1E | C3D_4N_10C | C3D_4N_1F |
| 0 | 4.43699 | 5.27465 | 4.94572 | 5.17902 | 5.16265 | 4.76144 |
| 3 | 39% | 1% | 21% | 10% | 12% | 30% |
| 4.9 | 82% | 0% | 75% | 69% | 71% | 79% |

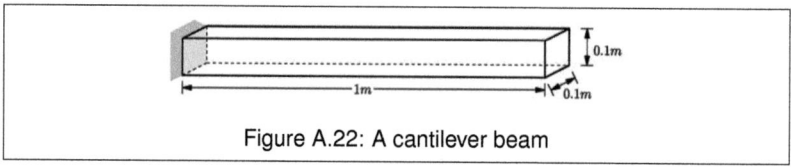

Figure A.22: A cantilever beam

The 200-th eigenvalue of the element C3D_8N_1I converges to the exact value although the formulation is instable. That means, modal analysis of a beam is not necessarily a sufficient test to verify stability, even if higher eigenmodes are checked. An indicator for the instability of C3D_8N_1I is that the error is not strictly monotonic decreasing.

The 1st and 200-th natural frequencies of the 2nd order tetrahedra are plotted in figure A.25. While the nodal integrated elements converge to the wrong solution, all stable formulations converge to the accurate values.

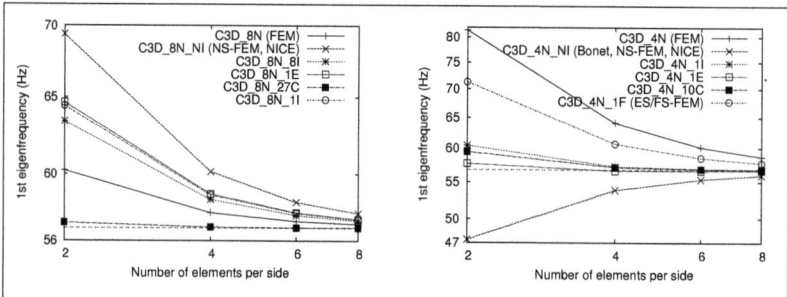

Figure A.23: Modal analysis of a beam: 1st natural frequency, $\nu = 0$. Left: 1st order hexahedra. Right: 1st order tetrahedra.

## A.8. Numerical efficiency

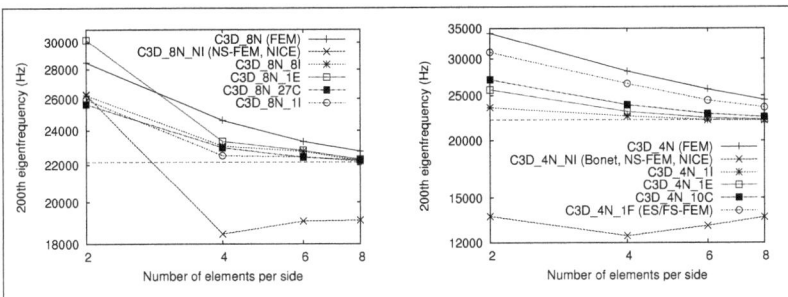

Figure A.24: Modal analysis of a beam: 200-th natural frequency, $\nu = 0$. Left: 1st order hexahedra. Right: 1st order tetrahedra.

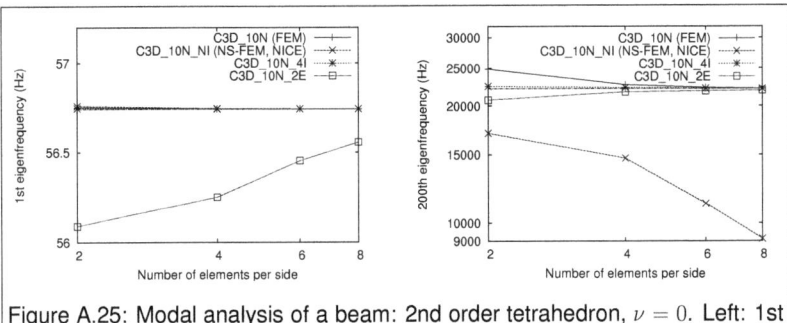

Figure A.25: Modal analysis of a beam: 2nd order tetrahedron, $\nu = 0$. Left: 1st natural frequency. Right: 200-th natural frequency.

## A.8 Numerical efficiency

Objective quantities for the efficiency can hardly be obtained. Numerical efficiency can be partially measured by the number of used integration points. This is at least true for explicit solution methods in which the time to assemble the restoring force vector and stiffness matrix grows with the number of material law evaluations, i.e. with the number of integration points. For implicit methods, the structure and sparsity of the stiffness matrix and the time used for assembling system matrices also play a significant role. Furthermore, the time to evaluate a constitutive relation (plasticity, nonlinearity) and the number of degrees of freedom must be balanced with the mentioned factors. The way how data structures are handled by the codes also varies a lot, for example one could aim at pre-computing constant quantities; focus on parallel or distributed environments, etc. Therefore, the computing time is not an objective quantity.

Figure A.26 compares the number of integration points for individual meshes. Notice, this number is proportional to the number of elements in standard FEM. For the other methods this number depends on the actual mesh topology. Then, the number of strain energy integration points must not be mistaken for the number of (Gaussian) integration points which are used to compute geometric volume integrals being used to determine the assumed gradient. More precisely, the number of material evaluation points is considered which is the number of support points of the assumed gradient field in this study (or the number of smoothing cells in SFEM).

Figure A.27 compares the sparsity of the stiffness matrix for individual meshes. In particular, the number of nonzero stiffness components is related to standard FEM. The sparsity is even more dependent on the actual mesh topology than the number of integration points. Obviously, the continuity condition decreases sparsity. This is, in particular, the case when nodal averages are applied. Furthermore, the simplified averaging of equation (4.108) is used in the plot. The approach using dual multipliers behaves worse.

Figure A.28 plots the used CPU time for the assembly of the restoring force vector with respect to the mesh size (cantilever beam example in section A.2). The times required to assemble the restoring force vector and stiffness matrix turn out to be more or less proportional to the number of integration points. Figure A.29 plots the relative displacement error with respect to the CPU time which is needed to create the restoring force vector. The relation between CPU time and displacement error may be different for each example.

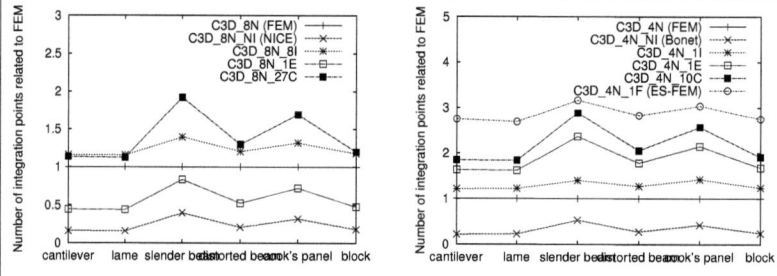

Figure A.26: Relative number of integration points of different examples (FEM=1). Left: brick elements. Right: simplex elements.

## A.8. Numerical efficiency

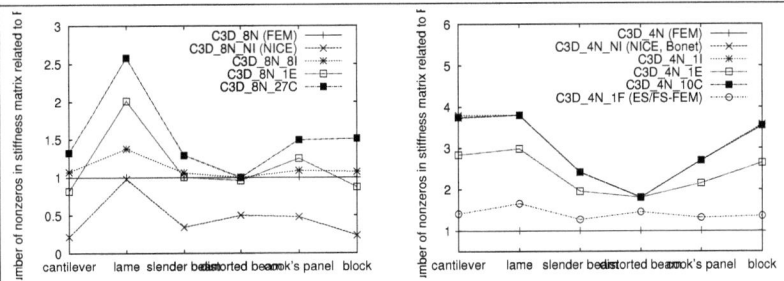

Figure A.27: Relative sparsity of different examples (FEM=1). Left: brick elements. Right: simplex elements.

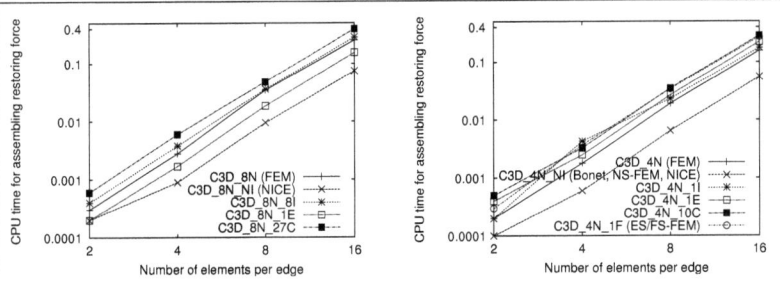

Figure A.28: CPU time for assembling the restoring force vector with respect to the number elements per edge. Left: brick elements. Right: simplex elements.

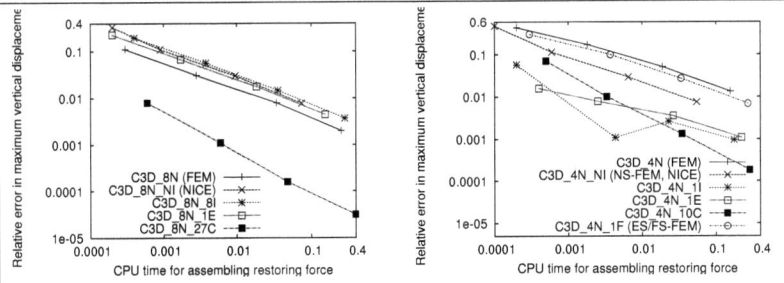

Figure A.29: Relative error in vertical displacement with respect to time needed for restoring force assembly. Left: brick elements. Right: simplex elements.

# Appendix B

# Variations on the contact interface

## Variations of parameterized quantities in contact frame

This paragraph serves to develop directional derivatives of the parameterized quantities following standard textbooks [141, 273].

For problems not involving contact, one would have

$$D_{\delta\hat{\phi}^{(i)}}[\hat{\phi}^{(j)}] = 0, \forall\, i \neq j \tag{B.1}$$

stating that the fields $\hat{\phi}^{(1)}$ and $\hat{\phi}^{(2)}$ are independent. In contact situations, however, both functions are coupled through the contact frame, i.e. the fields are dependent on $\hat{\xi}$. When computing their variations and subsequent directional derivatives, it must be emphasized that any perturbed quantity based on $\bar{\mathbf{X}}^{(i)}(\hat{\xi})$ includes implicitly the identification of $\hat{\xi}(\hat{\mathbf{X}}^{(i)})$. The variation of this identification must also be considered leading to a term containing $\delta\hat{\xi}^{\beta}$ as will be shown.

- The variation of a point $\hat{\mathbf{x}} = \hat{\phi}(\hat{\mathbf{X}})$ being located on the contact frame can be obtained from

$$\delta\hat{\mathbf{x}} = \left.\frac{d}{d\epsilon}\right|_{\epsilon=0} [\hat{\phi}(\hat{\mathbf{X}}) + \epsilon\delta\hat{\phi}] = \frac{\partial\hat{\phi}}{\partial\hat{\mathbf{X}}}\delta\hat{\mathbf{X}} + \delta\hat{\phi} = \hat{\mathbf{F}}(\hat{\mathbf{X}}) \cdot 0 + \delta\hat{\phi} \tag{B.2}$$

- The variation of the mapped quantity

$$\hat{\mathbf{x}}^{(i)} = \hat{\phi}^{(i)}(\hat{\mathbf{X}}^{(i)}) = \hat{\phi}^{(i)}(\bar{\mathbf{X}}^{(i)}(\hat{\xi}^{(i)}))$$

is

$$\begin{aligned}\delta\hat{\mathbf{x}}^{(i)} &= \left.\frac{d}{d\epsilon}\right|_{\epsilon=0}[\hat{\phi}^{(i)}(\bar{\mathbf{X}}^{(i)}(\hat{\xi}^{(i)})) + \epsilon\delta\hat{\phi}^{(i)}] \\ &= \frac{\partial\hat{\phi}^{(i)}}{\partial\bar{\mathbf{X}}^{(i)}} \cdot \frac{\partial\bar{\mathbf{X}}^{(i)}}{\partial\hat{\xi}^{(i)\beta}}\delta\hat{\xi}^{(i)\beta} + \delta\hat{\phi}^{(i)}\end{aligned}$$

since it depends on the parametrization of the contact frame. Using equations (7.12) and (7.13), one obtains

$$\delta\hat{\mathbf{x}}^{(i)} = \delta\hat{\phi}^{(i)} + \hat{\tau}_{\beta}^{(i)}\delta\hat{\xi}^{(i)\beta} \tag{B.3}$$

Notice, in feasible contact, i.e. $g(\hat{\xi}) = 0$, it is assumed that the surfaces $C^{(1)}$ and $C^{(2)}$ are described by the same tangential planes, i.e. $\hat{\tau}_\beta^{(1)} = \hat{\tau}_\beta^{(2)} = \hat{\tau}_\beta$ and $\xi^{(1)} = \xi^{(2)}$.

- The variation of the gap function (7.24) becomes

$$\delta \hat{g} = \delta \left( \hat{\mathbf{x}}^{(2)} - \hat{\mathbf{x}}^{(1)} \right) \cdot \hat{\nu} + \left( \hat{\mathbf{x}}^{(2)} - \hat{\mathbf{x}}^{(1)} \right) \cdot \delta \hat{\nu}$$
$$= \left( \delta \hat{\phi}^{(2)} - \delta \hat{\phi}^{(1)} + \hat{\tau}_\beta^{(2)} \delta \hat{\xi}^{(2)\beta} - \hat{\tau}_\beta^{(1)} \delta \hat{\xi}^{(1)\beta} \right) \cdot \hat{\nu} + \left( \hat{\mathbf{x}}^{(2)} - \hat{\mathbf{x}}^{(1)} \right) \cdot \delta \hat{\nu} \quad \text{(B.4)}$$

Due to the perpendicularity condition $\hat{\tau}_\beta^{(i)} \cdot \hat{\nu} = 0$ and and with $\hat{\nu} \cdot \delta \hat{\nu} = 0$ [273], this can be simplified to

$$\delta \hat{g} = \hat{\nu} \cdot \left( \delta \hat{\phi}^{(2)} - \delta \hat{\phi}^{(1)} \right) \quad \text{(B.5)}$$

- The variation of the glide path $\hat{\xi}^\beta$ may be obtained implicitly starting with the perpendicularity condition [141]

$$\left( \hat{\mathbf{x}} - \hat{\mathbf{x}}^{(i)} \right) \cdot \hat{\tau}_\alpha(\hat{\mathbf{x}}) = 0 \quad \text{(B.6)}$$

which holds since $\hat{\mathbf{x}}^{(i)}$ is assumed to be an orthogonal projection of $\hat{\mathbf{x}}$ onto $C^{(i)}$. Computing its directional derivative yields

$$0 = \delta \left( \hat{\mathbf{x}} - \hat{\mathbf{x}}^{(i)} \right) \cdot \hat{\tau}_\alpha(\hat{\mathbf{x}}) + \left( \hat{\mathbf{x}} - \hat{\mathbf{x}}^{(i)} \right) \cdot \delta \hat{\tau}_\alpha(\hat{\mathbf{x}}) \quad \text{(B.7)}$$

The first term writes

$$\delta \left( \hat{\mathbf{x}} - \hat{\mathbf{x}}^{(i)} \right) = \delta \hat{\phi} - \delta \hat{\phi}^{(i)} - \hat{\tau}_\beta^{(i)} \delta \hat{\xi}^{(i)\beta} \quad \text{(B.8)}$$

For the second term, the tangential variation is obtained from (7.13)

$$\delta \hat{\tau}_\alpha = \delta \hat{\Psi}_{t,\alpha}(\hat{\xi}) = \hat{\Psi}_{t,\alpha,\beta}(\hat{\xi}) \delta \hat{\xi}^\beta + \delta \hat{\Psi}_{,\alpha} \quad \text{(B.9)}$$

and the mapping $\hat{\mathbf{x}}$ onto $C^{(i)}$ is rewritten as

$$\hat{\mathbf{x}} - \hat{\mathbf{x}}^{(i)} = \hat{g}^{(i)} \hat{\nu} \quad \text{(B.10)}$$

Equation (B.7), therefore, yields

$$0 = \left( \delta \hat{\phi} - \delta \hat{\phi}^{(i)} - \hat{\tau}_\beta^{(i)} \delta \hat{\xi}^{(i)\beta} \right) \cdot \hat{\tau}_\alpha + \hat{g}^{(i)} \hat{\nu} \cdot \left( \hat{\Psi}_{t,\alpha,\beta}(\hat{\xi}) \delta \hat{\xi}^\beta + \delta \hat{\Psi}_{,\alpha} \right) \quad \text{(B.11)}$$

Defining the symmetric matrix

$$A_{\alpha\beta}^{(i)} := \hat{\tau}_\beta^{(i)} \cdot \hat{\tau}_\alpha + \hat{g}^{(i)} \hat{\nu} \cdot \hat{\Psi}_{,\alpha,\beta} \quad \text{(B.12)}$$

one obtains

$$A_{\alpha\beta}^{(i)} \delta \hat{\xi}^{(i)\beta} = \left( \delta \hat{\phi} - \delta \hat{\phi}^{(i)} \right) \cdot \hat{\tau}_\alpha + \hat{g}^{(i)} \hat{\nu} \cdot \delta \hat{\Psi}_{,\alpha} \quad \text{(B.13)}$$

In case of perfect sliding, $\hat{g}^{(i)} = 0$, (B.13) may be simplified to

$$\delta \hat{\xi}^{(i)\beta} = \left( \delta \hat{\phi} - \delta \hat{\phi}^{(i)} \right) \cdot \hat{\tau}^\beta \quad \text{(B.14)}$$

Using the definition of $\delta \hat{\xi}^\beta$ to be the relative variation of $C^{(1)}$ with respect to $C^{(2)}$

$$\delta \hat{\xi}^\beta := \delta \hat{\xi}^{(1)\beta} - \delta \hat{\xi}^{(2)\beta} \quad \text{(B.15)}$$

one obtains from (B.14)

$$\delta \hat{\xi}^\beta = \left( \delta \hat{\phi}^{(2)} - \delta \hat{\phi}^{(1)} \right) \cdot \hat{\tau}^\beta \quad \text{(B.16)}$$

# Appendix C

# Spatial discretization of the contact boundary

## C.1 Exact integration domains

In the original application of distance fields to collision detection [95] a finite element mesh based on first order tetrahedra is used. Then the intersection of the one body's boundary and the second body's volume is determined. The resulting interface is triangulated, see figure C.1 on the left. The contact constraints are enforced using a quadratic penalty function. The integration of the contact tractions on the interface can be done analytically: The integration domains are first order triangles with linearly interpolated displacements. The gap function is interpolated using the same ansatz functions.

Beside the disadvantages of the employed penalty method, the resulting algorithm is computationally expensive. The collision detection requires the parametrization of triangles within local element coordinate systems of individual tetrahedra, the identification of the actual cutting plane boundary, a triangulation and the integration for each triangular integration cell. When comparing the inaccuracy of the penalty method with the complexity of the collision detection and integration, other methods may be better suited. In many contact methods [89, 99, 140, 201, 230, 274] it is common to represent the contact boundary by a point cloud, see figure C.1 on the right. Each point serves as a numerical integration point and represents a finite part of the contact interface. An intersection test of a point and a finite element is very simple, but one has to approximate a suitable sampling density in order to stay accurate.

Noteworthy is the approach of [120, 213] for thin walled structures, where two surface triangles are tested on intersection. Furthermore, a simple formula for the signed spanned volume of both triangles is presented which can be used directly for collision detection and as a constraint function. The strategy is, however, difficult to apply to distance fields and fails for surface triangles which lie completely in the interior of a body.

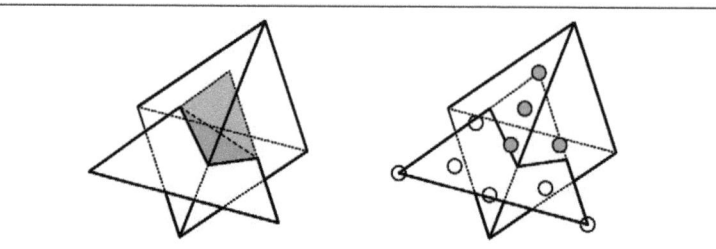

Figure C.1: Exact integration domain versus point cloud for the intersection of a triangle and a tetrahedron. Left: Triangulation of the exact intersection. Right: Intersecting discrete points which are sampled on the triangle.

## C.2 Integration strategies

Some strategies to define the integration points on the contact interface are described in the following paragraphs. Let a contact pair denote a pair of two bodies, the contactor and the target.

**Node-to-segment** The contact frame is identical with the boundary of the contactor body. Nodes on the contact frame are associated to points on the target. When using closest point projection, these points are projection points on the target's boundary, see figure C.2(a). When using distance fields, these points are those points on the boundary or in the interior of the target which have the same deformed coordinate as the contactor nodes.

Node-to-segment integration [230] is simple, but error-prone. Intersections may not be detected if a contactor node slides off a target segment or if the target mesh is of finer detail than the contactor boundary. These problems can be improved by a two-pass strategy, i.e. one considers the same contact pair twice by exchanging the contactor and the target, see figure C.2(c). The sampling density may still be too rough. Then one may add support for topological features such as finite element edges, leading to edge-to-edge and edge-to-segment search algorithms. In specific situations, the two-pass approach leads to overconstrained systems. Herein, some discrete constraint functions are linearly dependent or even identical for both sides. This may happen if both contacting boundaries are planes.

A further issue with node-to-segment integration is the representation of curved interfaces as pointed out in [144, 218]. Assume the discretization of curved surfaces by linear elements, see figure C.3. Exact enforcement of the nodal contact constraints, with the contactor being the right body, may lead to interpenetrations as in figure C.3(a). Let the contactor be the left body or assume the application of the two-pass approach. Then exact enforcement of the nodal constraints leads to gaps between both interfaces, see figure C.3(b). A strategy to improve the representation of curved interfaces is to interpolate the geometry of contact boundaries by $C1$-continuous functions [52, 61, 62, 131, 249].

**Segment-to-segment** Compared with node-to-segment, the accuracy of the contact tractions is improved by application of a $C0$-continuous interpolation of the contact tractions. They may be interpolated on the boundary of the contactor, figure C.2(b) [144, 216, 218, 219], or independently on an intermediate surface, figure C.2(d) [73, 187, 220, 221]. The problem of finding a sufficiently high sampling density of the integration points remains the same as in node-to-surface integration. If bodies with different mesh refinements are used, the interpolation of the contact tractions on the rougher boundary may eliminate essential information. Using an independent mesh on the intermediate surface complicates the problem by switching from a nonconform mapping between two surfaces to a pair of nonconform mappings involving three interfaces.

Simple segment-to-segment approaches may be used to eliminate overconstraint in two-pass node-to-segment integration, figure C.2(c). The issue of curved interfaces can be improved by enforcing the contact constraints in a weak sense, see figure C.3(c). The enforcement of the average of positive and negative gap function values leads to a deformed configuration with only small regions that overlap or exhibit gaps. Since the discrete distance field is only defined for non-negative values, this benefit is not available.

## C.3 Mortar method

The subsequent paragraphs follow the approach of [187] who applied the mortar method to contact problems using an intermediate surface.

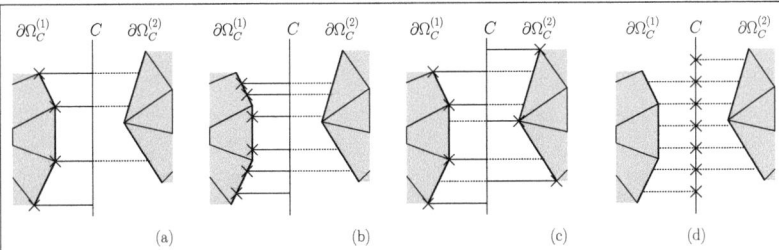

Figure C.2: Integration point strategies on the contact frame. (a) one-sided node-to-segment. (b) segment-to-segment on one boundary. (c) both-sided node-to-segment/node-based segment-to-segment. (d) segment-to-segment on the contact frame.

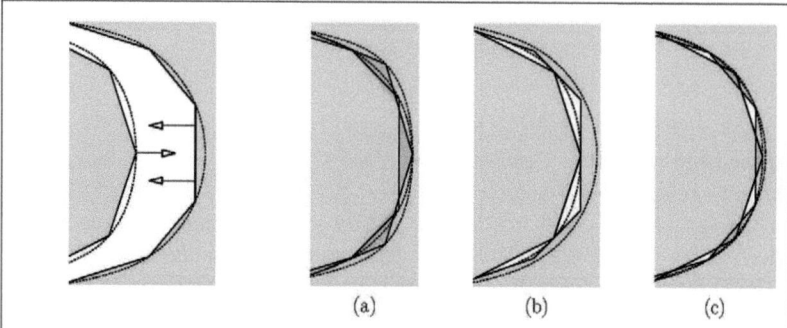

Figure C.3: Node-to-segment contact on curved interfaces. (a) one-sided node-to-segment. (b) both-sided node-to-segment. (c) segment-to-segment.

The contact frame as well as the bodies are discretized using finite sets of elements such that

$$\Omega^{(i)} \approx \Omega^{(i)h} = \bigcup_e \Omega_e^{(i)}, \qquad C^{(i)} \approx C^{(i)h} = \bigcup_e C_e^{(i)} \tag{C.1}$$

with the discretized deformed coordinates

$$\phi^{(i)} \to \phi^{(i)h}, \quad \hat{\phi}^{(i)} \to \hat{\phi}^{(i)h} \tag{C.2}$$

The displacement fields of the bodies, measured in the global frame and in the contact frame, are generally nonconform, but represent the same dynamics. Hence, the discretized displacements of the two bodies must be equal when measured in the global frame $\mathbf{x}^{(i)}$ and in the contact frame $\hat{\mathbf{x}}$. The condition $\phi^{(i)} = \hat{\phi}^{(i)}$ is weakly enforced by associating it with a spatially dependent Lagrange multiplier field $\lambda^{(i)}$ such that the constraint will be weakly enforced through the mortar integral

$$G_m^{(i)}(\phi^{(i)h}, \hat{\phi}^{(i)h}, \lambda^{(i)h}) = \int_{\partial_C \Omega^{(i)h}} \left( \phi^{(i)h}(\mathbf{X}^{(i)}) - \hat{\phi}^{(i)}(\hat{\mathbf{X}}^{(i)}) \right) \lambda^{(i)h}(\hat{\mathbf{X}}^{(i)}) d\Gamma = 0 \tag{C.3}$$

Therefore, the contact surfaces become interfaces coupling the mortar side, i.e. the displacement fields belonging to the finite element structure, with the non-mortar side, i.e. the equivalent fields defined in the contact frame. Adding the equations for both bodies, taking the first variation, adding the result to the virtual work in equation (7.9) and applying the substitution (7.20) yields a constrained variational problem which states:

Find $\phi^h, \hat{\phi}^h, \lambda^h$ such that

$$\begin{aligned} 0 &= G_{dyn}(\phi^h, \delta\phi^h) + \hat{G}_C(\hat{\phi}^h, \delta\hat{\phi}^h) + G_m(\delta\phi^h, \delta\hat{\phi}^h, \lambda^h) \\ 0 &= G_m(\phi^h, \hat{\phi}^h, \delta\lambda^h) \end{aligned} \tag{C.4}$$

## C.4. Node-to-element integration

**Discretization** The contact quantities are expanded in terms of finite element shape functions. Assume that the multipliers, deformations fields and their variations may be interpolated as

$$\lambda^{(i)h}(\mathbf{X}^{(i)}) = \sum_{A^{(i)}=1}^{n^{(i)}} M_{A^{(i)}}^{(i)}\left(\hat{\xi}^{(i)}(\mathbf{x}^{(i)})\right) \lambda_{A^{(i)}}^{(i)}$$

$$\phi^{(i)h}(\mathbf{X}^{(i)}) = \sum_{C^{(i)}=1}^{n^{(i)}} N_{C^{(i)}}^{(i)}\left(\xi^{(i)}(\mathbf{x}^{(i)})\right) \mathbf{x}_{C^{(i)}}^{(i)}$$

$$\hat{\phi}^{(i)h}(\mathbf{X}^{(i)}) = \sum_{E^{(i)}=1}^{\hat{n}^{(i)}} \hat{N}_E^{(i)}\left(\hat{\xi}^{(i)}(\mathbf{x}^{(i)})\right) \hat{\mathbf{x}}_E^{(i)} \tag{C.5}$$

Therein, $M_{A^{(i)}}^{(i)}$ denotes the interpolation function of the multiplier on the interface, $N_{C^{(i)}}^{(i)}$ is the finite element shape function being isoparametric and $\hat{N}_E^{(i)}$ is an isoparametric shape function interpolating the contact frame coordinates on the interface. That means, $\hat{\xi}^{(i)} \in \mathbb{R}^2$ and $\xi^{(i)} \in \mathbb{R}^3$.

If the discretization of the contact frame is identical with the mesh of one of the boundaries, the multipliers $\lambda_{A^{(i)}}^{(i)}$ can be replaced by the contact traction $\hat{\mathbf{t}}_{A^{(i)}}$ and the contact integral itself takes the character of the mortar integral [218].

The variation for the discrete multipliers $\lambda_{A^{(i)}}^{(i)}$ leads to the discrete equation

$$0 = \sum_{C^{(i)}} n_{AC^{(i)}}^{(i)} \mathbf{x}_{C^{(i)}}^{(i)} - \sum_E \hat{n}_{AE} \hat{\mathbf{x}}_E^{(i)}, \quad n_{AC}^{(i)} = \int_{\partial_C \Omega^{(i)}} M_A N_C^{(i)} d\Gamma, \quad \hat{n}_{AE} = \int_{C^h} M_A \hat{N}_E d\Gamma \tag{C.6}$$

where the integration domain becomes $\partial_C \Omega^{(i)} \approx C^{(i)} \approx C^h$. The Lagrange multipliers are stored at the nodes of the contact frame and, therefore, the matrix $\hat{n}_{AE}$ is quadratic. For a given spatial configuration vector $\mathbf{x}^{(i)}$ one can solve for $\hat{\mathbf{x}}^{(i)}$ requiring the inverse of $\hat{n}_{AE}$. In the standard mortar method, the interpolation function of the multipliers is the finite element shape function, i.e. $M_A = \hat{N}_A$, and $\hat{n}_{AE}$ is positive definite. The matrix is sparse but its inverse is dense leading to an expensive operation. Therefore, a dual multiplier space is often used [272]. Herein, the interpolation function $M_A$ is chosen such that it obeys the biorthogonality condition

$$\int_C M_A \hat{N}_B d\Gamma = \delta_{AB} \int_C \hat{N}_B d\Gamma \tag{C.7}$$

More details on dual multipliers are presented in section 4.4.3 on page 103. If the Lagrange multipliers are defined on the finite element structure instead of the contact frame, then one is chosing $M$ to enforce the biorthogonality of $n_{AC}^{(i)}$ computing $\mathbf{x}^{(i)}$ explicitly from a given $\hat{\mathbf{x}}^{(i)}$.

## C.4 Node-to-element integration

Let the contact frame $C$ be identical to the boundary $\partial_C \Omega^{(1)}$ of the contactor $\Omega^{(1)}$. Then one has to find the intersection of the contactor boundary with the target

$\Omega^{(2)}$. If there is contact, the resulting interface may partially lie on the boundary (feasible contact) or in the interior (infeasible contact) of the target. The mortar method involves only one integral, i.e. the contact integral which has to map the contact tractions and displacements of the contactor side of the interface to the generally nonconform target side of the interface. The contact tractions will be interpolated using the finite element shape functions of the contactor. On the target side, the contact tractions are surface loads which are applied to material points in the interior or on the boundary of finite elements.

The numerical integration scheme influences the accuracy of the interpolation of contact tractions, the accuracy and robustness, but also the efficiency of collision detection. In this work, integration points are coincident with finite element nodes on the contactor's boundary.

A general contact strategy is assumed, i.e. no contact pairs are defined by the user a-priori. Instead, all boundaries may be in contact with all bodies in the system. As a result, no assumption about the mortar side can be done in order to reduce overconstraint in a two-pass node-to-element strategy. This would require a clear definition of contact pairs such that the algorithm can merge the integration points on the non-mortar side into the mortar integral.

Using the collision response presented in section 3.8.5 improves some problems with two-pass node-to-element integration being mentioned earlier in section C.2. The issue of curved interfaces is circumvented by enforcing constraints on the relative normal velocity which is independent from the actual penetration depth. Overconstraint issues do not exist in the asynchronous collision algorithm where the individual discrete contact constraints are applied sequentially.

With $M_A(\xi_B) = \delta_{AB}$ the mortar condition (C.6) simplifies to

$$\hat{\mathbf{x}}_A^{(1)} = \mathbf{x}_A^{(1)} \tag{C.8}$$

$$\hat{\mathbf{x}}_A^{(2)} = \sum_B N_B^{(2)}(\xi^{(2)}(\hat{\mathbf{x}}_A^{(1)})) \mathbf{x}_B^{(2)} \tag{C.9}$$

which are used to express the variations of the gap function and the glide path in the contact integral (7.30) with respect to the virtual displacements in the global frame.

The gap function (7.61) requires the evaluation of two distance fields, i.e. the distance $d_A^{(1)}$ of the point $\hat{\mathbf{x}}_A^{(1)}$ to the contactor's boundary and the distance $d_A^{(2)}$ of the point $\hat{\mathbf{x}}_A^{(1)}$ to the target's boundary. Only the latter must be evaluated since $d_A^{(1)} = 0$. It should be noted that this simplification can not be applied to the variation of the gap (7.64), i.e. $\delta d_A^{(1)} \neq 0$, but the gradient of $d_A^{(1)}$ is assumed to be identical with the negative gradient of $d_A^{(1)}$, $\gamma_A^{(1)} = -\gamma_A^{(2)}$.

# Appendix D

# Collision detection

During the simulation, the contact conditions must be satisfied at discrete points in time. Depending on the temporal discretization of the contact constraints, this may require an iterative solution process involving the evaluation of the constraints at predictor configurations. For the solution algorithm one generally requires information on the activity of the constraints, the current residuum at the predictor configuration and eventually the first derivative or even the second derivative of the residuum.

Given a predictor coordinate vector $\mathbf{x}$ one requires the following steps:

1. **Global search**  One has to identify the local element coordinates $\xi^{(i)}$ on the contactor and target side for each given $\mathbf{x}$. This is equivalent to the inverse mapping $\phi^{-1}$ with $\mathbf{x} = \phi(\mathbf{X}(\xi))$. A brute force approach would compare all finite elements with $\mathbf{x}$ which unnecessarily increases numerical complexity. Instead, a global collision phase is done before the actual contact detection. In this phase, the set of possible collision candidates is reduced to a reasonably small number by means of very fast methods. These are potential contact pairs where each member is sufficiently close to the other.

2. **Local search**  The local search does the accurate collision detection, i.e. it performs the inside-outside test for the specified finite element geometry and finds the local coordinates $\xi^{(i)} = \xi^{(i)}(\mathbf{x})$.

3. **Generation of constraint equations**  For each positive detection, the discrete constraints are evaluated and temporarily stored (including discrete gradients, etc.).

4. **Computation of response**  Given the active set of constraints the response is computed.

If an iterative solution procedure is used, one repeats the process until the desired accuracy is obtained. Since the number of constraints may be large, it is advisable to integrate points 3 and 4 with the local detection phase: Once a positive interpenetration of a node with some finite element was found the response will be applied immediately before the next local intersection test takes place. This will reduce the number of temporary data objects.

## D.1 Global contact search

### D.1.1 Bounding volume hierarchies

Plain collision detection algorithms require a large amount of geometrical intersection tests, checking if any of the polygons used to model the surface of one entity touch or penetrate any member on the other body. To improve the efficiency, hierarchical representations of entities are generated to localize the regions where the actual collision appears or to delimit the domain where the exact collision test must be performed. Such representations approximate the topology of an object at different levels of detail. These include bounding volume hierarchies like sphere trees [25, 101, 210, 212, 246], shell trees [130], OBB-trees (Oriented Bounding Boxes) [77], AABB trees [246, 260] and hierarchies of k-DOPs (Discrete Orientation Polytopes) [128], as well as spatial partitioning like octrees (octant trees) [265], bucket trees [67], kd-trees [15] and position code algorithms [49, 50, 208], or totally different approaches like spatial hashing [254] or image based methods [255]. Surveys may be found in [160, 255]. Most of these algorithms involve two or more phases of detection at varying levels of accuracy.

The performance of the individual methods depends on the represented object geometry and behavior. Obviously, the requirements for the representation of a convex rigid body are different from an arbitrarily shaped deformable body consisting of a very large number of finite elements. Continuing the list of examples, in small deformation analysis an algorithm with large computational cost for the creation of the data structure and fast intersection time may be more efficient than a procedure with small computing times for coordinate updates and only reasonable intersection times. Furthermore, the number of levels must be balanced against the complexity of the bounding volume. The more accurate the bounding volume, the less levels of accuracy are required, but the more time is needed to build the data structure and to perform the intersection test.

A very simple bounding volume, being suitable for finite elements, are axis-aligned bounding boxes (AABBs). They require two coordinates to be saved which describe the minimal and maximal distribution of a three-dimensional object along the coordinates axis'. Intersection tests can be performed easily using the separating axis theorem [77] for convex polytopes, see figure D.1. When creating an AABB as a convex boundary approximation to a finite element, a safety margin (halo) may be applied [50]. Depending on the element's velocity and of the size of the margin, the next coordinate update of the hierarchical data structure is, therefore, not required at all time steps.

## D.1. Global contact search

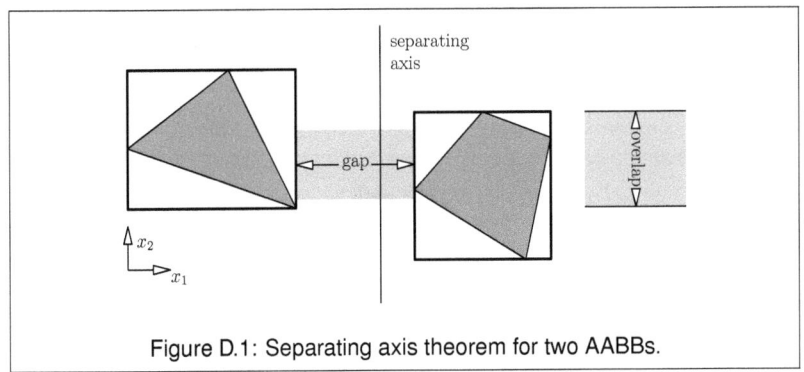

Figure D.1: Separating axis theorem for two AABBs.

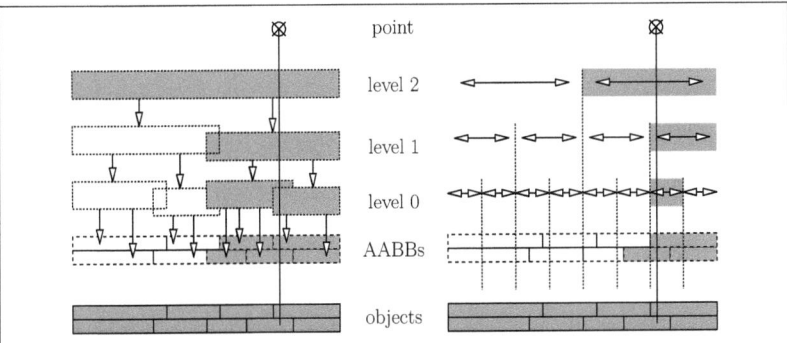

Figure D.2: AABB tree (left) and binary tree (right) in one dimension. The highlighted areas are the bounding volumes to be tested on intersection with the given point. The AABBs below level 0 are bounding volumes representing a single finite element. The highlighted AABBs are the objects to be eventually tested on intsersection.

A further general design decision is if the bounding volumes within individual levels may overlap or if they describe mutually exclusive domains, see figure D.2 which compares an AABB-tree with an octree (binary tree) in one dimension. For the first case, only one bounding volume is required per element, but a coordinate update requires the update of all bounding volumes at higher levels and intersection tests need more comparisons among individual bounding volumes. For the second case, multiple bounding volumes may be required for one element, but coordinate updates and intersection tests require less comparisons.

### D.1.2 Position codes

An approach to global collision detection based on position codes is presented in [49, 50, 269]. Therein, the complete space is subdivided into mutually exclusive cells. The spatial cooordinates of each cell are constant in time. Objects in the structure are then associated to the intersecting cells. When the objects' coordinates are changing, one checks if an update of the associated cells is necessary at all before the actual update takes place.

To each cell one can assign a position code, i.e. a unique integer number which identifies the cell. This code is dependent on the number of levels $l$ and the spatial dimension $n = 3$. Each level contributes $2^n$ further subdivisions into octants, i.e. $2^{(n \cdot l)}$ possible codes exist. Therefore, position code algorithms are related to spatial partition methods, in particular to octrees [265]. Using a plain octree, see figure D.2, one traverses from the top level to the bottom in order to find the appropriate cell (level 0) for a given coordinate. On each level, one has to decide on 8 suboctants (in three dimensions). On the contrary, by using position codes one creates a one-dimensional ordered space. Then one can use optimized search algorithms which are based on plain order conditions, for example binary trees or hash tables.

In synchronous time stepping schemes, the main question in collision detection is: What are the finite element nodes which intersect with a given finite element? In this case, the cells contain the appropriate finite element nodes. One assumes that all cells are assigned to a sorted set in ascending order (one actually creates a sorted set of nodes which are ordered with respect to their position code). For the given element one creates a bounding volume and then computes the smallest and the largest position code of all cells which intersect with the bounding volume. Subsequently, one searches for the cell with smallest position code. Starting at this cell one iterates through all cells within the sorted set until the cell with largest position code is reached. For each iterate one obtains the list of contained finite element nodes for which the local intersection test will be performed.

There exist approaches which reduce the number of required iterates by changing the ordering of the cells. The aim is to reduce the average index distance, i.e. the distance in the sorted set between the minimal and maximal position codes of a given finite element. The position code is interpreted as a space filling curve.

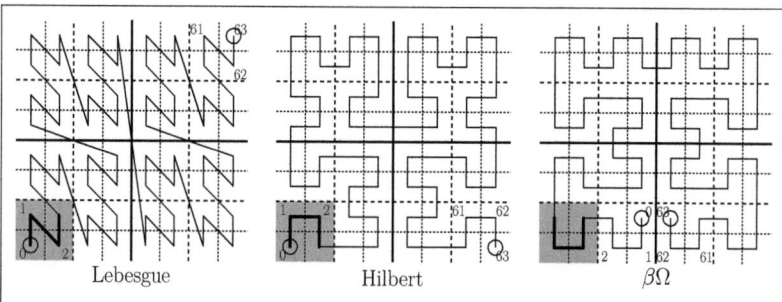

Figure D.3: Space-filling curves in two dimensions. Highlighted is the base shape.

## D.1. Global contact search

Improvements over linear position codes [208] were found through Lebesgue curves (standard octree scheme); even more efficient are Hilbert curves and the slightly better $\beta\Omega$ pattern [268], see figure D.3. Both patterns are circular whereby $\beta\Omega$ is constructed from two competing rules which define the rotation of the base "U". The contact detection must be adopted: First, the level is identified where the bounding volume does not exceed the level's cell bounds, see figure D.4. Within this level, there are at most four cells intersected - one for each AABB's vertex in two dimensions. The AABB is cut by at most one separating plane of the following finer level, splitting the search domain into eight coherent octants (or four adjacent quadrants).

### D.1.3 Application in dynamics

The advantage of cell based approaches, such as position codes and octrees, is that the data structures must be updated locally only if a contactor point moves from one cell to the other. The opposite is true for AABB trees, where one needs to update the complete hierarchy if an AABB on the base level changes its coordinates. In explicit time stepping schemes, the coordinates change only little and an update of the position code is usually required at a few times only. Additionally, the AABB enclosing a target element can be enlarged by a safety distance, forming the halo box [269]. Then one can determine the bounding position codes or, respectively, octants for the halo. The associated search needs not be repeated at each time step; it takes only place whenever the target element's coordinate changes (since the last search) become larger than the safety distance.

The examples of this chapter were tested with a modified octree. The base cell size is determined by the average element size. The origin and dimension of the domain to be subdivided is defined prior the simulation and remains fixed. Then one creates only those cells which intersect with contactor nodes or the AABBs of target elements. The cells are sorted according to an octree hierarchy (Lebesgue position codes). Each cell is associated to a list of points and target

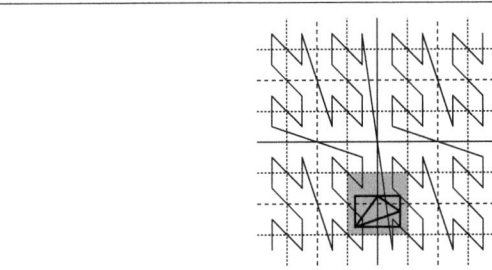

Figure D.4: Contact detection using a Lebesgue curve. Highlighted are the four cells to check.

AABBs which intersect the cell. Each contactor point and target AABB obtains a list of associated cells. After each time step, one checks if a point has changed its position code and moves it to the appropriate cell. If a target AABB has changed the position codes of its vertices, then one has to determine which cells must be added to and removed from the list of associated cells. These operations may include the creation of new cells. After the octree update one has to delete all cells, which do not contain any points or AABBs. Although quite memory consuming, the actual collision test is very efficient. Since the points and elements are stored at each cell, one can very efficiently determine lists of contact candidates given the questions

1. For a given target element, which contactor points intersect the same cells?

2. For a given contactor point, which target elements are located at the same position code?

The global collision test includes the test on intersection of contactor points with associated target AABBs.

## D.2 Local contact search

For each collision candidate pair obtained in the global detection one has to decide if the contactor point is inside of the target element. To do so one has to determine the local target finite element coordinate $\xi^{(2)}$ of the point coordinate $\hat{x}^{(1)}$. Two strategies are proposed:

1. The target element is subdivided into first order simplices. The space of coordinates of each tetrahedron is linear with respect to the local coordinate system of the simplex $\xi^S$ which in turn can be obtained by a linear transformation from the local coordinates $\xi^{(2)}$ of the target finite element. The local simplex element $\xi^S$ can be obtained by solving a linear equation. For each tetrahedron, an intersection test takes place. If no tetrahedron of the target element intersects with the contactor point, the intersection test has failed.

2. The local finite element coordinate can be iteratively obtained by solving the system of equation given by (7.68). This strategy is applied in the examples. The number of iterations is limited to $5$. After obtaining the local element coordinate, the intersection test takes place. A point with local coordinate $\xi$ intersects a tetrahedron if

$$\xi_1 \geq 0, \quad \xi_2 \geq 0, \quad \xi_3 \geq 0, \quad 1 - \xi_1 - \xi_2 - \xi_3 \geq 0 \qquad (D.1)$$

It intersects a hexahedron if

$$-1 \leq \xi_\alpha \leq 1, \quad \alpha = \{1, 2, 3\} \qquad (D.2)$$

## D.2. Local contact search

The local finite element coordinate found during the local collision test can be directly used to evaluate the distance field gradient given by equation (7.50).

# Bibliography

[1] R. Abedi, B. Petracovici, and R.B. Haber. A space-time discontinuous galerkin method for linearized elastodynamics with element-wise momentum balance. *Computer Methods in Applied Mechanics and Engineering*, 195(25-28):3247 – 3273, 2006.

[2] B. O. Almroth, P. Stern, and F. A. Brogan. Automatic choice of global shape functions in structural analysis. *AIAA J.*, 16:525–528, 1978.

[3] H. Andersen. RATTLE: A "velocity" version of the SHAKE algorithm for molecular dynamics calculations. *Journal of Computational Physics*, 52(1):24–34, October 1983.

[4] F. Armero. Assumed strain finite element methods for conserving temporal integrations in non-linear solid dynamics. *International Journal for Numerical Methods in Engineering*, 74(12):1795–1847, 2008.

[5] J. C. Baez and J. W. Gilliam. An algebraic approach to discrete mechanics. *Lett. Math. Phys.*, 31:205–212, 1994.

[6] E. Barth and B. Leimkuhler. Symplectic methods for conservative multibody systems. *Fields Institute Communications*, 10:25–43, 1996.

[7] T. J. Barth and J. A. Sethian. Numerical schemes for the Hamilton-Jacobi and level set equations on triangulated domains. *J. Comput. Phys.*, 145:1–40, September 1998.

[8] K. J. Bathe. *Finite Element Procedures*. Prentice Hall, 2nd edition, 1995.

[9] S. Beissel and T. Belytschko. Nodal integration of the element-free Galerkin method. *Computer Methods in Applied Mechanics and Engineering*, 139(1-4):49 – 74, 1996.

[10] T. Belytschko and R. Mullen. Mesh partitions of explicit-implicit time integration. In K. Mathe, J. Oden, and W. Wunderlich, editors, *Formulations and Computational Algorithms in Finite Element Analysis*, pages 673–690. MIT press, New York, 1976.

[11] T. Belytschko and R. Mullen. Explicit integration of structural problems. *Finite Elements in Nonlinear Mechanics*, 2:669–720, 1977.

[12] T. Belytschko and R. Mullen. Stability of explicit-implicit mesh partitions in time integration. *International Journal for Numerical Methods in Engineering*, 12:1575–1586, 1978.

[13] T. Belytschko and S. Ong. Hourglass control in linear and nonlinear problems. *Computer Methods in Applied Mechanics and Engineering*, 43:251–276, 1984.

[14] M. Benes and K. Matous. Asynchronous multi-domain variational integrators for nonlinear hyperelastic solids. *Computer Methods in Applied Mechanics and Engineering*, 199(29-32):1992 – 2013, 2010.

[15] J. L. Bentley. Multidimensional binary search trees used for associative searching. *Communications of the ACM*, 18:509–517, 1975.

[16] C. Bernardi, C. Y. Maday, and A.T. Patera. A new nonconforming approach to domain decomposition: The mortar element method. *Nonlinear Partial Differential Equations and Their Applications*, pages 13–51, 1992.

[17] Jeffrey J. Biesiadecki and Robert D. Skeel. Dangers of multiple time step methods. *J. Comput. Phys.*, 109:318–328, December 1993.

[18] T. C. Bishop, R. D. Skeel, and K. Schulten. Difficulties with multiple time stepping and fast multipole algorithm in molecular dynamics. *Journal of Computational Chemistry*, 18(14):1785–1791, 1997.

[19] S. Blanes. High order numerical integrators for differential equations using composition and processing of low order methods. *Appl. Numer. Math*, 37:289–306, 2001.

[20] S. Blanes and C. Budd. Explicit Adaptive Symplectic (Easy) Integrators: A Scaling Invariant Generalisation of the Levi-Civita and KS Regularisations. *Celestial Mechanics and Dynamical Astronomy*, 89:383–405(23), August 2004.

[21] J. Bonet and A.J. Burton. A simple average nodal pressure tetrahedral element for incompressible and nearly incompressible dynamic explicit applications. *Communications in Numerical Methods in Engineering*, 14:437–449, 1998.

[22] J. Bonet, H. Marriott, and O. Hassan. An averaged nodal deformation gradient linear tetrahedral element for large strain explicit dynamic applications. *Communications in Numerical Methods in Engineering*, 17(8):551–561, 2001.

[23] C. L. Bottasso. A new look at finite elements in time: a variational interpretation of Runge-Kutta methods. *Applied Numerical Mathematics*, 25(4):355–368, 1997.

# Bibliography

[24] N. Bou-Rabee and J. E. Marsden. Hamilton-Pontryagin Integrators on Lie Groups Part I: Introduction and Structure-Preserving Properties. *Found. Comput. Math.*, 9:197–219, March 2009.

[25] G. Bradshaw and C. O'Sullivan. Sphere-tree construction using dynamic medial axis approximation. *ACM SIGGRAPH Symposium on Computer Animation*, pages 33–40, July 2002.

[26] M. Broccardo, M. Micheloni, and P. Krysl. Assumed-deformation gradient finite elements with nodal integration for nearly incompressible large deformation analysis. *International Journal for Numerical Methods in Engineering*, 78:1113–1134, 2009.

[27] R. Brown, P. Bryant, and H. D. I. Abarbanel. Computing the Lyapunov spectrum of a dynamical system from an observed time series. *Phys. Rev. A*, 43(6):2787–2806, Mar 1991.

[28] Luigi Brugnano, Felice Iavernaro, and Donato Trigiante. On the existence of energy-preserving symplectic integrators based upon gauss collocation formulae. Technical Report arXiv:1005.1930, CERN, May 2010.

[29] C. Bucher. Stabilization of explicit time integration by modal reduction. In *Trends in Computational Structural Mechanics*, Barcelona, 2001. CIMNE.

[30] C. Bucher and S. Wolff. SLangTNG - a numerical tool for linear algebra, stochastics and structural analysis. http://tng.tuxfamily.org, 2007-2011.

[31] M. P. Calvo and J. M. Sanz-Serna. The development of variable-step symplectic integrators with application to the two-body problem. *SIAM J. Sci. Comput.*, 14:936–952, July 1993.

[32] B. Cano and J. M. Sanz-Serna. error Growth in the Numerical Integration of Periodic Orbits, with Application to Hamiltonian and Reversible systems. *SIAM J. Numer. Anal.*, 34:1391–1417, August 1997.

[33] A. S. L. Chan and K. M. Hsiao. Nonlinear analysis using a reduced number of variables. *Comp. Meth. in Appl. Mech. and Eng.*, 52:899–913, 1985.

[34] Anil B. Chaudhary and Klaus-Jürgen Bathe. A solution method for static and dynamic analysis of three-dimensional contact problems with friction. *Computers & Structures*, 24(6):855 – 873, 1986.

[35] J. Chung and G. M. Hulbert. Time integration algorithm for structural dynamics with improved numerical dissipation: the generalized-alpha method. *Journal of Applied Mechanics, Transactions ASME*, 60(2):371–375, 1993.

[36] F. Cirak and M. West. Decomposition Contact Response (DCR) for Explicit Finite Element Dynamics. *International Journal for Numerical Methods in Engineering*, 64(8):1078–1110, 2005.

[37] S. Cirilli, E. Hairer, and B. Leimkuhler. Asymptotic error analysis of the adaptive verlet method. *BIT Numerical Mathematics*, 39:25–33, 1999.

[38] R. Courant, K. Friedrichs, and H. Lewy. On the partial difference equations of mathematical physics. *IBM J. Res. Dev.*, 11:215–234, March 1967.

[39] A. Czekanski and S. A. Meguid. Analysis of dynamic frictional contact problems using variational inequalities. *Finite Elements in Analysis and Design*, 37:861–879, 2001.

[40] A. Czekanski and S. A. Meguid. On the use of variational inequalities to model impact problems of elasto-plastic media. *Int. J. Impact Eng.*, 2:1485–1511, 2006.

[41] M. Van Daele and G. Vanden Berghe. Geometric numerical integration by means of exponentially-fitted methods. *Applied Numerical Methematics*, 57:415–435, 2007.

[42] K. Y. Dai, G. R. Liu, and T.T. Nguyen. An n-sided polygonal smoothed finite element method (nsfem) for solid mechanics. *Finite Elements in Analysis and Design*, 43:847–860, 2007.

[43] W. J. T. Daniel. A study of the stability of subcycling algorithms in structural dynamics. *Computer Methods in Applied Mechanics and Engineering*, 156(1-4):1 – 13, 1998.

[44] W. J. T. Daniel. A partial velocity approach to subcycling structural dynamics. *Computer Methods in Applied Mechanics and Engineering*, 192(3-4):375 – 394, 2003.

[45] E. A. de Souza Neto, D. Peric, M. Dutko, and D. R. J. Owen. Design of simple low order finite elements for large strain analysis of nearly incompressible solids. *Int. J. Solids Struct.*, 33:3277–3296, 1996.

[46] E. A. de Souza Neto, F. M. Andrade Pires, and D. R. J. Owen. F-bar based linear triangles and tetrahedra for finite element analysis of nearly incompressible solids. Part I: formulation and benchmarking. *International Journal for Numerical Methods in Engineering*, 62:353–383, 2005.

[47] P. Deuflhard. A study of extrapolation methods based on multistep schemes without parasitic solutions. *Z. angew. Math. Phys.*, 30:177–189, 1979.

[48] P. Deuflhard, R. Krause, and S. Ertel. A contact-stabilized newmark method for dynamical contact problems. *Int. J. Numer. Meth. Engng.*, 73:1274–1290, 2008.

[49] R. Diekmann, J. Hungershöfer, M. Lux, L. Taenzer, and J.-M. Wierum. Efficient contact search for finite element analysis. In *Proc. ECCOMAS*, 2000.

# Bibliography

[50] R. Diekmann, J. Hungershöfer, M. Lux, L. Taenzer, and J.-M. Wierum. Using space filling curves for efficient contact searching. In *Proc. IMACS*, 2000.

[51] C.R. Dohrmann, M. W. Heinstein, J. Jung, S. W. Key, and W. R. Witkowski. Node-based uniform strain elements for three-node triangular and four-node tetrahedral meshes. *International Journal for Numerical Methods in Engineering*, 47(9):1549–1568, 2000.

[52] N. El-Abbasi and S. A. Meguid amd A. Czekanski. On the modelling of smooth contact surfaces using cubic splines. *Int. J. Num. Meth. Eng.*, 50(4):953–967, 2001.

[53] T. Elguedj, Y. Bazilevs, V. M. Calo, and T. J. R. Hughes. B and F projection methods for nearly incompressible linear and non-linear elasticity and plasticity using higher-order NURBS elements. *Comput. Methods Appl Mech. Engrg.*, 197:2732–2762, 2008.

[54] V. Emel'yanenko. An explicit symplectic integrator for cometary orbits. *Celestial Mechanics and Dynamical Astronomy*, 84:331–341, 2002.

[55] R. D. Engle, R. D. Skeel, and M. Drees. Monitoring energy drift with shadow Hamiltonians. *J. Comput. Phys.*, 206:432–452, July 2005.

[56] J. Erickson, D. Guoy, M. Sullivan, and A. Üngör. Building spacetime meshes over arbitrary spatial domains. *Eng. with Comput.*, 20:342–353, August 2005.

[57] C. Bucher et al. Slang - the structural language. University of Innsbruck, Austria; Bauhaus University Weimar, Germany; http://www.uni-weimar.de/cms/bauing/forschung/institute/ism/slang.html, 1992-2006.

[58] R. C. Fetecau, J. E. Marsden, M. Ortiz, and M. West. Nonsmooth Lagrangian Mechanics and Variational Collision Integrators. *SIAM J. Appl. Dyn. Syst.*, 2(3):381–416, 2003.

[59] S. Fisher and M. C. Lin. Deformed distance fields for simulation of non-penetrating flexible bodies. In *Proceedings of the Eurographic workshop on Computer animation and simulation*, pages 99–111, New York, NY, USA, 2001. Springer-Verlag New York, Inc.

[60] D. P. Flanagan and T. Belytschko. A uniform strain hexahedron and quadrilateral with orthogonal hourglass control. *International Journal for Numerical Methods in Engineering*, 17:679–706, 1981.

[61] B. Flemisch, J. M. Melenk, and B. I. Wohlmuth. Mortar methods with curved interfaces. *Appl. Numer. Math.*, 54(3-4):339–361, 2005.

[62] B. Flemisch and B. Wohlmuth. Stable Lagrange multipliers for quadrilateral meshes of curved interfaces in 3D. *Comput. Methods Appl. Mech. Engrg.*, 196(8):1589–1602, 2007.

[63] M. Focardi and P. M. Mariano. Convergence of asynchronous variational integrators in linear elastodynamics. *Internat. J. Numer. Methods Engrg.*, 75:755–769, 2008.

[64] W. Fong, E. Darve, and A. Lew. Stability of asynchronous variational integrators. In *Proceedings of the 21st International Workshop on Principles of Advanced and Distributed Simulation*, PADS '07, pages 38–44, Washington, DC, USA, 2007. IEEE Computer Society.

[65] W. Fong, E. Darve, and A. Lew. Stability of asynchronous variational integrators. *Journal of Computational Physics*, 227(18):8367 – 8394, 2008.

[66] K. Gallivan, G. Grimme, and P. Dooren. A rational lanczos algorithm for model reduction. *Numerical Algorithms*, 12:33–63, 1996. 10.1007/BF02141740.

[67] F. Ganovelli, J. Dingliana, and C. O'Sullivan. Buckettree: Improving collision detection between deformable objects. *Spring Conference in Computer Graphics (SCCG)*, 2000.

[68] B. Garcia-archilla, J. M. Sanz-serna, and R. D. Skeel. Long-time-step methods for oscillatory differential equations. *SIAM J. Sci. Comput*, 20:930–963, 1999.

[69] M. Gates, K. Matous, and M. T. Heath. Asynchronous multi-domain variational integrators for non-linear problems. *Internat. J. Numer. Methods Engrg.*, 76(29-32):1353–1378, 2008.

[70] W. Gautschi. Numerical integration of ordinary differential equations based on trigonometric polynomials. *Numer. Math.*, 3:381–397, 1961.

[71] M. W. Gee, C. R. Dohrmann, S. W. Key, and W. A. Wall. A uniform nodal strain tetrahedron with isochoric stabilization. *International Journal for Numerical Methods in Engineering*, 78:429 – 443, 2008.

[72] C. Glocker. Concepts for modeling impacts without friction. *Acta Mechanica*, 168(1-2):1–19, 2004.

[73] José. González, K. C. Park, and Carlos A. Felippa. Partitioned formulation of frictional contact problems using localized lagrange multipliers. *Commun. Numer. Meth. Engng*, 22(4):319–333, 2005.

[74] M. Gonzalez. *Energy and force stepping integrators in Lagrangian mechanics*. PhD thesis, California Institute of Technology, 2011.

# Bibliography

[75] M. Gonzalez, B. Schmidt, and M. Ortiz. Energy-stepping integrators in Lagrangian mechanics. *Int. J. Numer. Meth. Engng*, 82:205–241, 2010.

[76] M. Gonzalez, B. Schmidt, and M. Ortiz. Force-stepping integrators in Lagrangian mechanics. *Int. J. Numer. Meth. Engng*, 84:1407–1450, 2010.

[77] S. Gottschalk, M. C. Lin, and D. Manocha. OBBTree: A hierarchical structure for rapid interference detection. *Computer Graphics*, 30(Annual Conference Series):171–180, 1996.

[78] A. Gravouil and A. Combescure. Multi-time-step explicit-implicit method for non-linear structural dynamics. *International Journal for Numerical Methods in Engineering*, 50:199–225, 2001.

[79] H. Grubmüller, H. Heller, A. Windemuth, and K. Schulten. Generalized verlet algorithm for efficient molecular dynamics simulations with long-range interactions. *Mol. Simulation*, 6:121–142, 1991.

[80] E. Hairer. Backward analysis of numerical integrators and symplectic methods. *Annals of Numerical Mathematics*, 1:107–132, 1994.

[81] E. Hairer. Variable time step integration with symplectic methods. *Applied Numerical Mathematics: Transactions of IMACS*, 25(2–3):219–227, 1997.

[82] E. Hairer. Symmetric projection methods for differential equations on manifolds. *BIT Numerical Mathematics*, 40:726–734, 2000.

[83] E. Hairer. Important aspects of geometric numerical integration. *J. Sci. Comput.*, 25:67–81, October 2005.

[84] E. Hairer. Geometric numerical integration - lecture notes. University of Geneve, 2010.

[85] E. Hairer and C. Lubich. Asymptotic expansions and backward error analysis for numerical integrators. In R. de la Llave, L.R. Petzold, and J. Lorenz, editors, *Dynamics of Algorithms*, volume 118, pages 91–106. Springer, Berlin, 2000.

[86] E. Hairer, C. Lubich, and G. Wanner. *Geometric Numerical Integration*, volume 31 of *Springer Ser. Comput. Math.* Springer-Verlag, 2002.

[87] E. Hairer and Ch. Lubich. Long-time energy conservation of numerical methods for oscillatory differential equations. *SIAM J. Numer. Anal.*, 38:414–441, 2000.

[88] Ernst Hairer. Global modified Hamiltonian for constrained symplectic integrators. *Numerische Mathematik*, 95:325–336, 2003. 10.1007/s00211-002-0428-7.

[89] J. O. Hallquist. *LS-DYNA theoretical manual*. Livemore Software Technology Corporation, 1998.

[90] W. R. Hamilton. On a general method in dynamics. Part I, 1834, 247-308; Part II, 1835, 95-144.

[91] D. Harmon, E. Vouga, B. Smith, R. Tamstorf, and E. Grinspun. Asynchronous contact mechanics. In *SIGGRAPH '09 (ACM Transactions on Graphics)*, New York, NY, USA, 2009. ACM.

[92] S. Hartmann, E. Ramm, S. Brunssen, and B. Wohlmuth. A primal-dual active set strategy for unilateral non-linear dynamic contact problems of thin-walled structures. In *III European Conference on Computational Mechanics*, pages 321–321. Springer Netherlands, 2006.

[93] Z.C. He, G.R. Liu, Z.H. Zhong, S.C. Wu, G.Y. Zhang, and A.G. Cheng. An edge-based smoothed finite element method (ES-FEM) for analyzing three-dimensional acoustic problems. *Computer Methods in Applied Mechanics and Engineering*, 199:20–33, 2009.

[94] B. Heidelberger, M. Teschner, R. Keiser, M. Muller, and M. Gross. Consistent penetration depth estimation for deformable collision response. *Proceedings of Vision, Modeling, Visualization VMV*, 04:339–346, November 16-18 2004.

[95] G. Hirota. *An improved finite element contact model for anatomical simulations*. PhD thesis, The University of North Carolina at Chapel Hill, 2002.

[96] G. Hirota, S. Fisher, A. State, C. Lee, and H. Fuchs. An implicit finite element method for elastic solids in contact. In *Computer Animation, 2001. The Fourteenth Conference on Computer Animation. Proceedings*, pages 136–147, 2001.

[97] P. G. Hjorth and N. Nordkvist. Classical Mechanics and Symplectic Integration - lecture notes. University of Hawaii, 2005.

[98] M. Hochbruck, C. Lubich, and H. Selhofer. Exponential integrators for large systems of differential equations. *SIAM J. Sci. Comput.*, 19:1552–1574, 1998.

[99] D. Hoffmann. *Das Augmented-Lagrange-Verfahren bei Reibkontaktproblemen unter transienter Belastung*. PhD thesis, Universität Karlsruhe, Fak. f. Bauingenieur- und Vermessungswesen, Germany, 2003.

[100] W. Huang and B. Leimkuhler. The Adaptive Verlet Method. *SIAM J. Sci. Comput.*, 18:239–256, January 1997.

[101] P. M. Hubbard. Collision detection for interactive graphics applications. *IEEE Transactions on Visualization and Computer Graphics*, 1(3):218–230, 1995.

# Bibliography

[102] S. Hüeber and B. I. Wohlmuth. A primal-dual active set strategy for nonlinear multibody contact problems. *Comput. Methods Appl. Mech. Engrg.*, 194:3147–3166, 2005.

[103] T. J. R. Hughes. Equivalence of finite elements for nearly incompressible elasticity. *J. Appl. Mech., Trans. ASME*, 44:181–183, 1977.

[104] T. J. R. Hughes, J. A. Cottrell, and Y. Bazilevs. Isogeometric analysis: CAD, finite elements, NURBS, exact geometry and mesh refinement. *Comput. Methods Appl. Mech. Engrg.*, 194:4135–4195, 2005.

[105] T. J. R. Hughes and W. Liu. Implicit-explicit finite elements in transient analysis: implementation and numerical examples. *Journal of Applied Mechanics*, 45:375–378, 1978.

[106] T. J. R. Hughes and W. Liu. Implicit-explicit finite elements in transient analysis: stability theory. *Journal of Applied Mechanics*, 45:371–374, 1978.

[107] T. J. R. Hughes, R.L. Taylor, and J.L. Sackman. Finite element formulation and solution of contact-impact problems in continuum mechanics-iii. *Report No. UC SESM 75-7*, July 1975.

[108] Thomas J. R. Hughes. Generalization of selective integration procedures to anisotropic and nonlinear media. *International Journal for Numerical Methods in Engineering*, 15:1413–1418, 1980.

[109] G. M. Hulbert and J. Chung. Explicit time integration algorithms for structural dynamics with optimal numerical dissipation. *Computer Methods in Applied Mechanics and Engineering*, 137(2):175 – 188, 1996.

[110] P. Hut, J. Marino, and S. Mcmillan. Building a better leapfrog. *Astrophysical Journal*, 443(2 PART 2):L93–L96, 1995.

[111] S. R. Idelsohn and A. Cardona. A reduction method for nonlinear structural dynamic analysis. *Comp. meth. in Appl. Mech. and Eng.*, 49:253–279, 1985.

[112] B. M. Irons. Applications of a theorem on eigenvalues to finite element problems. Technical report, University of Wales, Department of Civil Engineering, Swansea, 1970.

[113] J. A. Izaguirre. Longer time steps for molecular dynamics. Technical report, University of Illinois at Urbana-Champaign, Champaign, IL, USA, 1999.

[114] C. Jiun-Shyan, W. Cheng-Tang, Y. Sangpil, and Y. Yang. A stabilized conforming nodal integration for galerkin mesh-free methods. *International Journal for Numerical Methods in Engineering*, 50(2):435–466, 2001.

[115] E.R. Johnson and T.D. Murphey. Dangers of two-point holonomic constraints for variational integrators. In *American Control Conference, 2009. ACC '09.*, pages 4723 –4728, june 2009.

[116] G. R. Joldes, A. Wittek, and K. Miller. Improved linear tetrahedral element for surgical simulation. In *Proceedings of MICCAI Workshop*, pages 54–66, Copenhagen, Oct. 2006.

[117] K. Kale and A. Lew. Parallel asynchronous variational integrators. *Internat. J. Numer. Methods Engrg.*, 70:291–321, 2007.

[118] C. Kane, J. E. Marsden, and M. Ortiz. Symplectic-energy-momentum preserving variational integrators. *J. of Math. Phys.*, 40(7):3353–3371, July 1999.

[119] C. Kane, J. E. Marsden, M. Ortiz, and M. West. Variational integrators and the newmark algorithm for conservative and dissipative mechanical systems. *Internat. J. Numer. Methods Engrg.*, 49:1295–1325, 1999.

[120] C. Kane, E. A. Repetto, M. Ortiz, and J. E. Marsden. Finite element analysis of nonsmooth contact. *Computer Methods in Applied Mechanics and Engineering*, 180:1–26, 1999.

[121] H. Kantz and T. Schreiber. *Nonlinear Time Series Analysis.* Cambridge University Press, Cambridge, 2004.

[122] R. K. Kapania and C. Byun. Reduction methods based on eigenvectors and ritz vectors for nonlinear transient analysis. *Computational Mechanics*, 11:65–82, 1993.

[123] K. Karhunen. Über lineare methoden in der wahrscheinlichkeitsrechnung. *Annals of Academic Science Fennicae, series A1 Math. & Physics*, 37:3–79, 1946.

[124] G. Kerschen, J.-C. Golinval, A. F. Vakakis, and L. A. Bergman. The method of proper orthogonal decomposition for dynamical characterization and order reduction of mechanical systems: An overview. *Nonlinear Dynamics*, 41:147–169, 2005.

[125] H. B. Khenous, P. Laborde, and Y. Renard. On the discretization of contact problems in elastodynamics. In *Lecture Notes in Applied and Computational Mechanics*, volume 27, pages 31–38. Springer-Verlag, 2006.

[126] C. Kim, R. D. Lazarov, and J. E. Pasciak. Multiplier spaces for the mortar finite element method in three dimensions. *SIAM J. Numer. Anal.*, 39(2):519–538, 2001.

[127] T. Y. Kim, J. Dolbow, and T. Laursen. A mortared finite element method for frictional contact on arbitrary interfaces. *Computational Mechanics*, 39(3):223–235, 2007.

# Bibliography

[128] J. T. Klosowski, M. Held, J. S. B. Mitchell, H. Sowizral, and K. Zikan. Efficient collision detection using bounding volume hierarchies of $k$-DOPs. *IEEE Transactions on Visualization and Computer Graphics*, 4(1):21–36, 1998.

[129] R. H. Krause and B. I. Wohlmuth. Multigrid methods for mortar finite elements. In *Multigrid Methods VI*, volume 14, pages 136–142, Berlin, 2000. Springer–Verlag.

[130] S. Krishnan, M. Gopi, M. Lin, D. Manocha, and A. Pattekar. Rapid and accurate contact determination between spline models using ShellTrees. *Computer Graphics Forum*, 17(3), 1998.

[131] L. Krstulovic-Opara, P. Wriggers, and J. Korelc. A c1-continuous formulation for 3d finite deformation frictional contact. *Computational Mechanics*, 29:27–42, 2002.

[132] P. Krysl, S. Lall, and J. E. Marsden. Dimensional model reduction in nonlinear finite element dynamics of solids and structures. *Int. J. Numer. Meth. Eng.*, 51:479–504, 2001.

[133] P. Krysl and B. Zhu. Locking-free continuum displacement finite elements with nodal integration. *International Journal for Numerical Methods in Engineering*, 76:1020–1043, 2008.

[134] W. Krätzig and P. Nawrotzki. Computational concepts in structural stability. *Archives of Computational Methods in Engineering*, 3:81–119, 1996. 10.1007/BF02736131.

[135] S. Lall, P. Krysl, and J. E. Marsden. Structure-preserving model reduction for mechanical systems. *Physica D*, 184(1–4):304–318, 2003.

[136] S. Lall, J. E. Marsden, and S. Glavaki. A subspace approach to balanced truncation for model reduction of nonlinear control systems. *International Journal of Robust and Nonlinear Control*, 12(6):519–535, February 2002.

[137] S. Lall and M. West. Discrete variational hamiltonian mechanics. *Journal of Physics A: Mathematical and General*, 39(19):5509, 2006.

[138] B. P. Lamichhane and B. I. Wohlmuth. Higher order dual Lagrange multiplier spaces for mortar finite element discretizations. *CALCOLO*, 39:219–237, 2002.

[139] B. P. Lamichhane and B. I. Wohlmuth. A quasi-dual Lagrange multiplier space for serendipity mortar finite elements in 3D. *M2AN*, 38:73–92, 2004.

[140] T. A. Laursen. *Formulation and Treatment of Frictional Contact Problems Using Finite Elements*. PhD thesis, Department of Mechanical Engineering, Stanford University, 1992.

[141] T. A. Laursen. *Computational Contact and Impact Mechanics*. Springer, 1st edition, 2002.

[142] T. A. Laursen and V. Chawla. Design of energy conserving algorithms for frictionless dynamic contact problems. *International Journal for Numerical Methods in Engineering*, 40:863–886, 1997.

[143] T. A. Laursen and G. R. Love. Improved implicit integrators for transient impact problems - geometric admissibility within the conserving framework. *International Journal for Numerical Methods in Engineering*, 53:245–274, 2002.

[144] T. A. Laursen, M. A. Puso, and M. W. Heinstein. Practical issues associated with mortar projections in large deformation contact/impact analysis. *5th World Congress on Computational Mechanics*, Jul 2002 7-12.

[145] T. A. Laursen, B. Yang, and M. A. Puso. Implementation of frictional contact conditions in surface-to-surface mortar based computational frameworks. In *ECCOMAS 2004 Proceedings*, 2004.

[146] D. J. Lawson. Generalized runge–kutta processes for stable systems with large lipschitz constants. *SIAM J. Numer. Anal.*, 4:372–380, 1967.

[147] R. Lehoucq and R. Koteras. Estimating the critical time-step in explicit dynamics using the lanczos method. *International Journal for Numerical Methods in Engineering*, 69:2780–2788, 2007.

[148] B. Leimkuhler. An efficient multiple time-scale reversible integrator for the gravitational n-body problem. *Appl. Numer. Math.*, 43:175–190, October 2002.

[149] B. Leimkuhler and S. Reich. A reversible averaging integrator for multiple time-scale dynamics. *Journal of Computational Physics*, 171(1):95 – 114, 2001.

[150] B. J. Leimkuhler and R. D. Skeel. Symplectic numerical integrators in constrained hamiltonian systems. *Journal of Computational Physics*, 112(1):117 – 125, 1994.

[151] R. I. Leine, U. Aeberhard, and C. Glocker. Hamilton's principle as variational inequality for mechanical systems with impact. *Journal of Nonlinear Science*, 19:633–664, 2009.

[152] E. Lens, A. Cardona, and M. Géradin. Energy preserving time integration for constrained multibody systems. *Multibody System Dynamics*, 11:41–61, 2004.

[153] M. Leok. Generalized galerkin variational integrators, August 18 2005. Preprint.

# Bibliography

[154] A. Lew. *Variational time integrators in computational solid mechanics.* PhD thesis, California Institute of Technology, 2003.

[155] A. Lew, J. E. Marsden, M. Ortiz, and M. West. Asynchronous variational integrators. *Archive for Rational Mechanics & Analysis*, 167(2):85–146, 2003.

[156] A. Lew, J. E. Marsden, M. Ortiz, and M. West. An overview of variational integrators. In *Finite Element Methods: 1970's and Beyond. Theory and engineering applications of computational methods*, pages 1–18. International Center for Numerical Methods in Engineering (CIMNE), Barcelona, 2004.

[157] A. Lew, J. E. Marsden, M. Ortiz, and M. West. Variational time integrators. *Internat. J. Numer. Methods Engrg.*, 60(1):153–212, 2004.

[158] A. Lew and M. Ortiz. Asynchronous variational integrators. In *Geometry, Mechanics and Dynamics*, pages 91–110. Springer, 2002.

[159] S. Leyendecker, J. E. Marsden, and M. Ortiz. Variational integrators for constrained dynamical systems. *ZAMM*, 88:677–708, 2008.

[160] M. C. Lin and S. Gottschalk. Collision detection between geometric models: A survey. In *In Proc. of IMA Conference on Mathematics of Surfaces*, pages 37–56, 1998.

[161] T. R. Littell, R. D. Skeel, and M. Zhang. Error analysis of symplectic multiple time stepping. *SIAM J. Numer. Anal.*, 34:1792–1807, October 1997.

[162] G. R. Liu. A generalized gradient smoothing technique and the smoothed bilinear form for galerkin formulation of a wide class of computational methods. *International Journal of Computational Methods*, 5(2):199–236, 2008.

[163] G. R. Liu. A G space theory and a weakened weak (W2) form for a unified formulation of compatible and incompatible methods: Part I theory. *International Journal for Numerical Methods in Engineering*, 81:1093 – 1126, 2009.

[164] G. R. Liu, K. Dai, and T. Nguyen. A smoothed finite element method for mechanics problems. *Computational Mechanics*, 39(6):859–877, May 2007.

[165] G. R. Liu, Y. Li, K. Y. Dai, M. T. Luan, and W. Xue. A linearly conforming radial point interpolation method for solid mechanics problems. *International Journal of Computational Methods*, 3:401–428, 2006.

[166] G. R. Liu, T. T. Nguyen, K. Y. Dai, and K. Y. Lam. Theoretical aspects of the smoothed finite element method (SFEM). *International Journal for Numerical Methods in Engineering*, 71:902–930, 2007.

[167] G. R. Liu, T. Nguyen-Thoi, and K. Y. Lam. An edge-based smoothed finite element method (es-fem) for static, free and forced vibration analyses of solids. *Journal of Sound and Vibration*, 320:1100–1130, 2008.

[168] G. R. Liu, T. Nguyen-Thoi, and K. Y. Lam. A novel alpha finite element method ($\alpha$FEM) for exact solution to mechanics problems using triangular and tetrahedral elements. *Comput. Methods Appl. Mech. Engrg.*, 197:3883–3897, 2008.

[169] G. R. Liu, T. Nguyen-Thoi, H. Nguyen-Xuan, and K. Y. Lam. A node-based smoothed finite element method (NS-FEM) for upper bound solutions to solid mechanics problems. *Comput. Struct.*, 87(1-2):14–26, 2009.

[170] G. R. Liu, H. Nguyen-Xuan, and T. Nguyen-Thoi. A theoretical study on the smoothed FEM (S-FEM) models: Properties, accuracy and convergence rates. published online in Wiley Interscience, 2010. DOI: 10.1002/nme.2941.

[171] G. R. Liu, H. Nguyen-Xuan, T. Nguyen-Thoi, and X. Xu. A novel galerkin-like weakform and a superconvergent alpha finite element method (S$\alpha$FEM) for mechanics problems using triangular meshes. *J. Comput. Phys.*, 228(11):4055–4087, 2009.

[172] G. R. Liu, X. Xu, G. Y. Zhang, and T. Nguyen-Thoi. A superconvergent point interpolation method (sc-pim) with piecewise linear strain field using triangular mesh. *International Journal for Numerical Methods in Engineering*, 77:1439–1467, 2008.

[173] G. R. Liu and G. Y. Zhang. Upper bound solution to elasticity problems: A unique property of the linearly conforming point interpolation method (lc-pim). *International Journal for Numerical Methods in Engineering*, 74:1128–1161, 2008.

[174] G. R. Liu and G. Y. Zhang. A novel scheme of strain-constructed point interpolation method for static and dynamic mechanics problems. *International journal for applied mechanics*, 1:233–258, 2009.

[175] G. R. Liu, G. Y. Zhang, K. Y. Dai, Y. Y. Wang, Z. H. Zhong, G. Y. Li, and X. Han. A linearly conforming point interpolation method (LC-PIM) for 2D solid mechanics problems. *International Journal of Computational Methods*, 2:645–665, 2005.

[176] M. Loeve. Fonctions aléatoires du second ordre. In *Processus stochastiques et mouvement Brownien*, Paris, 1948. Gauthier-Villars.

[177] M. López-Marcos, J. M. Sanz-Serna, and R. D. Skeel. Explicit symplectic integrators with maximal stability intervals. In D. F. Griffiths and G. A. Watson, editors, *Numerical Analysis, A. R. Mitchell 75th Birthday Volume*, pages 163–176. World Scientific, Singapore, June 1996.

# Bibliography

[178] M. A. Lopez-Marcos, J. M. Sanz-Serna, and R. D. Skeel. Explicit symplectic integrators using hessian-vector products. *SIAM J. Sci. Comput*, 18:223–238, 1997.

[179] G. R. Love and T. A. Laursen. Improved implicit integrators for transient impact problems - dynamic frictional dissipation within an admissible conserving framework. *Computer Methods in Applied Mechanics and Engineering*, 192:2223–2248, 2003.

[180] Ch. Lunk and B. Simeon. Solving constrained mechanical systems by the family of newmark and $\alpha$-methods. *ZAMM - Journal of Applied Mathematics and Mechanics / Zeitschrift für Angewandte Mathematik und Mechanik*, 86(10):772–784, 2006.

[181] A. M. Lyapunow. *Stability of Motion*. Academic Press, New York, 1966.

[182] Q. Ma and J. A. Izaguirre. Long time step molecular dynamics using targeted langevin stabilization. In *Proceedings of the 2003 ACM symposium on Applied computing*, SAC '03, pages 178–182, New York, NY, USA, 2003. ACM.

[183] H. Mang and G. Hofstetter. *Festigkeitslehre*. Springer, Wien, New York, 2 edition, 2004.

[184] D. Marchal, F. Aubert, and C. Chaillou. Collision between deformable objects using fast-marching on tetrahedral models. In *SCA '04: Proceedings of the 2004 ACM SIGGRAPH/Eurographics symposium on Computer animation*, pages 121–129, New York, NY, USA, 2004. ACM Press.

[185] J. E. Marsden and M. West. Discrete mechanics and variational integrators. *Acta Numerica*, 10:357–514, 2001.

[186] Breen D. Mauch S. A fast marching method of computing closest points. *http://www.cco.caltech.edu/sean/closestpoint/closept.html*, dec 2006.

[187] T.W. McDevitt and T.A. Laursen. A mortar-finite element formulation for frictional contact problems. *Int. J. Num. Meth. Eng.*, 48:1525–1547, 2000.

[188] M. Meyer and H. G. Matthies. Efficient model reduction in nonlinear dynamics using the karhunen-loève expansion and dual-weighted-residual methods. *Computational Mechanics*, 31:179–191, 2003.

[189] S. Mikkola. Practical symplectic methods with time transformation for the few-body problem. *Celestial Mechanics and Dynamical Astronomy*, 67:145–165, 1997.

[190] S. Mikkola and K. Tanikawa. Explicit symplectic algorithms for time-transformed hamiltonians. *Celestial Mechanics and Dynamical Astronomy*, 74:287–295, 1999.

[191] K. Modin. On explicit adaptive symplectic integration of separable hamiltonian systems. *J. Mult. Body Mech.*, 222(4):289–300, 2008.

[192] B. Moran, M. Ortiz, and F. Shih. Formulation of implicit finite element methods for multiplicative finite deformation plasticity. *International Journal for Numerical Methods in Engineering*, 29:483–514, 1990.

[193] J. J. Moreau. Unilateral contact and dry friction in finite freedom dynamics. In *Nonsmooth Mechanics and Applications, CISM Courses and Lectures*, volume 302, pages 1–82, Linköping, Sweden, 1988. Springer-Verlag.

[194] J. J. Moreau. Numerical aspects of the sweeping process. *Comput. Methods Appl. Mech. Engrg.*, 177:329–349, 1999.

[195] J. Moser and A. P. Veselov. Discrete versions of some classical integrable systems and factorization of matrix polynomials. *Comm. Math. Phys*, 139(2):217–243, 1991.

[196] H. M. Mourad, J. Dolbow, and K. Garikipati. An assumed-gradient finite element method for the level set equation. *International Journal for Numerical Methods in Engineering*, 64(8):1009–1032, 2005.

[197] J.C. Nagtegaal, D.M. Parks, and J.R. Rice. On numerical accurate finite element solutions in the fully plastic range. *Computer Methods in Applied Mechanics and Engineering*, 4:153–177, 1974.

[198] P. Nawrotzki and C. Eller. Numerical stability analysis in structural dynamics. *Computer Methods in Applied Mechanics and Engineering*, 189(3):915 – 929, 2000.

[199] M. O. Neal and T. Belytschko. Explicit-explicit subcycling with noninteger time step ratios for structural dynamic systems. *Comput. Struct.*, 31(6):871–80, 1989.

[200] N. M. Newmark. A method of computation for structural dynamics. *ASCE J. of the Engineering Mechanics Division*, pages 67–94, 1959.

[201] N.G.Bourago and V.N.Kukudzhanov. A survey on contact algorithms (extended version). Technical report, The Institute for Problems in Mechanics of RAS, Moscow, 2003.

[202] N. Nguyen-Thanh, T. Rabczuk, H. Nguyen-Xuan, and S. Bordas. A smoothed finite element method for shell analysis. *Computer Methods in Applied Mechanics and Engineering*, 198:165–177, June 2008.

[203] T. Nguyen-Thoi, G. R. Liu, K. Y. Lam, and G. Y. Zhang. A face-based smoothed finite element method (FS-FEM) for 3D linear and geometrically non-linear solid mechanics problems using 4-node tetrahedral elements. *International Journal for Numerical Methods in Engineering*, 78:324 – 353, 2008.

# Bibliography

[204] T. Nguyen-Thoi, G. R. Liu, and H. Nguyen-Xuan. An n-sided polygonal edge-based smoothed finite element method (nES-FEM) for solid mechanics. published online in Wiley Interscience, 2010. DOI: 10.1002/cnm.1375.

[205] H. Nguyen-Xuan, T. Rabczuk, S. Bordas, and J. F. Debongnie. A smoothed finite element method for plate analysis. *Computer Methods in Applied Mechanics and Engineering*, 197:1184–1203, 2008.

[206] R. E. Nickel. Nonlinear dynamics by mode superposition. *Comp. Meth. in Appl. Mech. and Eng.*, 7:107–129, 1976.

[207] E. Noether. Invariante Variationsprobleme. *Kgl. Ges. Wiss. Nachr. Göttingen Math. Physik*, 2:235–257, 1918.

[208] M. Oldenburg and L. Nilsson. The position code algorithm for contact searching. *International Journal for Numerical Methods in Engineering*, 37(3):359–386, 1994.

[209] S. Osher and J. A. Sethian. Fronts propagating with curvature-dependent speed: Algorithms based on Hamilton-Jacobi formulations. *Journal of Computational Physics*, 79:12–49, 1988.

[210] C. O'Sullivan and J. Dingliana. Real-time collision detection and response using sphere-trees. *15th Spring Conference on Computer Graphics*, pages 83–92, April 1999.

[211] P. Oswald and B. Wohlmuth. On polynomial reproduction of dual fe bases. *Thirteenth International Conference on Domain Decomposition Methods*, 2001.

[212] I. J. Palmer and R. L. Grimsdale. Collision detection for animation using sphere-trees. *Computer Graphics Forum, 14(2)*, pages 105–116, jun 1995.

[213] A. Pandolfi, C. Kane, J.E. Marsden, and M. Ortiz. Time-discretized variational formulation of non-smooth frictional contact. *Int. J. Numer. Meth. Engng.*, 53:1801–1829, 2002.

[214] M. Preto and S. Tremaine. A class of symplectic integrators with adaptive time step for separable hamiltonian systems. *The Astronomical Journal*, 118(5):2532, 1999.

[215] M. A. Puso, J. S. Chen, E. Zywicz, and W. Elmer. Meshfree and finite element nodal integration methods. *International Journal for Numerical Methods in Engineering*, 74(3):416–446, 2008.

[216] M. A. Puso, T. L. Laursen, and J. Solberg. A 3d frictional segment-to-segment contact method for large deformations and quadratic elements. *ECCOMAS 2004*, 2004.

[217] M. A. Puso and J. Solberg. A stabilized nodally integrated tetrahedral. *International Journal for Numerical Methods in Engineering*, 67(6):841–867, 2006.

[218] M.A. Puso and T.A. Laursen. A mortar segment-to-segment contact method for large deformation solid mechanics. *Computer Methods in Applied Mechanics and Engineering*, 193:601–629, 2004.

[219] M.A. Puso and T.A. Laursen. A mortar segment-to-segment frictional contact method for large deformations. *Computer Methods in Applied Mechanics and Engineering*, 193:4891–4913, 2004.

[220] G. Rebel and K. C. Park. Application of the localised lagrange multiplier method to a 3d contact patch test. *Proc. 2002 AIAA SDM Conference, Paper No. AIAA-2002-1577*, April 22-26 2002.

[221] G. Rebel, K. C. Park, and C. A. Felippa. A contact formulation based on localized lagrange multipliers: formulation and application to two-dimensional problems. *International Journal for Numerical Methods in Engineering*, 54(2):263–297, 2002.

[222] S. Reese, M. Küssner, and B. D. Reddy. A new stabilization technique for finite elements in non-linear elasticity. *International journal for numerical methods in engineering*, 44:1617–1652, 1999.

[223] S. Reese, P. Wriggers, and B. D. Reddy. A new locking-free brick element technique for large deformation problems in elasticity. *Computers and Structures*, 75:291–304, 2000.

[224] S. Reich. Symplectic integration of constrained hamiltonian systems by composition methods. *SIAM J. Numer. Anal.*, 33:475–491, April 1996.

[225] S. Reich. Backward error analysis for numerical integrators. *SIAM J. Numer. Anal.*, 36:1549–1570, July 1999.

[226] J. Remke and H. Rothert. Eine modale Reduktionsmethode zur geometrisch nichtlinearen statischen und dynamischen Finite-Element-Berechnung. *Archive of Applied Mechanics*, 63:101–115, 1993. 10.1007/BF00788916.

[227] McLachlan R.I., Perlmutter M., and Quispel G.R.W. On the nonlinear stability of symplectic integrators. *Bit Numerical Mathematics*, 44:99–117(19), 2004.

[228] G. Rowlands. A numerical algorithm for Hamiltonian systems. *J. Comput. Phys.*, 97:235–239, November 1991.

# Bibliography

[229] J.-P. Ryckaert, G. Ciccotti, and J. J. C. Berendsen. Numerical integration of the cartesian equations of motion of a system with constraints: Molecular dynamics of n-alkanes. *Journal of Computational Physics*, 23:327–341, 1977.

[230] R.G. Sauve and G.D. Morandin. Simulation of contact in finite deformation problems - algorithm and modelling issues. *International Journal of Mechanics and Materials in Design*, 1:287–316, 2004.

[231] J. M. A. Scherpen. Balancing for nonlinear systems. *Syst. Control Lett.*, 21:143–153, August 1993.

[232] T. Schlick, M. Mandziuk, R. D. Skeel, and K. Srinivas. Nonlinear resonance artifacts in molecular dynamics simulations. *Journal of Computational Physics*, 140(1):1 – 29, 1998.

[233] J. Sethian. Level set methods and fast marching methods: Evolving interfaces in computational geometry, 1998.

[234] J. A. Sethian. A fast marching level set method for monotonically advancing fronts. In *Proc. Nat. Acad. Sci*, pages 1591–1595, 1996.

[235] J. A. Sethian. Fast marching methods. *SIAM Rev.*, 41:199–235, June 1999.

[236] Y. Shibberu. Time-discretization of hamiltonian dynamical systems. *Computers and Mathematics with Applications*, 28:123–145, 1994.

[237] Y. Shibberu. Time-discretization of hamiltonian dynamical systems. *Computers & Mathematics with Applications*, 28(10-12):123 – 145, 1994.

[238] Y. Shibberu. How to regularize a symplectic-energy-momentum integrator, July 2005. url: http://www.rose-hulman.edu/~shibberu/DTH_Dynamics/DTH_Dynamics.htm.

[239] A. Signorini. Questioni di elasticità non linearizzata e semilinearizzata (issues in non linear and semilinear elasticity). *Rendiconti di Matematica e delle sue applicazioni*, 5(18):95–139, 1959.

[240] J. C. Simo, R. L. Taylor, and K. S. Pister. Variational and projection methods for the volume constraint in finite deformation elasto-plasticity. *Computational Methods in Applied Mechanical Engineering*, 51:177–208, 1985.

[241] J.C. Simo and S. Rifai. A class of mixed assumed strain methods and the method of incompatible modes. *International Journal for Numerical Methods in Engineering*, 29:1595–1638, 1990.

[242] R. D. Skeel. Integration schemes for molecular dynamics and related applications. In M. Ainsworth, J. Levesley, and M. Marletta, editors, *The Graduate Student's Guide to Numerical Analysis '98*, volume 26 of *Springer Ser. Comput. Math.*, pages 119–176. Springer-Verlag, 1999.

[243] R. D. Skeel and D. J. Hardy. Practical construction of modified hamiltonians. *SIAM J. Sci. Comput.*, 23:1172–1188, April 2001.

[244] R. D. Skeel, G. Zhang, and T. Schlick. A family of symplectic integrators: Stability, accuracy, and molecular dynamics applications. *SIAM J. Sci. Comput.*, 18:203–222, January 1997.

[245] Robert D. Skeel and K. Srinivas. Nonlinear stability analysis of area-preserving integrators. *SIAM Journal on Numerical Analysis*, 38(1):129–148, 2000.

[246] A. Smith, Y. Kitamura, H. Takemura, and F. Kishino. A simple and efficient method for accurate collision detection among deformable objects in arbitrary motion. *Proc. of the IEEE Virtual Reality Annual International Symposium*, pages 136–145, 1995.

[247] P. Smolinski. Subcycling integration with non-integer time steps for structural dynamics problems. *Computers & Structures*, 59(2):273 – 281, 1996.

[248] P. Smolinski and S. Sleith. Explicit multi-time step methods for structural dynamics. In *New Method in Transient Analysis*, pages 1–4. ASME, 2nd edition, 1992.

[249] M. Stadler and G.A. Holzapfel. Subdivision schemes for smooth contact surfaces of arbitrary mesh topology in 3d. *Int. J. Num. Meth. Eng.*, page in press, 2004.

[250] E. Stein and M. Rüter. Finite Element Method for Elasticity with Error-controlled Approximation and Model Adaptivity. In E. Stein, R. de Borst, and T.J.R. Hughes, editors, *Encyclopedia of Computational Mechanics*, volume 2, chapter 2, page 60. Wiley, 2007.

[251] A. Stern. Discrete Hamilton–Pontryagin mechanics and generating functions on Lie groupoids. *J. Symplectic Geom.*, 8(2):225–238, 2010.

[252] A. Stern and E. Grinspun. Implicit-explicit variational integration of highly oscillatory problems. *Multiscale Model. Simul.*, 7(4):1779–1794, 2009.

[253] D. Stoffer. Variable steps for reversible integration methods. *Computing*, 55(1):1–22, 1995.

[254] M. Teschner, B. Heidelberger, M. Mueller, D. Pomeranets, and M. Gross. Optimized spatial hashing for collision detection of deformable objects, 2003.

[255] M. Teschner, S. Kimmerle, B. Heidelberger, G. Zachmann, L. Raghupathi, A. Fuhrmann, M. Cani, F. Faure, N. Magnenat-Thalmann, W. Strasser, and P. Volino. Collision detection for deformable objects. *Computer Graphics Forum 24(1)*, 2005.

# Bibliography

[256] S. Thite. Adaptive spacetime meshing for discontinuous galerkin methods. *Comput. Geom. Theory Appl.*, 42:20–44, January 2009.

[257] S. P. Timoshenko and J. N. Goodier. *Theory of Elasticity*. McGraw-Hill, New York, 3 edition, 1970.

[258] M. Tuckerman, B. J. Berne, and G. J. Martyna. Reversible multiple time scale molecular dynamics. *The Journal of Chemical Physics*, 97(3):1990–2001, 1992.

[259] M. Hochbruck V. Grimm. Error analysis of exponential integrators for oscillatory second-order differential equations. *J. Phys. A:Math. Gen.*, 39:5495–5507, 2006.

[260] G. van den Bergen. Efficient collision detection of complex deformable models using aabb trees. *J. Graph. Tools*, 2(4):1–13, 1997.

[261] A. Varga. Enhanced modal approach for model reduction. *Mathematical and Computer Modelling of Dynamical Systems,*, 1:91 – 105, 1995.

[262] L. Verlet. Computer "Experiments" on Classical Fluids. I. Thermodynamical Properties of Lennard-Jones Molecules. *Physical Review*, 159(1):98–103, 1967.

[263] A. P. Veselov. Integrable discrete-time systems and difference operators. *Funct. Anal. Appl.*, 22(2):83–93, 1988.

[264] S. P. Wang and E. Nakamachi. The inside–outside contact search algorithm for finite element analysis. *International Journal for Numerical Methods in Engineering*, 40(19):3665–3685, 1997.

[265] M. S. Warren and J. K. Salmon. A parallel hashed oct-tree n-body algorithm. In *Supercomputing*, pages 12–21, 1993.

[266] J. M. Wendlandt and J. E. Marsden. Mechanical integrators derived from a discrete variational principle. *Phys. D*, 106:223–246, August 1997.

[267] M. West. *Variational integrators*. PhD thesis, California Institute of Technology, 2004.

[268] J.-M. Wierum. Logarithmic path-length in space-filling curves. In *14th Canadian conference on computational geometry*, pages 22–26, 2002.

[269] J.-M. Wierum. *Anwendung diskreter raumfüllender Kurven : Graphpartitionierung und Kontaktsuche in der Finite-Elemente-Simulation*. PhD thesis, Paderborn, Univ., Fakultät für Elektrotechnik, Informatik und Mathematik, 2004.

[270] E. L. Wilson, M.-W. Yuan, and J. M. Dickens. Dynamic analysis by direct superposition of ritz vectors. *Earthquake Eng. and Structural Dynamics*, 10:813–821, 1982.

[271] E.L. Wilson, E.L. Taylor, W.P. Doherty, and J. Ghaboussi. *Incompatible displacement models*, pages 43–57. Academic Press, 1973.

[272] B. Wohlmuth. A mortar finite element method using dual spaces for the Lagrange multiplier. *SIAM J. Numer. Anal.*, 38(3):989–1012, 2000.

[273] P. Wriggers. *Computational Contact Mechanics*. Springer, 2nd edition, 2006.

[274] P. Wriggers and M. Imhof. On the treatment of nonlinear unilateral contact problems. *Archive of Applied Mechanics*, 63:116–129, 1993.

[275] P. Wriggers, T. Vu Van, and E. Stein. Finite element formulation of large deformation impact-contact problems with friction. *Computers & Structures*, 37(3):319 – 331, 1990.

[276] X. Xua, G.R. Liu, and G.Y. Zhang. A point interpolation method with least square strain field (pim-lss) for solution bounds and ultra-accurate solutions using triangular mesh. *Computer Methods in Applied Mechanics and Engineering*, 198:1486–1499, 2009.

[277] Z. C. Xuan, T. Lassila, G. Rozza, and A. Quarteroni. On computing upper and lower bounds on the outputs of linear elasticity problems approximated by the smoothed finite element method. *International Journal for Numerical Methods in Engineering*, 83(2):174–195, 2010.

[278] H. Yoshida. Construction of higher order symplectic integrators. *Phys. Lett. A*, 150:262–268, 1990.

[279] G. Y. Zhang, G. R. Liu, Y. Y. Wang, H. T. Huang, Z. H. Zhong, G. Y. Li, and X. Han. A linearly conforming point interpolation method (lc-pim) for three-dimensional elasticity problems. *International Journal for Numerical Methods in Engineering*, 72:1524–1543, 2007.

[280] Z.-Q. Zhang and G. R. Liu. Temporal stabilization of the node-based smoothed finite element method and solution bound of linear elastostatics and vibration problems. *Computational Mechanics*, 46(2):229–246, July 2010.

[281] Ge Zhong and J. E. Marsden. Lie-poisson hamilton-jacobi theory and lie-poisson integrators. *Physics Letters A*, 133(3):134 – 139, 1988.

[282] A. Üngör and A. Sheffer. Tent-Pitcher: A meshing algorithm for space–time discontinuous Galerkin methods. In *Proc. 9th Internat. Meshing Roundtable*, pages 111–122, Sandia National Laboratories, 2000.

Die VDM Verlagsservicegesellschaft sucht für wissenschaftliche Verlage abgeschlossene und herausragende

## Dissertationen, Habilitationen, Diplomarbeiten, Master Theses, Magisterarbeiten usw.

für die kostenlose Publikation als Fachbuch.

Sie verfügen über eine Arbeit, die hohen inhaltlichen und formalen Ansprüchen genügt, und haben Interesse an einer honorarvergüteten Publikation?

Dann senden Sie bitte erste Informationen über sich und Ihre Arbeit per Email an *info@vdm-vsg.de*.

**Sie erhalten kurzfristig unser Feedback!**

VDM Verlagsservicegesellschaft mbH
Dudweiler Landstr. 99
D - 66123 Saarbrücken
**www.vdm-vsg.de**

Telefon +49 681 3720 174
Fax +49 681 3720 1749

Die VDM Verlagsservicegesellschaft mbH vertritt

Printed by Books on Demand GmbH, Norderstedt / Germany